Biomedical Engineering

Sang C. Suh • Varadraj P. Gurupur
Murat M. Tanik
Editors

Biomedical Engineering

Health Care Systems, Technology and Techniques

Springer

Editors
Sang C. Suh
Texas A&M University - Commerce
Department of Computer Science
Commerce, TX
USA
Sang_Suh@tamu-commerce.edu

Varadraj P. Gurupur
The University of Alabama
at Birmingham
Department of Neurology
3rd Avenue South 1530
35294 Birmingham, AL
USA

Murat M. Tanik
The University of Alabama
at Birmingham
Department of Electrical
and Computer Engineering
3rd Avenue South 1530
35294 Birmingham, AL
USA

ISBN 978-1-4899-8938-3   ISBN 978-1-4614-0116-2 (eBook)
DOI 10.1007/978-1-4614-0116-2
Springer New York Dordrecht Heidelberg London

Springer is part of Springer Science+Business Media (www.springer.com)

Foreword

That health care creates and relies on massive information systems is of no surprise to anyone. This has long been understood to be inherent to the management of a myriad of subsystems involved, from electronic medical records to billing and every thing in-between. Less understood, albeit still not a surprise, the aggregation of these subsystems is often an ad hoc exercise. This volume collects a treasure trove of design exercises that span the wide span of those component systems. From long reaching transdisciplinary education planning, to the specifics of imaging systems, and the delicate balancing act of devising protocols for interoperability in heterogeneous environments, this collection offers a cross section of applications to health care through the use of formal analytical tools. As a consequence, this collection also has the effect of offering a cross-section of those methodological formalisms.

I could not help but reading these reports as a preparation for the gathering storm heading our way with the personalization of medicine. The volume and heterogeneity of data sources, and the complexity of the medical decision they inform will require that the ad hoc aggregation be raised to an entirely new level. The only way to prepare for the higher abstraction of semantic modeling of those future systems, while retaining their practical use as tools for design, is to be as thorough as possible with the components over which we have direct control. The case-studies and analysis collected here provide a comprehensive overview of where current technology is. Accordingly, this is just the sort of foundation from which we need to follow Alan Kay's often quoted recommendation that the best way to predict what the information systems for the personalized medicine future will be like is indeed to invent them.

Professor/Director   Jonas S. Almeida, PhD
Division of Informatics
Department of Pathology
University of Alabama at Birmingham

Preface

No book is the sole product of its authors. There are always supporting and facilitating individuals who deserve major credit. If a book is composed of an integrated set of chapters aimed at conveying a single focus, the number of contributing individuals is significantly increased. Therefore, we would like to thank all contributing authors for their dedication and patience, and in some cases, enduring multiple edits and revisions to maintain a common format and seamless presentation. Without their dedication and hard work this book would not have been completed. A special thanks goes to one of the editors, Dr. Varadraj P. Gurupur, who conceived the idea and mobilized us as well as many of the contributors. As key next generation leader, Dr. Sang C. Suh negotiated the contract and played a significant supporting role throughout the process. We also extend our thanks to Society for Design and Process Science (SDPS: www.sdpsnet.org). Many of the authors are members of SDPS and had presented some of the materials in these chapters at prior SDPS conferences. Therefore, we would like to offer our sincere gratitude to all of you; the chapter contributors, Springer editors, numerous reviewers, and the SDPS Board of Directors. More importantly, we extend our gratitude to the SDPS family and friends, who are the main force and energy behind all our community oriented efforts.

The planning for this book germinated during the SDPS-2010 conference when we invited the SDPS family and friends to attend the Society's 15th Anniversary celebration in Dallas, Texas. Friday, June 11, 2010, the last day of the conference, was completely dedicated to the Next Generation activities. The day started with the Next Generation Keynote Address of Dr. Aditya Mathur, Department Head of Computer Science at Purdue University. The next was an informal panel of our distinguished seniors: Dr. C.V.Ramamoorthy and Dr. Oktay Sinanoglu and others who offer guidance and inspiration to next generation participants. This event immediately followed with panel participants from the next generation, namely Dr. Hironori Washizaki, Dr. Herwig Unger's representative, Dr. Yong Zeng, Dr. John N. Carbone, Dr. Sang Suh, Dr. Ali Dogru, Dr. Radmila Juric, Dr. Fan Yang, and Dr. Peng Han. All these interactions are characteristic of the traditional openness and sincerity of past SDPS conferences but with a goal to develop ideas for the next generation. The afternoon began with Jim Brazell, President, Ventureramp, Inc., who conducted the Next Generation Facilitation Workshop. During this workshop many ideas for the future of SDPS were extended. Dr. Gurupur suggested the idea of this book and followed through with the support of Dr. Suh to complete the project. You are holding in your hands the fruits of a year-long effort of hard work and dedicated concentration. This book represents the first volume of a series of volumes to be produced on many topics concerning transformative integration—a subject dear to the SDPS family, friends, and supporters. Therefore, we wish to diverge a moment to present a short history of SDPS in this first volume.

HISTORY AND THE FUTURE

Let's start with a recollection of how SDPS vision has started. During the summer of 1989 Dr. Tanik and Dr. Bernd Krämer met at Naval Postgraduate School, where Dr. Richard Hamming, a founder of coding theory provided intellectual support for *The Mathematical Foundations for Abstract Design* book Dr. Tanik was working on. The book eventually got published in 1991 under the name Fundamentals of Computing for Software Engineers. During the 1994 Systems Integration Conference in Brazil organized by one of the founders of SDPS Dr. Fuad Sobrinho, fate brought together Dr. David Gibson and Dr. Tanik. They had just arrived at the airport and were waiting for the conference bus. They did not know each other but Dr. Tanik initiated a conversation and started to expound the plans for establishing a new society which had emerged from a small group earlier that summer. This new society he described would focus on transdisciplinary science. An early version of the star-shaped logo was already designed. Fundamental nature of process was suggested by Dr. R. T. Yeh and the name "Abstract Design and Process" was in the works. Dr. Gibson informed Dr. Tanik that he should visit with Dr. George Kozmetsky and explain these ideas to him upon their return. Eventually, Dr. Gibson, Dr. Kozmetsky, and Dr. Tanik met at IC²Institute in Austin. The beginnings of SDPS were already taking shape at that time, including the original name, original star-shaped logo, bylaws, and basic concepts. The logo represents information integration through communication. Little did Dr. Tanik know, but both he and Dr. Kozmetsky had a mutual friend, Dr. C.V. Ramamoorthy. After some discussions during that first meeting, Dr. Kozmetsky said, "Tell Ram to call me," indicating Dr. C.V. Ramamoorthy. It was this connection along with the support of many people that triggered the launch of SDPS. The first SDPS conference was held in December, 1995 at the IC² facilities in Austin and Dr. Tanik, at that time was a member of the Department of Electrical and Computer Engineering under the chairmanship of Dr. Steven Szygenda. Eventually, SDPS was established by the founding board, composed of late Dr. Kozmetsky (chairman), Dr. C. V. Ramamoorthy, Dr. R. T. Yeh, Dr. A. Ertas, and Dr. M. Tanik. SDPS was incorporated in the State of Texas on September 6, 1995 as a non-profit organization and Dr. R. T. Yeh was elected as the first and founding president of SDPS. The current Society for Design and Process Science (SDPS) board is composed of Dr. C. V. Ramamoorthy, Dr. R. T. Yeh (honorary member), and Dr. B. Krämer, and Dr. M. Tanik.

The second impetus for furthering the transformative vision of SDPS happened in the year 2000. During SDPS-2000 Keynote Speech, Dr. Herbert A. Simon is quoted as saying:

...Today, complexity is a word that is much in fashion. We have learned very well that many of the systems that we are trying to deal with in our contemporary science and engineering are very complex indeed. They are so complex that it is not obvious that the powerful tricks and procedures that served us for four centuries or more in the development of modern science and engineering will enable us to understand and deal with them. We are learning that we need a science of complex systems, and we are beginning to construct it...

With this direction and our accumulated experience, the SDPS governing board in 2000 established the Software Engineering Society (SES) and a publishing arm of SDPS. Neither SES nor the publishing arm of SDPS fully served their intended purposes partly due to our self-reliant attitudes. Eventually, our publishing arm split to a smaller entity at the end of 2008 by a SDPS board decision. As we completed our fifteen years, it is time to rededicate ourselves to the notion of worldwide, sustainable, and transformative goals of SDPS and SES. This is where we need the energy, creativity, and high aspirations of the next generation. Furthermore, SDPS should play a leading role in the development of the processes for next generation transformative knowledge dissemination. The time for knowledge dissemination with classical book methods and traditional journal publications is passing. The next generation would play a critically leading role in this arena with new dissemination initiatives. This edited volume is a small step in this direction.

In the year 2000, then SDPS President, Dr. Tanik, delivered the following message in his opening address:

During the last five years, SDPS greatly enjoyed your support and intellectual participation. What makes SDPS successful is the combination of all your individual efforts and participation. That means SDPS is really YOU...nothing more nothing less. SDPS = YOU. Actually for us, YOU is also an acronym that stands for Young-Organized-Unspoiled. In other words, SDPS is Young and does not carry the burden of old societies. SDPS is organized worldwide with truly international and dedicated members. SDPS is Unspoiled, meaning pure, refreshing, sparkling, uncontaminated, and whole. We are now ready to expand our organization, as we stand poised to make a great impact in the new millennium. Hence, the collective knowledge and intellectual energy each of YOU contribute at every conference is the driving force behind the ascent of SDPS.

SDPS stayed the course for another decade and persevered, but it is now time to advance to new horizons. We have carried the light of Dr. Kozmetsky, who had established one of the first, if not the first, comprehensive and practical transdisciplinary center complete with an educational component: IC^2 Institute. Dr. Kozmetsky also provided a direction for expansion of SDPS fifteen year ago. In his address published in Transactions of SDPS Vol-1 No-1 (1996) he emphasized that

...We need to develop a step-by-step generational road map that sets forth the opportunities to commercialize our scientific and technological breakthroughs as well as their timing. Equally important, the road map must set forth the commercialization process including their global initiatives.

On this vision, we now recognize that the time has come for us to actively seek collaboration from our colleagues in many disciplines in order to implement our goals. Along these lines, thanks to the founding leaders of SDPS such as Nobel Laureate Dr. Herbert A. Simon, distinguished professors Dr. C.V. Ramamoorthy and Dr. R. T. Yeh, as well as dedicated worldwide membership, including various other Nobel Laureates such as Dr. Ilya Prigogine and Dr. Steven Weinberg, we are extending the vision of integration of business, engineering, and science processes towards transformative goals. Again this book provides a first small step towards these goals for the next generation.

The book composed of twenty two chapters organized in two sections. The focus of first section composed of twelve chapters is medical processes.These include: Medical Devices Design Based on EBD: A Case Study; The Great Migration: Information Content to Knowledge Using Cognition Based Frameworks; Design and Process Development for Smart Phone Medication Dosing Support System and Educational Platform in HIV/AIDS-TB Programs in Zambia; Image-Based Methodologies and their Integration in a Cyber-Physical System for Minimally Invasive and Robotic Cardiac Surgeries; Implementing a Transdisciplinary Education and Research Program; JIGDFS: A Secure Distributed File System for Medical Image Archiving; Semantic Management of the Submission Process for Medicinal Products Authorization; Sharing Healthcare Data by Manipulating Ontological Individuals; Sharing Cancer Related Genes Research Resources on Peer-to-Peer Network; IHCREAD: An automatic Immunohistochemistry Image Analysis Tool; A Systematic Gene Expression Explorer Tool for Multiple and Paired Chips; and On Solving Some Heterogeneous Problems of Healthcare Information Sharing and Interoperability Using Ontology Computing. The focus of the second section composed of eleven chapters is techniques and technologies. These include: Clinical Applications and Data Mining; Image Processing Techniques in Biomedical Engineering; Probabilistic Techniques in Bioengineering; A Complete Electrocardiogram (ECG) Methodology for Assessment of Chronic Stress in Unemployment; Coherence of Gait and Mental Workload; Efficient Design of LDPC Code for Wireless Sensor Network; Toward Multi-Service Electronic Medical Records Structure; The New Hybrid Method for Classification of Patients by Gene Expression Profiling; New Musculoskeletal Joint Modeling Paradigms; and The Role of Social Computing for Health Services Support.

The National Institutes of Health (NIH) states that a transformative research project is based on creative ideas and projects that have the potential to transform a field of science. This has also been the goal of SDPS since the beginning by following Dr. Simon's lead in integrative research on complex systems and unstructured problems. The current motto of SDPS, *Transformative research and education through transdisciplinary means,* is in line with the NIH and National Science Foundation (NSF) goals on transformative initiatives. We sincerely hope that this book will trigger the series of edited manuscripts lead by next generation SDPS leadership. The contributions by many authors from different disciplines will serve the goals of transformative research and education for the benefit of and civilizing effect on society.

March 2011   Tanik, Suh and Gurupur

Contents

Part I   Healthcare Systems

Part I
Healthcare Systems

Chapter 1
MEDICAL DEVICES DESIGN BASED ON EBD: A CASE STUDY

Suo Tan, Yong Zeng, and Aram Montazami

1 Introduction

The demands of designing safe and effective medical devices have significantly increased for the past decades. For example, Long [4] expressed a demand of improving current orthopedic medical devices by illustrating and discussing the uses, general properties, and limitations of orthopedic biomaterials. The emergence of fraudulent devices drove the need for regulations [5]. Countries all over the world have established their laws and regulations to systematically manage the medical devices in their respective markets. In the United States, for example, to assure the safety and effectiveness of medical devices in its market, the FDA (Food and Drug Administration) has established three regulatory classes based on the level of control: Class I General Controls (with or without exemptions), Class II General Controls and Special Controls (with or without exemptions), and Class III General Controls and Premarket Approval. For Class I and some Class II devices, simple controls will suffice for FDA clearance. Class III medical devices are subject to quality system requirements and stringent adverse event reporting and post-market surveillance [6]. For companies that produce Class III medical devices, a premarket approval (PMA) will be required before the devices can be marketed. This is because the risk to the user or patient determines that a mass of trials have to be done before approval [7]. All device manufacturing facilities are expected to be inspected every two years.

Although governments are regulating the medical devices to ensure safety, it is estimated that 44,000 to 98,000 people die in hospital each year because of preventable medical errors, according to a report released from the Institute of Medicine [8]. This is more than the number of people who die annually from motor vehicle accidents, breast cancer, or AIDS [2]. It makes hospital-based errors alone the eighth leading cause of death in the United States [9]. Among different preventable

Y. Zeng (✉)
Montreal, Quebec, Canada
e-mail: zeng@ciise.concordia.ca

S. Suh et al. (eds.), *Biomedical Engineering*,
DOI 10.1007/978-1-4614-0116-2_1, © Springer Science+Business Media, LLC 2011

medical errors, a large portion is related to their misuse [2]. Rousselle and Wicks [10] presented the importance of customizing every aspect of the preparation process to the type of device and the study endpoints for pathologist, in order to avoid any error that might result in irreparable loss. According to CDRH [11], user-related errors can be divided into user-interface issue, user-environment issue, and user issue. Those user-related errors are usually a result of the ignorance of human factors in the design of medical devices [2]. Hence, human factors are significant aspects that designers should take in consideration to improve the safety and performance of medical devices.

Human Factors Engineering (HFE) is the science and the methods used to make devices easier and safer to use [12]. HFE improves the usability, performance, reliability, and user satisfaction of a product. Meanwhile, it reduces operational errors, training expense, and operator stress. It is often interchanged with the term "ergonomics". The inclusion of human factors in medical devices design will definitely help the designer consider the users' characteristics and hospital environment settings in a more effective manner [2]. The consideration of such factors may introduce massive overhead in the development of medical devices, but it is significant and beneficial. Extensive research on HFE in the field of medical devices design has been conducted [13–18]. Lin et al. [13] highlighted the benefits of HFE by the comparison of two simulations, one with HFE while another not. The results supported the idea that HFE improved patient safety in clinical setting. Gosbee [14] presented a case study with vulnerabilities of human factors design and its analyses to illustrate the crucial role of HFE in patient safety. Ginsburg [16] conducted a human factors evaluation to inform hospital procurement decision-making in selecting a general-purpose infusion pump. The author initialized a human-factor heuristic assessment of pumps according to four sets of criteria, and followed by a task analysis from designers' and users' feedback to rate their usability. Then the Human Factors Engineer visited different clinical areas to get participants' rates of the pumps within preset scenarios and the usability error. Based on the usability rates and usability error, a better decision was made for hospital procurement. However, there were limitations for this study, no one pump won in every aspect, trade-offs must be made, and cost efficiency was not considered. This study was only for user-end assessment, not at the design phase. Malhotra et al. [15] proposed a cognitive approach for medical devices design. By obtaining information related to medical errors and patient safety through investigations, the device design process could be customized and modified for better patient safety management. In this way, the end-production could be creative and successful in market competition. Kools et al. [19] proposed another cognitive method to contribute to the design of effective health education information. By using pictures, the end user could better understand the medical devices that were inherently less clear. They illustrated an intuitive and reasonable way to reduce the chance of potential misuse of medical devices. This supported the idea Berman [20] presented. Fairbanks and Wears [18] highlighted the role of design in medical devices from several of previous work. Technology, manufacturing, regulations and rules, human factors, those conditions come together to enable a medical device to be designed safe and reliable. Martin *et al.* [17] gave a comprehensive review on HFE (ergonomics) by applicable methods in medical devices design.

This paper aims to use the Environment-Based Design (EBD) approach to design a medical device by analyzing design requirements in terms of product environments. A brief introduction is given in section two followed by a section describing a case study on a particular medical device design process, with human factors considered prior to the design stage based on the EBD model. Conclusions are given in Section 4.

2   Environment-Based Design (EBD)

Environment-Based Design (EBD) [21] is a design methodology based on the recursive logic of design, proposed by Zeng and Cheng [22], and the axiomatic theory of design modeling developed by Zeng [23]. The logic of design captures the fundamental nature of the design process, reveals the logic underlying design creativity and leading to solutions of a design problem. The axiomatic theory of design modeling allows for the development of design theories following logical steps based on mathematical concepts and axioms. The theory of "Environment-Based Design" has been derived in 2004 [21]. The underlying concept is that a design problem is implied in a product system and is composed of three parts: the environment in which the designed product is expected to work, the requirements on product structure, and the requirements on performances of the designed product. The requirements on product structure and performance are related to the product environment. In addition, the product environment includes three major environments: natural, built, and human (shown in Figure 1). According to the EBD, a design process includes three main activities: environment analysis, conflict identification, and solution generation (see Figure 2).

2.1   Environment Analysis

A design problem is often described in an informal plain language whereas any scientific method usually is founded on a certain formal structure. In order to bridge

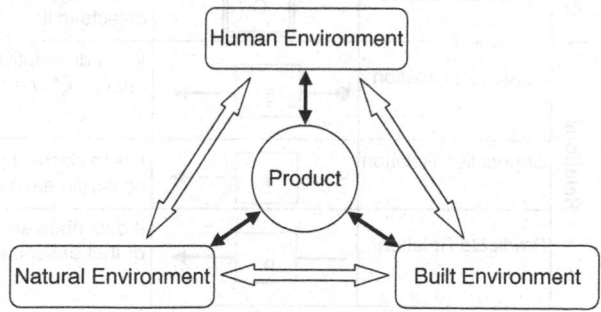

Figure 1. Three major environments for a product working in [2].

Figure 2. EBD: process flow [1].

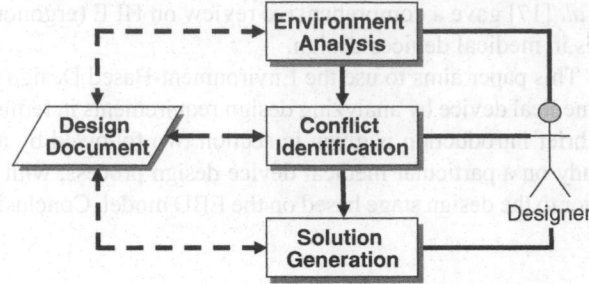

the gaps between the two representations, the Recursive Object Model (ROM) was proposed by Zeng [24] to conduct linguistic analysis. Using the two kinds of objects and three relations between objects shown in Table 1, a design problem can be transformed into a ROM representation of the design problem.

However, based on the transformation only, it is not sufficient for the designer to extract all the possible environment components and their relations. S/he has to make sure everything in the diagram is clear. This can be completed by object analysis. Related questions need to be asked to clarify the meaning of unclear objects by following the rules defined by Wang and Zeng [25]. The ROM diagram should be updated by iterative question-and-answer process.

In order to further verify the completeness of the extracted environment components and their relations, a roadmap was proposed as guidance for requirements modeling [3]. In this roadmap, requirements (structural or performance) were categorized by two criteria in terms of different partitions of product environment. One criterion classifies the product requirements by partitioning product environment in terms of the product life cycle whereas the other classifies them by partitioning the product environment into natural, built, and human environments. For their

Table 1. Elements of ROM diagram [24].

	Type	Graphic Representation	Description
Objects	Object	O	Everything in the universe is an object
	Compound Object	O	It is an object that includes at least two objects in it.
Relations	Constraint Relation	ξ	It is a descriptive, limiting, or particularizing relation of one object to another.
	Connection Relation	$/$	It is to connect two objects that do not constrain each other.
	Predicate Relation	ρ	It describes an act of an object on another or that describes the states of an object.

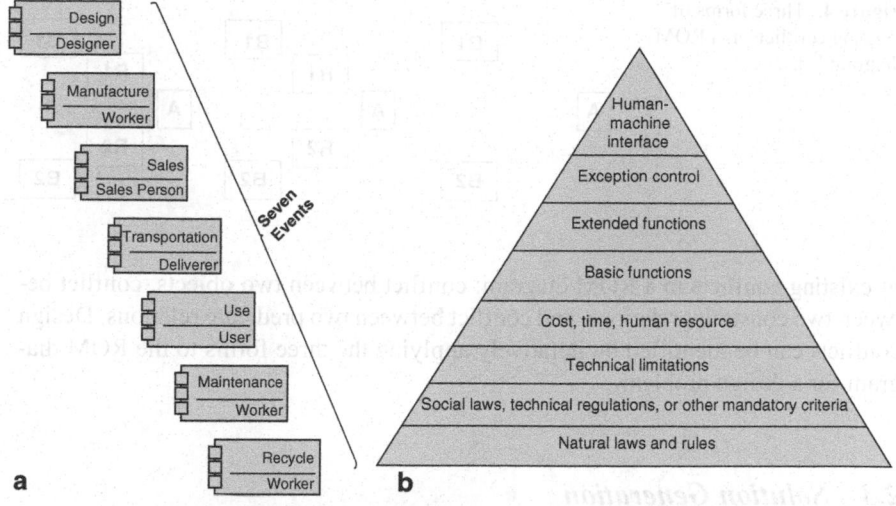

Figure 3. Two approaches to requirements classification [3].

case, Chen and Zeng [3] divided the life cycle into seven kinds of events, which were design, manufacture, sales, transportation, use, maintenance, and recycling (see Figure 3(a)). The requirements were fit into the seven events. At any time point of product life, one or some of these seven events might occur simultaneously or alternatively [3]. On the other hand, as illustrated in Figure 3(b), they classified the product requirements into eight levels: natural laws and rules, social laws and regulations, technical limitations, cost, time and human resources, basic functions, extended functions, exception control, and human-machine interface. In this pyramid-like model, requirements at the lower levels have higher priority in the development of a design solution while those products meeting the requirements at the highest level are said to be called high usability products. Among those requirements, the highest four levels of product requirements come from the human environment. They pertain to the purposes of the human use of the product [25]; the lowest level of product requirements comes from the natural environment; and the rest are the result of the built environment. Although product life cycle varies from product to product, by following the roadmap, the implicit requirements can be found, which gives the designer a very detailed "whole picture" about a design problem.

2.2   Conflict Identification

Identifying conflicts based on the "whole picture" is much easier than finding them from the natural language descriptions. As shown in Figure 4, there are three forms

Figure 4. Three forms of existing conflicts in a ROM diagram [1].

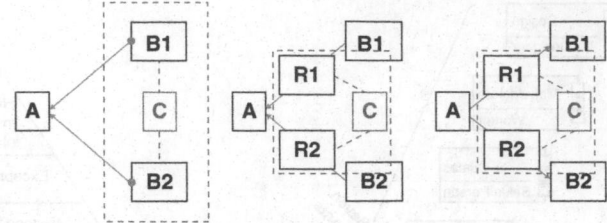

of existing conflicts in a ROM diagram: conflict between two objects, conflict between two constraint relations, and conflict between two predicate relations. Design conflicts can be identified by iteratively applying the three forms to the ROM diagram for a design problem.

2.3 Solution Generation

Before giving solutions to the identified conflicts, the designer should analyze them first. The principle is to find out the dependences among them as one conflict may be resulted from others. Tracing back to the root where the most basic conflict exists, and handling the root conflict first may eliminate some consequent conflicts that depend on it. Effective solutions can be generated in this manner. With the generated solutions, the ROM diagram, which represents the design problem, is updated again. Repeat the process until no more undesired conflicts exist.

The EBD is a logical and recursive process that aims to provide designers the right direction for solving a design problem. The three activities can be carried out simultaneously for an experienced designer. The next section will show the effectiveness of the EBD by applying it to analyze the design of a medical device design.

3 Case Study

Novatek International Inc. aims to provide innovative, leading edge and user-friendly LIMS (Laboratory Information Management System software) and other solutions that target the pharmaceutical, biotech and other health-care industries. In a collaborative research between Concordia and Novatek, the EBD was used for the design of algorithms and approaches to solving a few critical problems in Novatek's products. This section will provide a case study to show how the EBD can be used to support the development of a new product in Novatek. The description of the product is given as follows:

> Design an automated system to rapidly read various commercially available lateral flow tests using image recognition technology. The system should be effective and cost-efficient. It is for laboratory research. The solution should include a software kit which is compatible with the Novatek LIMS.

In the following subsections, we will follow the methods provided by the EBD to design the product.

3.1   Environment Analysis

From the design description, the corresponding ROM diagram is drawn as in Figure 5. To highlight the semantics implied in the problem, the ROM diagram is further simplified with a better layout as shown in Figure 6. It is clearly shown from the ROM diagram that the ultimate purpose of this problem is "design a system to read tests" in short (see the bold arrowed lines in Figure 6). Furthermore, we can observe five critical environment components and relations: system, read, tests, kit and LIMS, as they are the most constrained objects. However, there are also other environment components and relations. As previously described, the less obvious components can be found by object analysis. According to the rules of object analysis presented by [25], we can ask questions regarding the unclear constraints of the object "system" such as "What do you mean by *effective and cost-efficient*?", "What kind of *laboratory*?", "What do you mean by *automated*?".

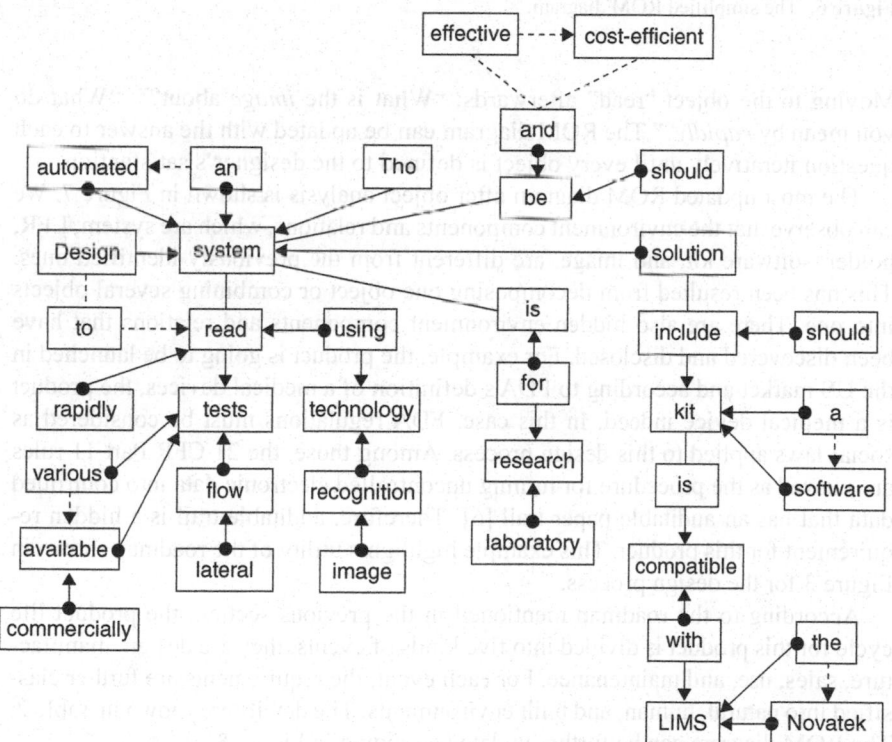

Figure 5. An initial ROM from the description.

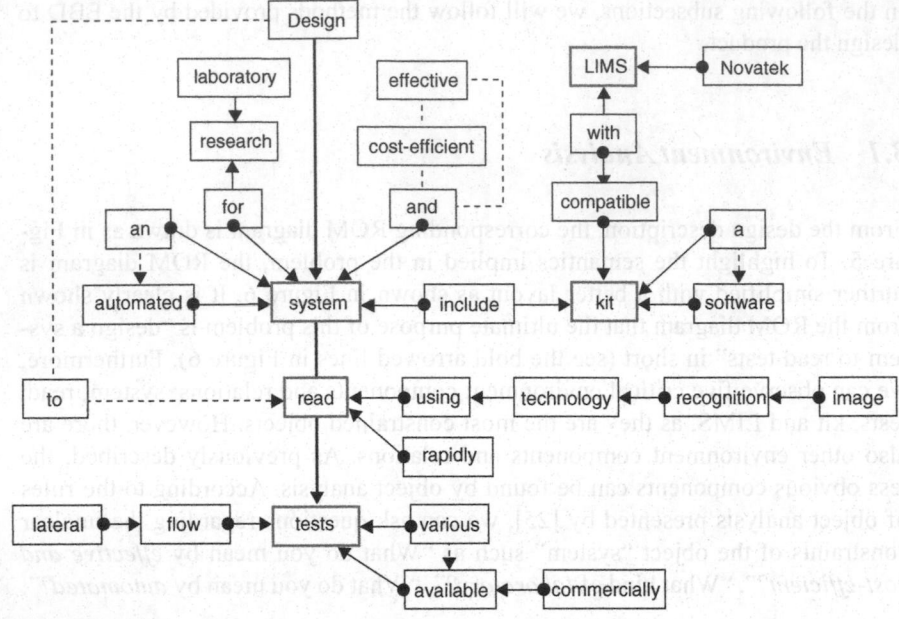

Figure 6. The simplified ROM diagram.

Moving to the object "read" afterwards: "What is the *image* about?", "What do you mean by *rapidly*?" The ROM diagram can be updated with the answer to each question iteratively until every object is defined to the designer's satisfaction.

The most updated ROM diagram after object analysis is shown in Figure 7. We can observe that the environment components and relations, which are system, LFR, holder, software kit, and image, are different from the previously identified ones. This has been resulted from decomposing one object or combining several objects into one. There are also hidden environment components and relations that have been discovered and disclosed. For example, the product is going to be launched in the US market and according to FDA's definition of a medical devices, the product is a medical device indeed. In this case, FDA regulations must be considered as social laws applied to this design process. Among those, the 21 CFR Part 11 rules are viewed as the procedure for turning uncontrolled electronic data into controlled data that has an auditable paper trail [6]. Therefore, auditable trail is a hidden requirement for this product. This example highlights utility of the roadmap shown in Figure 3 for the design process.

According to the roadmap mentioned in the previous section, the product life cycle for this product is divided into five kinds of events; they are design, manufacture, sales, use, and maintenance. For each event, the requirements are further classified into natural, human, and built environments. The details are shown in Table 2. The ROM diagram can be further updated as shown in Figure 8.

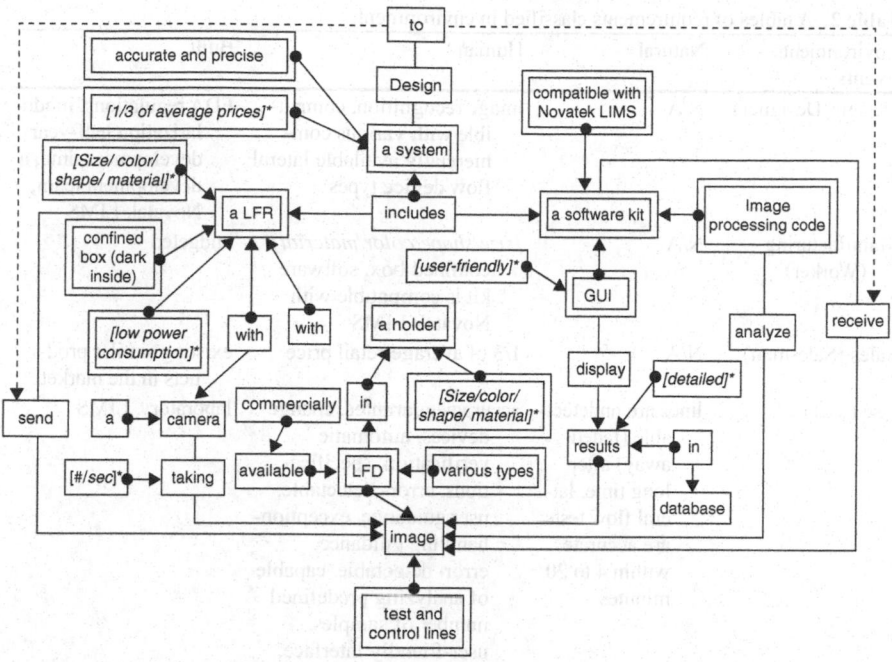

Figure 7. The updated ROM diagram after object analysis. LFR: lateral flow reader; LFD: lateral flow device; GUI: graphic user interface. (*[ ]**: Due to intellectual property protection, only conceptual descriptions are given.).

3.2   Conflict Identification and Solution Generation

As can be observed from Figure 8, the major environment components and relations are system, a LFR, a software kit, image, LFD, holder, results, confined box, module and GUI. There are six critical conflicts as listed in Table 3. Conflict 1 is a conflict resulted from the most basic level (see Table 2), and it is the most critical conflict for the basic functions that should be solved first. We can either solve it by reading the test result within 20 minutes after a test or by another method that is not limited by the test time. Finally, *[line-restore technology]** and *[sophistic image processing code]** are adopted to handle the conflict. Since conflict 1 is the root of conflict 2, the solution to conflict 1 may eliminate conflict 2 as well. It introduces an increase of cost, but it is a must. For conflict 3, we can put an illuminant inside. But this illuminant device increases the cost and consumes additional power, which in fact introduces two additional foreseeable conflicts which are unacceptable. Finally, a *[smart solution]** is adopted for this conflict without introducing any new conflicts at all. For conflict 4, an adjustable holder was proposed. The solution is designed for various LFD types from several manufacturers. By refining the code and designing a small drawer, conflicts 5 and 6 are handled. For the LFR, a USB cable can serve as both a power supply and a data transmission channel when connected

Table 2. A tables of requirements classified in environment.

Environments Events	Natural	Human	Built
Design (Designer)	N/A	image recognition, compatible with various commercially available lateral flow device types	FDA regulations, modularization half-year development time, a development team, Novatek LIMS
Manufacturing (Worker)	N/A	*[size/shape/color/material]**, confined box, software kit is compatible with Novatek LIMS	budgeted
Sales (Salesman)	N/A	1/3 of average retail price	existing similar products in the market
Use (user)	lines are undetectable (faded away) after long time, lateral flow tests are accurate within 4 to 20 minutes	accuracy guarantee, change devices, automatic verification, specifications, error- detectable, user guidance, exception-handling guidance, error- detectable, capable of analyzing predefined number of samples, user-friendly interface, definable user access right, independent audit trail, extendibility, low power consumption, capable of detecting errors automatically	laboratory, LIMS
Maintenance (Technician)	N/A	software update automatically, replace the damaged module	warranty

to a computer. So far, the most critical conflicts are handled without bringing any undesired problem. We then repeat all the steps and add the solutions back to the ROM diagram, until there are no more unacceptable conflicts.

The product to be designed is a medical device, which requires higher level of "safety and effectiveness" by FDA. As the LFR itself is not harmful for any living beings, much attention has been focused on the effectiveness and efficiency of the system, which improves the software. Based on an updated ROM diagram, we can clearly figure out what kinds of improvements are needed and how they improve product quality. As can be seen in the most updated ROM (Figure 8), many efforts were made to ensure that interface issue, user-environment issue, and user issue are thoroughly considered.

With continuous communication with the company, the manufacture agent and customers, the EBD can be recursively applied in the design process to make it better. This is indeed confirmed by the actual process happened in the company. After

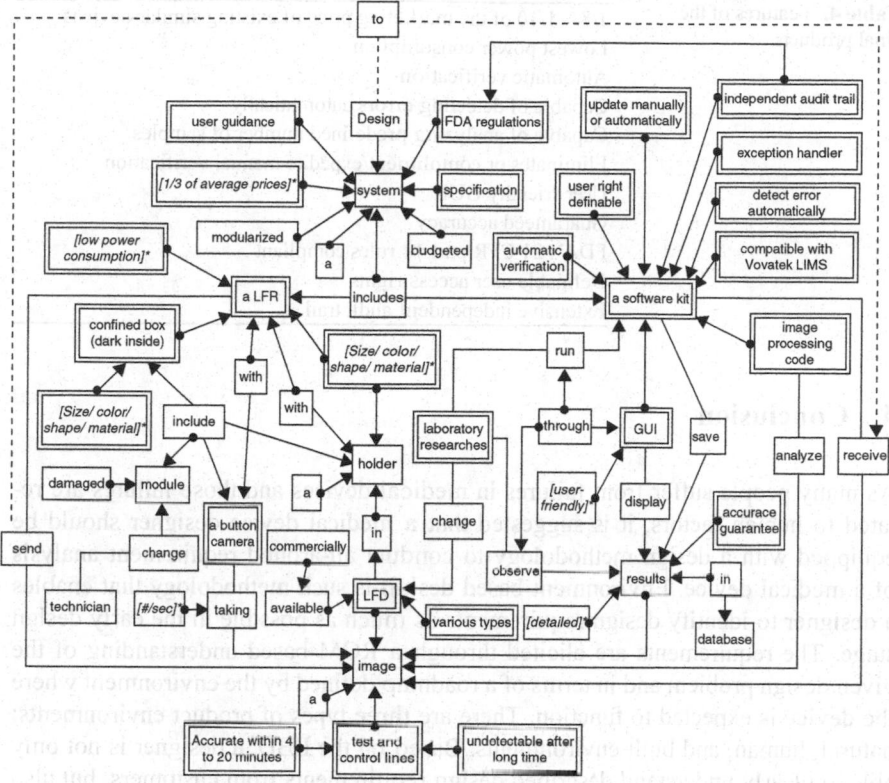

Figure 8. A further updated ROM diagram.

four prototypes, the product was finally released to the worldwide market. The design process in this paper does not dig too deep into the details due to intellectual property protection. However, it includes HMF as we can see from the Table 2. The effectiveness and efficiency is obvious. As a matter of fact, all the new product features at its release can be naturally developed following the process presented in this paper (see Table 4). Further improvements needed for the product can also be identified from the latest ROM diagram.

Table 3. Conflicts identified from Figure 8.

Number	Conflict	
1	Analysis the image with lines	The lines are undetectable
2	Accurate within 4 to 20 minutes	Accurate and prices guarantee
3	With a camera taking picture	Confined box (dark inside)
4	A holder	various LFD types
5	An image processing code	Image of various LFD types
6	Researcher change LFD	Confined box

Table 4. Features of the final products.

1/8 – 1/10 of the marketing price of existing similar devices
Lowest power consumption
Automatic verification
Capable of detecting errors automatically
Capable of analyzing predefined number of samples
Eliminates or compliment/expedite manual verification
User friendly GUI
Guaranteed accuracy
FDA's 21 CFR part 11 rules compliant
Definable user access rights
Extensive independent audit trail

4 Conclusion

As many people suffer from failures in medical devices and those failures are related to human factors, it is suggested that a medical device designer should be equipped with a design methodology to conduct all-around requirement analysis of a medical device. Environment-based design is such methodology that enables a designer to identify design requirements as much as possible in the early design stage. The requirements are elicited through a ROM based understanding of the given design problem and in terms of a roadmap defined by the environment where the device is expected to function. There are three types of product environments: natural, human, and built environments. Based on the EBD, a designer is not only able to clearly understand described design requirements from customers, but also able to discover hidden requirements. The presented case study shows how EBD is applied to a medical device design problem, which shows that the EBD provides an effective and efficient method for new product development.

The ongoing and future work of this research includes the development of tools and technologies that support the more effective application of the EBD to industries. Those technologies include a robust ROM analysis software systems that can automatically generate ROM diagrams from design text, semantics extraction algorithms that capture the right meaning of the current design problem, and a software tool supporting the entire EBD process.

Acknowledgements This work is partially supported by NSERC through its Discovery Grant program (Grant number RGPIN 298255).

References

1. Zeng, Y., *Environment-Based Design (EBD), The 2011 ASME International Design Engineering Technical Conferences (IDETC) and Computers and Information in Engineering Conference (CIE)*, Washington, DC, August 28–31, 2011, DETC2011–48263.

2. Chen, M., Z. Chen, L. Kong, and Y. Zeng, *Analysis of medical devices design requirements.* J. Integr. Des. Process Sci., 2005. **9**(4): p. 61–70.
3. Chen, Z.Y. and Y. Zeng, *Classification of product requirements based on product environment.* Concurrent Engineering, 2006. **14**(3): p. 219–230.
4. Long, P.H., *Medical devices in orthopedic applications.* Toxicologic Pathology, 2008. **36**(1): p. 85–91.
5. Rados, C. *Medical device and radiological health regulations come of age.* 2006 [cited Nov. 2010]; The Centennial Edition Available from: http://www.fda.gov/AboutFDA/WhatWeDo/History/ProductRegulation/MedicalDeviceandRadiologicalHealthRegulationsComeofAge/default.htm.
6. Pisano, D.J. and D.S. Mantus, eds. *Fda regulatory affairs: A guide for prescription drugs, medical devices, and biologics.* 2rd ed. 2008, Informa Healthcare: New York. 464.
7. Schuh, J.C.L., *Medical device regulations and testing for toxicologic pathologists.* Toxicologic Pathology, 2008. **36**(1): p. 63–69.
8. Kohn, L.T., J.M. Corrigan, and M.S. Donaldson, eds. *To err is human: Building a safer health system.* 1 ed. 1999, National Academies Press: Washington, D.C. 287.
9. Altman, D.E., C. Clancy, and R.J. Blendon, *Improving patient safety – five years after the iom report.* New England Journal of Medicine, 2004. **351**(20): p. 2041–2043.
10. Rousselle, S. and J. Wicks, *Preparation of medical devices for evaluation.* Toxicologic Pathology, 2008. **36**(1): p. 81–84.
11. CDRH. *Medical device use-safety: Incorporating human factors engineering into risk management.* 2000 [cited Nov. 2010]; Available from: http://www.fda.gov/downloads/MedicalDevices/DeviceRegulationandGuidance/GuidanceDocuments/ucm094461.pdf.
12. FDA. *Why is human factors engineering important for medical devices.* 2001 [cited Nov. 2010.
13. Lin, L., K.J. Vicente, and D.J. Doyle, *Patient safety, potential adse drug events, and medical device design: A human factors engineering approach.* Comput. Biomed. Res., 2001.ver **34**(4): p. 274–284.
14. Gosbee, *Human factors engineering and patient safety.* Qual. Saf. Health Care, 2002. **11**(4): p. 352–354.
15. Malhotra, S., A. Laxmisan, A. Keselman, J. Zhang, and V.L. Patel, *Designing the design phase of critical care devices: A cognitive approach.* J. of Biomedical Informatics, 2005. **38**(1): p. 34–50.
16. Ginsburg, G., *Human factors engineering: A tool for medical device evaluation in hospital procurement decision-making.* J. of Biomedical Informatics, 2005. **38**(3): p. 213–219.
17. Martin, J.L., B.J. Norris, E. Murphy, and J.A. Crowe, *Medical device development: The challenge for ergonomics.* Applied Ergonomics, 2008. **39**(3): p. 271–283.
18. Fairbanks, R.J. and R.L. Wears, *Hazards with medical devices: The role of design.* Annals of emergency medicine, 2008. **52**(5): p. 519–521.
19. Kools, M., M.W.J. van de Wiel, R.A.C. Ruiter, and G. Kok, *Pictures and text in instructions for medical devices: Effects on recall and actual performance.* Patient education and counseling, 2006. **64**(1): p. 104–111.
20. Berman, A., *Reducing medication errors through naming, labeling, and packaging.* J. Med. Syst., 2004. **28**(1): p. 9–29.
21. Zeng, Y., *Environment-based formulation of design problem.* J. Integr. Des. Process Sci., 2004. **8**(4): p. 45–63.
22. Zeng, Y. and G.D. Cheng, *On the logic of design.* Design Study, 1991. **12**(3): p. 137–141.
23. Zeng, Y., *Axiomatic theory of design modeling.* J. Integr. Des. Process Sci., 2002. **6**(3): p. 1–28.
24. Zeng, Y., *Recursive object model (ROM)-modelling of linguistic information in engineering design.* Comput. Ind., 2008. **59**(6): p. 612–625.
25. Wang, M. and Y. Zeng, *Asking the right questions to elicit product requirements.* Int. J. Comput. Integr. Manuf., 2009. **22**(4): p. 283–298.

Chapter 2
THE GREAT MIGRATION: INFORMATION CONTENT TO KNOWLEDGE USING COGNITION BASED FRAMEWORKS

John N. Carbone and James A. Crowder

1 Introduction

Research shows that generating new knowledge is accomplished via natural human means: mental insights, scientific inquiry process, sensing, actions, and experiences, while context is information, which characterizes the knowledge and gives it meaning. This knowledge is acquired via scientific research requiring the focused development of an established set of criteria, approaches, designs, and analysis, as inputs into potential solutions.

Transdisciplinary research literature clearly argues for development of strategies that transcend knowledge of any one given discipline and that enhance research collaboration. Increasingly, this cross-domain research is more commonplace, made possible by vast arrays of available web based search engines, devices, information content, and tools. Consequently, greater amounts of inadvertent cross-domain information content are exposed to wider research audiences. Researchers expecting specific results to queries end up acquiring somewhat ambiguous results and responses broader in scope. Therefore, resulting in a lengthy iterative learning process and query refinement, until sought after knowledge is discovered. This recursive refinement of knowledge and context occurs as user cognitive system interaction, over a period in time, where the granularity of information content results are analyzed, followed by the formation of relationships and related dependencies. Ultimately the knowledge attained from assimilating the information content reaches a threshold of decreased ambiguity and level of understanding, which acts as a catalyst for decision-making, subsequently followed by actionable activity or the realization that a research objective has been attained.

Additionally, underlying clinical decision-making is a great concern for handling ambiguity and the ramifications of erroneous inferences. Often there can be serious consequences when actions are taken based upon incorrect recommendations and can influence clinical decision-making before the inaccurate inferences can be

J. N. Carbone (✉)
Dallas, TX, USA
e-mail: John_N_Carbone@raytheon.com

S. Suh et al. (eds.), *Biomedical Engineering*,
DOI 10.1007/978-1-4614-0116-2_2, © Springer Science+Business Media, LLC 2011

detected and/or even corrected. Underlying the data fusion domain is the challenge of creating actionable knowledge from information content harnessed from an environment of vast, exponentially growing structured and unstructured sources of rich complex interrelated cross-domain data.

Therefore, this chapter addresses the challenge of minimizing ambiguity and fuzziness of understanding in large volumes of complex interrelated information content via integration of two cognition based frameworks. The objective is improving actionable decisions using a Recombinant Knowledge Assimilation (RNA) framework integrated with an Artificial Cognitive Neural Framework (ACNF) to recombine and assimilate knowledge based upon human cognitive processes which are formulated and embedded in a neural network of genetic algorithms and stochastic decision making towards minimizing ambiguity and maximizing clarity.

2   Knowledge

Nonaka and Takeuchi [1], when describing how Japanese companies innovate as knowledge creating organizations, described two types of knowledge: tacit and explicit. Tacit knowledge is personal and context-specific. Explicit knowledge is knowledge codified in books, journals and other documents for transmittal. Additionally, Nonaka [2] prescribed how dynamic organizational creation of knowledge needs to be strategically collected, understood, and managed across the entire company's organizational structure as intellectual capital. Knowledge theorist Polanyi and Sen [3], in describing what he called the "Tacit Dimension," used the idea of tacit knowledge to solve Plato's "Meno's paradox," that deals with the view that the search for knowledge is absurd, since you either already know it or you don't know what you are looking for, whereby you can not expect to find it. The author argued that if tacit knowledge was a part of knowledge then "we do know what to look for and we also have an idea of what else we want to know," therefore personal and context-specific knowledge must be included in the formalization of all knowledge.

Renowned fuzzy logic theorist Zadeh [4], described tacit knowledge as world knowledge that humans retain from experiences and education, and concluded that current search engines with their remarkable capabilities do not have the capability of deduction, that is the capability to synthesize answers from bodies of information which reside in various parts of a knowledge base. More specifically Zadeh describes fuzzy logic as a formalization of human capabilities: the capability to converse, reason and make rational decisions in an environment of imprecision, uncertainty, and incompleteness of information.

Tanik and Ertas [5] described, knowledge as generated through mental insights and the scientific inquiry process, usually stored in written form, assimilated through mental efforts, and disseminated through teachings and exposure in the context of a disciplinary framework. Kim et al. [6] used a case study to develop an organizational knowledge structure for industrial manufacturing. Specifically, a methodology was developed for capturing and representing organizational knowledge as a six-step procedure, which ranged from defining organizational knowledge to creation of a knowledge map for validation. The defined knowledge was extracted from the process as three types: prerequisite knowledge before process execution, used knowledge during execution, and produced knowledge after execution. Spender [7] stated that universal knowledge true at all times is the highest grade that knowledge can attain. Alternatively, Gruber [8] when describing social knowledge systems on the web and their relationship to semantic science and services, defined knowledge as "collective knowledge" that is collaborated upon.

Lastly, when describing how science integrates with information theory, Brillouin [9] defined knowledge succinctly as resulting from a certain amount of thinking and distinct from information which had no value, was the "result of choice," and was the raw material consisting of a mere collection of data. Additionally, Brillouin concluded that a hundred random sentences from a newspaper, or a line of Shakespeare, or even a theorem of Einstein have exactly the same information value. Therefore, information content has "no value" until it has been thought about and thus turned into knowledge.

3 Context

Dourish [10] expressed that the scientific community has debated definitions of context and it's uses for many years. He discussed two notions of context, technical, for conceptualizing human action relationship between the action and the system, and social science, and reported that "ideas need to be understood in the intellectual frames that give them meaning." Hence, he described features of the environment where activity takes place [11].

Alternatively, Torralba [12] derived context based object recognition from real-world from scenes, described that one form of performing the task was to define the 'context' of an object in a scene was in terms of other previously recognized objects and concluded, that there exists a strong relationship between the environment and the objects found within, and that increased evidence exists of early human perception of contextual information. Dey [13] presented a Context Toolkit architecture that supported the building of more optimal context-aware applications, because, he argued, that context was a poorly used resource of information in computing environments and that context was information which must be used to characterize the collection of states or as he called it the "situation abstraction" of a person, place or object relevant to the interaction between a user and the application. Similarly, when describing a conceptual framework for context–aware systems, Coutaz et al. [14] concluded that context informs recognition and mapping by providing a structured, unified view of the world in which a system operates. The authors provided a framework with an ontological foundation, an architectural foundation, and an approach to adaptation, which they professed, all scale alongside the richness of the environment. Graham and Kjeldskov [15] concluded that context was critical in the understanding and development of information systems. Winograd [16] noted that intention could only be determined through inferences based on context. Hong and Landay [17] described context as knowing the answers to the "W" questions (e.g. Where are the movie theaters?). Similarly, Howard and Qusibaty [18] described context for decision making using the interrogatory 5WH model (who, what, when, where, why and how). Lastly, Ejigu et al. [19] presented a collaborative context aware service platform, based upon a developed hybrid context management model. The goal was to sense context during execution along with internal state and user interactions using context as a function of collecting, organizing, storing, presenting and representing hierarchies, relations, axioms and metadata.

4   Organization of Knowledge & Context

Rooted in cognition, in 1957 Newell et al. [20] and Simon [21] together developed models of human mental processes and produced General Problem Solver (GPS) to perform "means-end analysis" to solve problems by successively reducing the difference between a present condition and the end goal. GPS organized knowledge into symbolic objects and related contextual information which were systematically stored and compared. Almost a decade later Sternberg [22] described a now well-known paradigm called the Sternberg Paradigm where, observations of participants were taken during experiments to determine how quickly the participants could compare and respond with answers based upon the size and level of understanding of their knowledge organized into numerical memory sets. Sternberg Paradigm is known for (1) organizing knowledge and modifying context while using a common process for describing the nature of human information processing and (2) human adaptation based upon changes in context.

Similarly, Rowley and Hartley [23] described the development of knowledge as the organization and processing required to convey understanding, accumulated learning, and experience. Object Oriented Design (OOD), as defined by Booch [24] and Rumbaugh et al. [25], organized knowledge and attributes describing details of objects in the form of general objects of information, using a bottom-up approach, iteratively building its components and attributes through a series of decisions. Booch's more generalized design decisions occurred via five basic phases which he described as part of the macro processes of OOD: Conceptualization which established the core requirements, analysis which developed the desired behavior via a model, design which included various architectural artifacts, and evolution which was the core component responsible for iterative bottom-up development, and lastly maintenance which managed the spiral delivery of functional capability. The details Booch described in the micro processes of his definition of OOD were the critical design mechanisms which fleshed out design details to take the conceptualization phase requirements to an implementable solution. The micro process components were, namely, identify and classify the abstraction of objects, identify the semantic representations of the objects and classes which define them, identify via specialized OOD notation the relationships between the objects, and finally the specification of the interfaces, the physical implementation of the defined classes and runtime objects.

More recently, Gruber [8] described the collection of knowledge and context on the web as "collective intelligence." Gruber based his opinion on Elgelbart's [26] principle which stated the need for creating combined human-machine interactive systems which can boost the collective intelligence of organizations and society via automated harvesting of collected knowledge for collective learning. Specifically, Gruber added that true collective intelligence can emerge if aggregated information from the people is recombined to create new knowledge. Van Ittersum et al. [27] organized knowledge and context as individual stand-alone knowledge components in agricultural systems which can be linked using a software infrastructure. Finally, Ejigu et al. [19] defined the organization of knowledge and context as a process of collection and storage. Their work proposed what they described as a neighborhood based context-aware service platform which was user collaborative in nature, that managed the reusability of context resources and reasoning axioms, and shared computational resources among multiple devices in the neighborhood space. They used a semantic ontology based hybrid model known as EHRAM as the core data source from which they systematically collected and stored information content, reasoned upon with their reasoning engine and then disseminated via their interface manager to the user. The main components of EHRAM context model were used to model the information content sources as a set of hierarchies (H), set of entities (E), set of entity relations (Re), set of attribute relations (Ra), set of axioms (A) and set of metadata (M). Hence, the information data source content was collected and stored as the EHRAM layered context representation structure.

5 Presentation of Knowledge & Context

Trochim [28] described Concept Maps to present knowledge and context as structured conceptualization used by groups to collaborate thoughts and ideas. Described was the typical case in which concept maps are developed via six detailed steps: the "Preparation," which included the selection of participants and development of the focus for conceptualizing the end goal, such as brainstorming sessions and developing metrics, (e.g. rating the focus), the "Generation" of specific statements which reflected the overarching conceptualization, the "Structuring" of statements which described how the statements are related to one another, the "Representation" of statements in the form of a presented visual concept map, which used multidimensional scaling [29] to place the statements in similar proximity to one another and cluster analysis [30] which determined how to organize the presentation into logical groups which made sense, the "Interpretation" of maps which was an exercise in consensus building once the representation had been created; and finally the "Utilization" of maps which was described as a process by which the groups within the process collectively determine how the maps might be used in planning or evaluation of related efforts. Stated was that concept mapping encouraged groups to stay on task which then resulted relatively quickly into an interpretable conceptual framework. It also expressed the framework entirely in the language of the participants and finally yielded a graphic or pictorial product. The product simultaneously presented all major ideas and their interrelationships and often improved group or organizational cohesiveness and morale. Graph theory, was shown to be used within many disciplines as an approach to visually and mathematically present knowledge and context relationships, [31].

In Software Engineering, many traditional tools exist: Entity Relationship Diagrams (ERD), Sequence Diagrams (SD), and State Transition Diagrams (STD) which each present different knowledge and context about database, and systems [32]. More recently, Universal Modeling Language (UML) [33] and semantic and ontology based software development tools, as well as, descriptive Resource Description Framework (RDF) language [34], and Web Ontology Language (OWL) [35] were used extensively to create, store, and present knowledge and context, using shapes, lines, and text as relationships between objects of information. However, Ejigu et al. [19] argued that ontology tools were only good at statically presenting knowledge of a domain and that they were not designed for scalable capturing and processing dynamic information in constantly changing environments.

6   Representation of Knowledge & Context

Dourish [11] concluded that representation of knowledge and context is an ethno methodological problem of encoding and representing social motivation behind action and that translating ideas between different intellectual domains can be exceptionally valuable and unexpectedly difficult. One reason is that ideas need to be understood within the intellectual frames that give them their meaning, and therefore need to be sensitive to the problems of translation between the frames of reference. Additionally, he describes four assumptions which represent context in systems, first, context as a form of information which can be encoded and represented in software systems just as other information content, second, context is delineable and therefore for a set of requirements, context can be defined as activities that an application supports and it can be done in advance, third, context is stable and hence can vary representation from one software application to another but does not vary from instance to instance of an event, it was specific to an activity or an event. Lastly, Dourish concluded, that most importantly context is separable from the action or activity, since context described the features of the environment where the activity takes place, separate from the activity itself. Dourish proposed an interactional model of context, where the central concern with representing context was with the questions, "how and why" during interactions, do people achieve and maintain a mutual understanding of the context for actions.

Clinical psychologists, Polyn and Kahana [36] described that cognitive theories suggest that recall of a known item representation is driven by an internally maintained context representation. They described how neural investigations had shown

that the recall of an item represented in the mind is driven by an internally maintained context representation that integrated information with a time scale. Howard
and Kahana [37] stated that by linking knowledge items and context representations
in memory, one could accomplish two useful functions. First, one could determine
whether a specific item occurred in a specific list (episodic recognition). Second, one
could use a state of context to cue item representations for recall (episodic recall).

Alternatively, Konstantinou et al. [38] concluded that a common knowledge representation formalism ought to allow inference extraction, and proposed "Relation-
al.OWL," based tool to automate structural representation of knowledge ontology
to database mapping. Additionally, Ejigu et al. [19] made the argument that context
and the organization of it was missing from systems and is instead in the "head"
of the user, and proposes an ontology based structure using RDF representation
of knowledge and context with metadata attributes. Similarly, Zouaq et al. [39],
proposed a Natural Language Processing (NLP) solution which enables structured
representations of documents. They proposed a knowledge puzzle approach using
ontology based learning objects, semantic maps, and grammatical maps, which represented structure of context on the basis of using text relations. Finally, similar to
Trochim [28], Novak and Canas [40] described the structure of concept maps as a
mechanism for structural representation of knowledge and context.

7   Frameworks for Knowledge and/or Context

Outlining the need for frameworks which can analyze and process knowledge and
context, Liao et al. [41] represented context in a knowledge management framework comprising processes, collection, preprocessing, integration, modeling and
representation, enabling the transition from data, information and knowledge to
new knowledge. The authors also indicated that newly generated knowledge was
stored in a context knowledge base and used by a rule-based context knowledge-
matching engine to support decision-making activities. Gupta and Govindarajan
[42] defined a theoretical knowledge framework and measured the collected increase of knowledge flow out of multinational corporations based upon "knowledge
stock" (e.g., the value placed upon the source of knowledge). Pinto [43] developed
a conceptual and methodological framework to represent the quality of knowledge
found in abstracts. Suh [44] concluded that collaborative frameworks do not provide
the contents which go in them, therefore, content was discipline specific, required
subject matter experts, and clear decision making criteria. Additionally, Suh noted
that processes promoting positive collaboration and negotiation were required to
achieve the best knowledge available, and were characterized by process variables
and part of what is defined as the Process Domain. Finally, Ejigu et al. [19] created
a framework for knowledge and context which collected and stored knowledge as
well as decisions in a knowledge repository that corresponded to a specific context
instance. Subsequently, the framework evaluated the knowledge and context via a
reasoning engine.

8   Collection of Knowledge & Context

Llinas et al. [45], observed that the synthesis of combining two bits of information into knowledge fusion requires knowledge and pedigree/historical information, which was context. Rowley and Hartley [23] describe knowledge as learning accumulation, hence, to accumulate knowledge and context "collective intelligence" was used as described by Gruber [8]. Therefore, not only is effort required to observe, select, and physically take hold of information, but also necessary is the understanding that collected knowledge and context has a historical relationship to existing information. Gruber [8] states that collective intelligence emerge if data collected from all people is aggregated and "recombined" to create new knowledge. To form an understanding of the relationship between different knowledge and contexts when assimilating knowledge, the associated relationships can be written symbolically as knowledge K_i and the associated context relationship R_j where, $K_i (R_j)$ represents a recombination of knowledge and context and finally represents the assimilation storage into the core domain repository. This is depicted in knowledge assimilation Figure 1. Figure 1 depicts a conceptual search space where a user would search for discipline specific knowledge and context within the Information Domain. The combined knowledge and context is then assimilated in the Temporary Knowledge Domain into a storage space shown on the right of the equation, the Knowledge Domain, to store knowledge and context which has reached a threshold level in the mind of the assimilator.

9   Storage of Knowledge and Context

Today, existing databases housing vast bits of information do not store the information content of the reasoning context used to determine their storage [19]. The knowledge collection and storage formula was therefore developed to include and store relationship context along with knowledge, recursively. This means that,

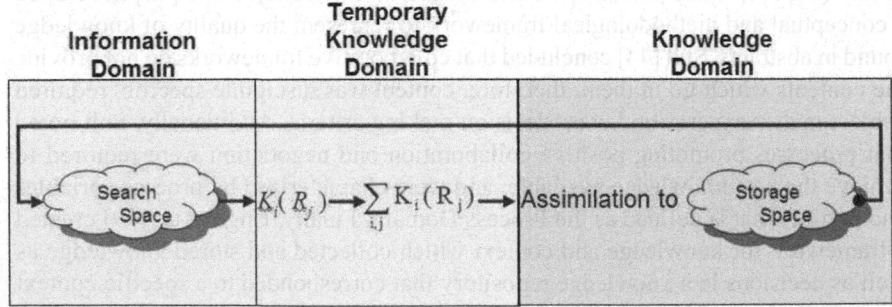

Figure 1. Recombinant Knowledge Assimilation Equation.

each act of knowledge and context pairing shown as in equation shown in Figure 1 $\sum_{i,j} K_i(R_j)$, recursively examined all of the previous relationships as they were re-combined into storage since they were all related and dependent on each other. Recursive refinement then occurred, per iteration of relationship pairing. Recursive refinement occurred when the user found what was looked for shown as $K_i(R_j)$, using interrogatives, (e.g. who, what when, where, why and how) [17–18]. The in-formation content contributing to finding the answer then has significant value and therefore, a higher degree of permanence in the mind of the stakeholder [46]. There-fore, the information content has reached a threshold where retaining the knowledge and context has become important.

The assimilation to storage can take physical and virtual form. Virtual storage can be described as the caching of a collection of temporary knowledge in the mind of the user per Ausubel et al. [47] along with a set of historical pedigree of precon-ceived/tacit or explicit knowledge and context per Nonaka [2] used to solve an issue at hand. Physical representations of assimilated stores are well known (e.g. librar-ies, databases, coin or philatelic collections.) However, whether virtual or physical, each unit of storage has a series of reasons or pedigree as to why it was collected and stored, or in the case of knowledge and context assimilation, why a knowledge and context relationship was created. For this result it is assumed that while knowl-edge and context are contemplated in the mind of the user [47], that knowledge and context are stored virtually until the point in time the user reaches the threshold where it is believed the virtual knowledge is of enough quality to become stored in a physical repository for someone else to see or use, or that a virtual memory constraint has been reached and thus the memory needs to saved physically so that it might not be lost if not captured.

10   Presentation of Knowledge and Context

Figure 2 represents a KRT. This approach for presentation of knowledge and con-text and was constructed to present five discrete attributes, namely, time, state, relationship distance, relationship value, and event sequence. The goal of a KRT is to map the dependencies of knowledge and related attributes as knowledge is

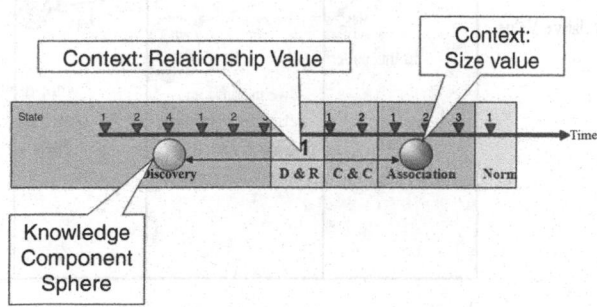

Figure 2. Knowledge Rela-tivity Thread.

developed from information content. In this figure, the timeline represented by the blue arrow from left to right, shows the events or state transitions in sequence and captures the decision points. During iteration of presentation of knowledge and context, intrinsic values were captured and placed close to each colored knowledge component. In Figure 2, these are represented as words under the cycles. The Basic Sentence Decomposition depicts how a KRT looks when it represents a sentence decomposed into pieces; in this case words. The red triangles, added next, depict a particular state for each iteration in the KRT development cycle. For emphasis, each colored sphere was built into the depiction and added in sequence to represent the fact that each word follows the other. Each icon represents each word of the sentence. The relative values in this Basic Sentence Decomposition between each sphere are perceived to be of the same value to each other. Therefore, the lines are the same distance as well. Since, this base representation depicted in Figure 2 can present time, state, and sequence, as well as, relationships, the challenge was addressed as described by Dourish [11] to create presentation of context which can visually capture and manage a continually renegotiation and redefinition of context as development of knowledge occurs over time.

Figure 3 shows a KRT presentation approach to comparing the knowledge and context between two distinct discipline abstracts. Specifically, for this example, Bioscience 1 abstract and Video Processing 1 abstract were compared to each other to find similarities, per the need as prescribed by Habermas [48] to have an original set of criteria to meet and by Ertas et al. [49] to find and integrate concepts and methods from other disciplines which share similar levels of analysis and finally by Trochim [28] which described the need to present knowledge and context so different groups can collaborate their different thoughts and ideas in a structured conceptualized manner. Therefore, a systematic approach was taken comparing and presenting the knowledge and context of each aggregated object to the other. As part of this enhanced systematic approach, each aggregated object in each ab-

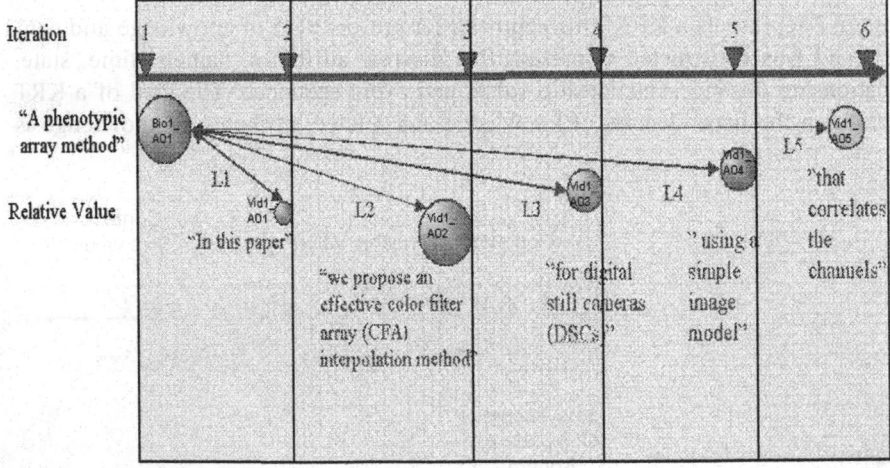

Figure 3. Comparing Research Journal Abstracts- KRT Approach.

stract is compared to each of the other aggregated objects in the other abstract. As this comparison occurred, the user captured each event in a log for every action and related reason which transpired during the systematic comparison. The details of the log are explained later in this paper. This logged information was used to help subsequent users gain a more complete understanding of the knowledge and context and thereby interpret a previous KRT collaboration presentation blueprint. The KRT visualization of this comparison shown in Figure 3 depicts the sequence of the aggregated objects that were compared. An important distinction about the observation of each comparison is that each was made from the perspective of the aggregated object being compared. This is defined conceptually as an analogy to Hibeller [50] where the concept of relating the motion of two particles is as a frame of reference and is measured differently by different observers.

Figure 3, is a snapshot in time, using simple length measures to show relative distance of a relationship which is described later in paragraph 0, for comparison of aggregated object 1 in the Bioscience 1 domain abstract or Bio1_AO1 compared to each aggregated object from Video Processing abstract 1 or Vid1_AO1 to Vid1_AO5. Iteration 1 shows Bio1_AO1="A phenotypic array method" solo. Iteration 2 shows Bio1_AO1 being compared to Vid1_AO1="In this paper." The relationship is not similar and therefore has little value and is presented by the smaller spheroid and distant relationship set namely to L1. By contrast, iteration 3 shows an equal size red spheroid showing an overlapping match was found (e.g. the word "array"). Meaning Bio1_AO1 has the word "array" in the text as does Vid1_AO2, thus presenting a change in relationship shown as a different length L2 as compared to Bio_AO1 and Vid1_AO1 (1). The reason why the relationship between Bio1_AO1 and Vid1_AO2 is not closer than L2 is that though the relationship has been found to be textually similar, until additional information content is gathered and understood as per Brillouin's [9] assertion that information has no value until it has been thought about, a final assertion can not be made that these two aggregated objects are exactly the same. Iteration 4 shows Bio1_AO1 compared to Vid1_AO3="for digital still cameras (DSCs)." The green spheroid is larger than the blue spheroid Vid1_AO1 because, at initial look, substantive information such as "digital still camera" presents additional information which might be relative to Bio1_AO1="A phenotypic array" when additional comparisons and knowledge and context are obtained. The distance of the relationship is therefore currently a bit further than that of Vid1_AO2 (L2), but closer than Vid1_AO1 which has little to no similarity, at this point, to Bio1_AO1. Lastly, Vid1_AO4 and Vid1_AO5 have similar attributes as Vid1_AO3 and therefore their knowledge and context relationship settings are similarly set.

11   Representation of Knowledge and Context

The representation of knowledge and context formula is introduced here and is presented by Equation (3-2). The independent results which follow are mathematical evaluations extended from Newton's law of gravitation shown in Equation (3-1). Newton's Law of Gravitation formula is,

$$F = G\frac{(M_1 M_2)}{r^2} \qquad \text{(Equation 3-1)}$$

where:

F is the magnitude of the gravitational force between the two objects with mass,

G is the universal gravitational constant,

M_1 is the mass of the first mass,

M_2 is the mass of the second mass, and

r is the distance between the two masses.

This equation was used as an analogy for the derivation of mathematical relationship between a basis made up of two objects of knowledge.

Abstracting Newton's Law of Gravitation as an analogy of Equation (3-1), representing relationships between two objects of knowledge using context, is written as Equation (3-2) shown below, which describes the components of the formula for representing relationships between two objects of knowledge using context:

$$A = B\frac{(I_1 I_2)}{c^2} \qquad \text{(Equation 3-2)}$$

A is the magnitude of the attractive force between the two objects of knowledge,

B is a balance variable,

I_1 is the importance measure of the first object of knowledge,

I_2 is the importance measure of the second object of knowledge, and

c is the closeness between the two objects of knowledge

Comparing the parameters of Equation (3-1) and Equation (3-2) F and A have similar connotations except F represents a force between two physical objects of mass M_1 and M_2 and A represents a stakeholder magnitude of attractive force based upon stakeholder determined importance measure factors called I_1, and I_2. As an analogy to F in Equation 3-1, A's strength or weakness of attraction force was also determined by the magnitude of the value. Hence, the greater the magnitude value, the greater the force of attraction and vice versa. The weighted factors represented the importance of the objects to the relationships being formed. The Universal Gravitational Constant G is used to balance gravitational equations based upon the physical units of measurement (e.g. SI units, Planck units). B represents an analogy to G's concept of a balance variable and is referred to as a constant of proportionality. For simplicity, no units of measure were used within equation (3-2) and the values for all variables only showed magnitude and don't represent physical properties (e.g. mass, weight) as does G. Therefore, an assumption made here is to set B to the value of 1.

For simplicity, all of these examples assume the same units and B was assumed to be one. The parameter c in Equation (3-2) is taken to be analogous to r in Equation (3-1). Stakeholder perceived context known as closeness c represented how closely two knowledge objects (KO) are related. Lines with arrows are used to present the closeness of the relationships between two pieces of knowledge presented as spheroids.

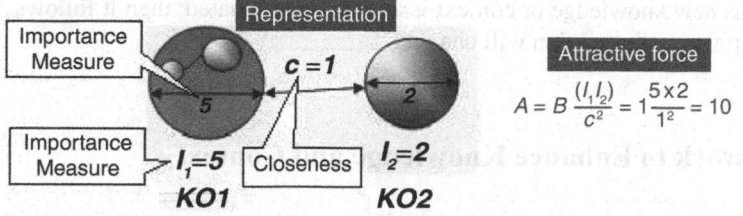

Figure 4. Representation of Knowledge Object and Context.

The representation of knowledge and context approach depicted in Figure 4 is a representative structure of knowledge and context as a snapshot in time for Bioscience 1 abstract. The first word of Bioscience 1 abstract is the word "A." "A" by itself has little meaning. However, it was still considered part of this abstract and was therefore marked as object of knowledge 1 (KO1) within the abstract. As the abstract was read and more information content was gained and understood, "A"'s knowledge value changed. Currently, all that is known at this juncture is that "A" described a singular entity and has foreshadowed that something will follow. Hence, that has some small value and creates cognitive structure in the mind of the "learner" per Ausubel et al. [47]. It is depicted in Figure 4 as knowledge object 1 (KO1) (e.g. red spheroid with the number 1) and mentally place only a small value on it for now because of our lack of knowledge. Next, as reading the abstract continued, the second word is found and marked as knowledge object 2 (KO2), "phenotypic." Figure 4, representing the knowledge and context of the mind of the learner now depicts KO1 and KO2, as related to each other. The word "A," or KO1 has a smaller spheroid than KO2, and therefore, structurally represents a smaller context of importance measure shown as a diameter, $I_1 < I_2$. The line distance between KO1 and KO2 structurally represents "closeness" or how closely related the objects are perceived to be to each other. The word "A," KO 1 has small relationship to KO 2. Hence, KO1's relationship to KO2 was characterized simply as residing within the same abstract and one of order sequence. Therefore, the knowledge objects remain further apart, shown as closeness or "c." Therefore, the snapshot in time shows a structural representation of knowledge relationship between two knowledge objects along with the context of magnitude importance value shown as the arrows representing the diameter magnitude of each knowledge object.

Using Equation (3-2), the value of the attraction force $A_{1\rightarrow2} = 5 \times 2$ divided by the relative closeness/perceived distance$^2 = 1$. Hence, the attraction force A in either direction was 10. The value of 10 is context which can be interpreted in relation to the scale. The largest possible value for attraction force A with the assumed important measure 1-10 scale is 100, therefore a force of attraction value of 10 was relatively small compared to the maximum. This means that the next stakeholder/ researcher understood that a previous stakeholder's conveyance was of small relative overall importance. However, the closeness value of 1 showed that the two objects were very closely related. Figure 4 therefore shows that when using Equation (3-2), if relationship closeness and/or perceived importance measure of the knowledge objects

change value, as new knowledge or context is added and evaluated, then it follows
that relationship force of attraction will change.

12   Framework to Enhance Knowledge and Context

The framework developed in this research to enhance knowledge and context is
shown in Figure 5 and was referred to as the Recombinant Knowledge Assimilation
(RNA). RNA and is made up of a combination of the organization of knowledge
and context, the presentation of knowledge and context, and the representation of
knowledge and context [19]. The three components make up the core pieces essential for building a knowledge and context framework [19, 41]. Cross discipline
domain research [21–22, 24, 27, 51] shows clearly that although all researchers use
their own flavor of unique rules, methodologies, processes and frameworks, they
use a core set of components for gathering, analyzing, organizing and disseminating
their work. Recently Liao et al. [41] and Ejigu et al. [19] defined these processes as:
collection, storage, presentation and representation.

13   RNA Flow Diagram

The RNA Flow Diagram shown in Figure 5 is shown to describe the flow of the
processes within the framework [19]. It is similar to the Liao et al. [41] framework
that collects, stores, presents and represents knowledge and context. The RNA flow
diagram comprised three major, discrete parts. First, "Content," which represents
all information content input into the flow diagram. Second, "Sub-Processes" for
synthesizing knowledge and context. Third, storage repositories known as pedigree
bins, where knowledge and context was stored during compilation. Compilation is
a path beginning from basic information content in the Information Domain, to the
Knowledge Domain, as described by Brillouin [9], where a set of initially "useless" information is "thought about" and turned into knowledge. This knowledge
becomes the collected pedigree knowledge and context, just as Gupta and Govindarajan [42] collected knowledge flow for measurement, for the next researcher, as
shown by the blue arrow leaving the Knowledge domain and feeding back into the
Information Domain in Figure 6.

In the RNA flow Diagram shown in Figure 5, each diamond shaped box represents a decision point. This is a critical point where a stakeholder of the process
contemplates the decision to be made using any previous knowledge components
acquired prior to making the decision as defined by Kim et al. [6]. Each red spheroid represents a sub-step within each of the larger components of the RNA process.
These red spheroids are used to identify an important portion of the process. Red
arrows signify action and green arrows represent "Yes" answers to a decision, hence
the red lines represent a stakeholder of the process performing an action such as,

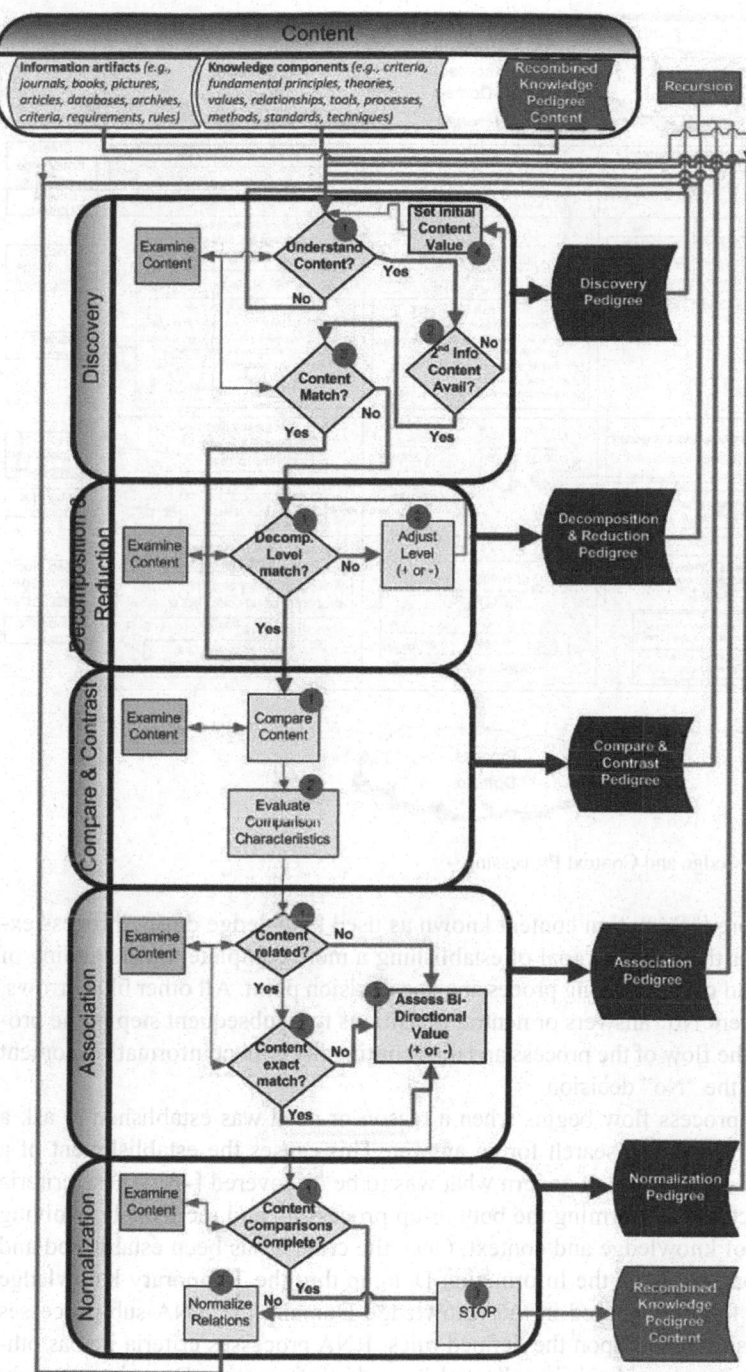

Figure 5.   RNA Flow Diagram.

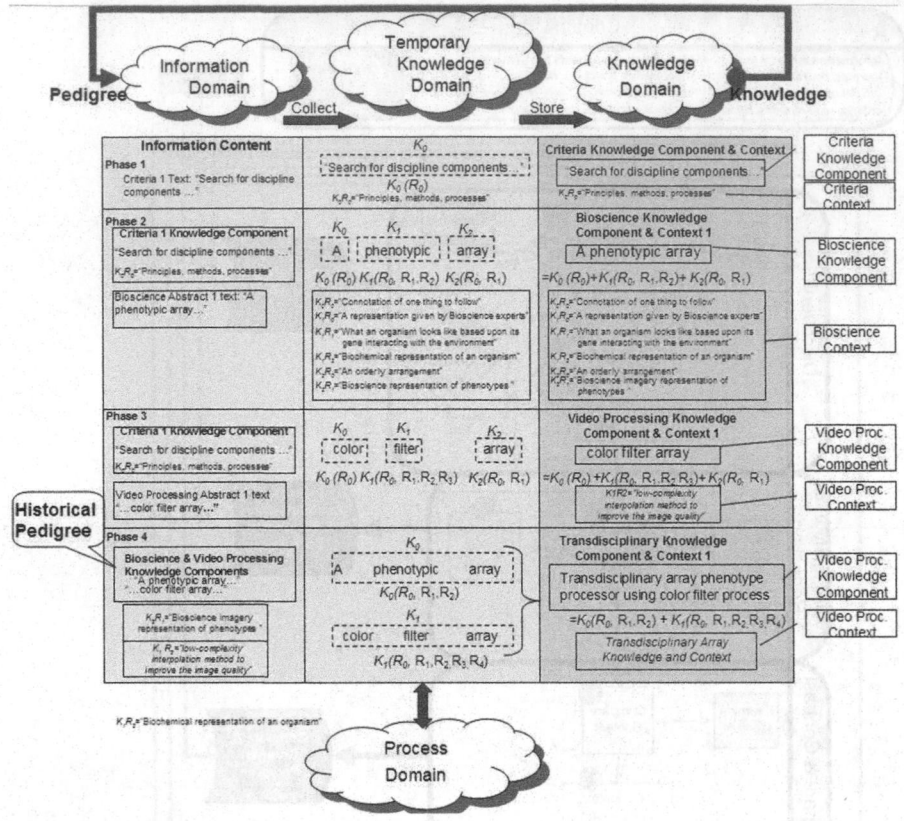

Figure 6. Knowledge and Context Processing.

collecting more information content known as used knowledge during process execution [6] for the eventual goal of establishing a more complete understanding of knowledge and context during processing at a decision point. All other blue arrows, represent either "No" answers or neutral transitions to a subsequent step in the process to track the flow of the process and thus continually collect information content used to make the "No" decision.

The RNA process flow begins when a reason or need was established to ask a question and to want to search for an answer. This causes the establishment of a set of criteria or rules which govern what was to be discovered [48]. These criteria govern the activity performing the bottom-up processing and recursively evolving the building of knowledge and context. Once the criteria has been established and understood passing from the Information Domain thru the Temporary Knowledge Domain and finally captured in the Knowledge Domain, the RNA sub-processes begin processing based upon the defined rules. RNA processes criteria just as other information content. Each is collected from the Information Domain, "thought about" [9] in the temporary Knowledge Domain and subsequently placed into the Knowledge Domain for use as shown in Figure 6.

The upper rounded box labeled "Content" represents all information content which can potentially be used when performing the steps of the RNA process to build knowledge components. This is the set of initially "useless" information built into knowledge, as described by Brillouin [9], and is represented by the information content under the Information Domain search space in Figure 6. Hence, when a stakeholder begins the process of examining information, it is the information content which was initially observed, using the senses, and then subsequently "thought about" and understood, via collecting, representing, presenting, and storing, until the stakeholder satisfies the desired threshold of understanding defined by the initiating criteria. The criteria were considered information content as well, since a set of criteria was established to setup rules to compare against until satisfied. The gathering and comparisons, shown by the red arrows in Figure 5, occur to the point where a stakeholder believes an understanding has been reached during each step in the process, just as Brillouin [9] defines knowledge as resulting from a certain amount of thinking. Therefore, the developed sub-Processes: Discovery, Decomposition and Reduction, Compare & Contrast, Association, and Normalization process information content based upon a set of initial criteria.

14   RNA Synthesis Sub-processes

The RNA common process contains five functional sub-processes, labeled Discovery, Decomposition and Reduction, Compare & Contrast, Association, and Normalization. These sub-processes synthesize knowledge and context within the framework down the left side of Figure 5. These sub-processes operate in the process domain [52] as shown in Figure 6. *Discovery* encompasses the review and understanding of existing knowledge and /or in the case of disciplines, the review of a discipline's fundamentals and/or First Principles. *Decomposition & Reduction* decomposes the domain knowledge into "bite size" digestible bits of information and reduces the representative domain knowledge to a core capability. *Compare & Contrast*, a cognitive examination process assimilating facts and information, comparing each to the other, looking for evolving associations, *Association* for establishing and assigning relationships between any two objects of information, and *Normalization* for functionally combining commonalities into a normalized form and validating the result. Finally, *recursion* is depicted as the blue domain knowledge feedback loops, which represents the iterative recursive refinement taking the knowledge gathered during each iteration and using it as input into the next iteration of the RNA process.

Since RNA's synthesis tasks, depicted in Figure 5, extend concepts from mature disciplines including Software Engineering. Specifically, recursion is shown by the feedback loops from each of the processes [24] [25]. Recursion is well suited for the goal of creating objects of information using a bottom-up approach, iteratively building its components and attributes through a series of decisions. Hence, RNA implements the mature bottom-up approach for developing knowledge and context

as discipline components, derived from discipline domain abstract readings and
the recursive nature of the process shown by the feedback loop in Figure 5 which
recombines knowledge and context.

15 Discovery

In the Discovery sub-process, the stakeholder must gather at least one additional
piece of information content to make a comparison. During the comparison process, the stakeholder was asking questions and developing answers, just as in the
Sternberg Paradigm [22]. However, the difference was that RNA developed and
retained empirical information during each specific step. Each question and answer
was developed and captured at each step. All thoughts regarding reasoning and the
information content used to develop the comparison were also captured at each
step. Consequently, the value the stakeholder placed upon each piece of information
content, shown in Discovery step 4, can be temporarily saved mentally or stored
physically to retain the context of the thoughts being developed. This was designated by all the dark blue arrows and boxes labeled (e.g. Discovery Pedigree). After
the first piece of Information Content has been observed, the flow diagram shows
that a stakeholder must have at least one other piece of information content in order
to form a comparison. Hence, the RNA process flow expands using a red arrow to
depict the setting of an initial value property for the first piece of content and then
continues back to Discovery step 1 to observe a 2nd piece of information content in
order to form a comparison.

Finally, if the stakeholder has found two pieces of content that was believed to be
an exact match and was exactly what has been searched for, then the flow diagram
resumes in the Association building block where a determination was made as to
the bi-directional value of force attraction of matching relationship pairs. If there
was not an exact match then the next Decomposition and Reduction building block
in the flow diagram was used to assist in determining whether there was simply an
inequality in the comparison, and the Decomposition and Reduction flow block as-
sists in rectifying that issue.

16 Decomposition & Reduction

The next step in the RNA Process was Decomposition and Reduction. This phase
extends and expands GPS [20], [53], used to solve problems by successively reducing the difference between a present condition and an end goal. This was important
because this section of the flow diagram was built so the stakeholder can establish
a comparison level by which one can create comparisons more easily. Therefore,
decomposition expands the RNA flow diagram as shown in box 2, and constitutes
the act of slicing the contextual bonds of a relationship between two pieces of in-

formation and comparing the logical context level to assess whether information content should be further sliced or whether information content should be aggregated instead. The process of decomposition and reduction to practice based upon knowledge and context is similar to the concept of graduated/granulated in fuzzy logic [4]. As expressed in the Decomposition definition above, a document can be sliced into paragraphs and paragraphs can be sliced into sentences.

However, this Decomposition and Reduction decision spot in the flow diagram is built so words can also be aggregated together into sentences, or so characters can be aggregated into words. Thus, the red arrow from the box labeled "Adjust Layer Up or Down" was created showing that the stakeholder decides whether the content being compared was at the same logical context level/OEA. As before, the capture of the reasoning and meanings behind the decisions to aggregate or decompose was gathered and the dark blue pedigree repository box was created to depict the pedigree storage. The flow diagram then was built to feed back, all pedigrees from all phases, into the information content repository each time new context, knowledge or information content is generated as output from the flow diagram.

The reasoning captured during decomposition can give valuable insight into the stakeholder context. For example, it is well known that words can have multiple definitions, and when they are aggregated together into sentence form they can portray different emphasis and meanings just by their sequence. Therefore, capturing this as pedigree provides the next evaluator of this information valuable reasoning context which could otherwise easily be misinterpreted. The detail log shown in Appendix A was created when abstracts were processed. The log describes details of state, sequence, and events which give insight into how the process was used to generate knowledge components from information content. Specifically, the Bioscience paper will be processed, and labeled pedigree will be shown, using the RNA flow chart below. The specific examples will show that the capture of the relationship pedigree along with the stakeholder weighting of relative relationships provides valuable insight into (e.g., who, what, when, where, how and why) relationships were developed and how the process contributes to the benefit of subsequent researchers evaluating the conveyed thoughts. Once the objects can be equated at the same contextual level, the OEA's can be passed to the next stage, Compare & Contrast.

17 Compare and Contrast

The Compare & Contrast building block was then added to capture the specific characteristics of the OEA relationship through a series of interrogatories. At this stage, simple interrogatories such as, Who, What, When, Where, How, and Why as well as more detailed questions can be asked based upon the context to determine relationship specifics. Hence, the box for comparing content was added to the flow diagram and then the "Evaluate Characteristics" box was added to designate the need to perform an analysis of the characteristics captured such that the next building block can be added called Association.

18 Association

The building block Association is where the critical analysis was performed for determining the value of the relationships formed during RNA. The decision box is added to designate the need to determine if, based upon the analysis captured during Compare and Contrast, the objects are related to one another. The flow diagram box is then added to designate the need to assess the value strength or weakness of the relationship bi-directionally. A value assessment of each object to the other is performed, based upon the context of the analysis. As in all the previous sub-processes, the iterative decisions and reasoning is captured in the created blue pedigree boxes for ultimate feed-back into the content repository box.

19 Normalization

The next building block added to the RNA flow diagram is the Normalization box representing evaluation of the overall content of the relationships developed under a set of rules governing what to discover. This is analogous to an automobile which is made up of many parts. Each part has an independent function. Each set of parts is related to each other based upon some specific context (e.g. Rim and Tire). However, the sum of all valued parts equals a car, but each part has a perceived value to the overall value of the car as well. An engine might be perceived as having more importance than the radio. Therefore, the Normalization building block was added to designate the need to evaluate all relationships created under the guise of a given criteria context to each other bi-directionally. If all comparisons are complete, then the RNA process flow diagram process stops and the Normalization pedigree are added to the content repository through the blue feedback pedigree box. The pedigree reasoning which was derived from normalizations of the all the relationships created under a certain criteria are related to each other to achieve a cohesive overall value chain of the relationships to each other and their importance to the overall context of the criteria.

In summary, the new RNA Common Process depiction in Figure 5, describes a process which can be generalized for use in a domain where knowledge assimilation is desired, by extending a bottom-up approach in OOD and applying concepts the natural language interrogatives found in 6WH. Therefore, RNA follows a path of creating knowledge and context in a natural manner combined with techniques described herein, for collecting, representing, presenting and storing.

20 Application of RNA to Journal Abstracts

The RNA common process was applied to research journal abstracts in Bioscience [54] and Video Processing [55]. The elements of the constructed RNA framework and sub-processes were applied to each journal abstract, yielding criteria knowledge component and context, knowledge component and context, and transdisciplinary knowledge component and context. This is depicted in by the four phases in Figure 6.

Additionally, the snapshot in time shown in Figure 6 depicts how the framework combined the use of RNA as a common process, the presentation approach for knowledge and context, and the representation approach for knowledge and context. Together the framework constructed and refined a sustainable blueprint of knowledge and context from abstract excerpts in Bioscience and Video Processing. Thus, via the log files and pedigree bin storage mechanisms, it was shown how a cohesive user collaborative [44] dependency trail of knowledge and context was created. The collaborative nature of the process showed how "collective intelligence" was created as defined by Gruber [8]. Therefore, the outcome satisfied the objective of locating reliable and relevant information out of an environment of rich domain specific Bioscience and Video processing abstracts. Finally, upon comparison of the two abstracts using the framework the outcome showed creation of transdisciplinary knowledge component and context.

21   Knowledge & Context Foundation Conclusion

A framework was constructed from the organization, presentation, and the representation of knowledge and context. The organization was derived from the concept of collection and storage, general problem solver, derived from Newell et al. [20] and Simon [21] who together developed models of human mental processes. Sternberg paradigm [22], and tenets of transdisciplinary engineering as defined by Tanik and Ertas [5]. The presentation was constructed from five discrete attributes, namely, time, state, relationship distance, relationship value, and event sequence from computer engineering and mathematics. The representation was derived by using Newton's law of gravitation as an analogy. Finally, the framework was applied to abstracts from research manuscripts and extracted disciplinary and transdisciplinary knowledge and components and therefore was able to as described by Ertas et al. [49], discover important knowledge within one discipline can be systematically discovered, and recombined into another, and via combined engineering visualization mechanisms and collaborative KRT blueprints satisfied Stokols [56], need to achieve a more complete understanding of prior research collaborations and sustain future ones. Finally, the framework satisfied the need as described by Liao et al. [41], enabling transition from data, information and knowledge to new knowledge.

Therefore, using RNA, disciplinary and transdisciplinary knowledge components and context were systematically discovered from tacit and explicit knowledge and context; a mechanism to dynamically interact with ever changing research knowledge, assimilating it to form explicit new knowledge while also retaining the causal pedigree. Thus, RNA was able to enhance transdisciplinary research knowledge and context and describe a foundation for using the mature physical science of n-dimensional relationships in a space and apply the domain to the development and management of complex interrelated knowledge and context.

22   Frameworks for Knowledge and/or Context Refinement

As the knowledge and context foundation described above depicts the process and tools for enhancing knowledge and context the Artificial Cognitive Neural Framework expounded upon in the following sections describe the mechanisms by which we apply additional refinement concepts and another formalization for the modular Decomposition and Reduction and Association sub-processes described in the RNA flow above. A formalization for increasing or decreasing levels of granularity and a formalization for increasing or decreasing the closeness of relationships occurs here as a hybrid, fuzzy-neural processing system using genetic learning algorithms. This processing system uses a modular artificial neural architecture. This architecture is based on a mixture of neural structures that add flexibility and diversity to overall system capabilities. In order to provide an artificially intelligent processing environment that is continually adaptable, we believe the system must possess the notion of artificial emotions that allow the processing environment to "react" in real-time as the systems outside the environment change and evolve recursively as recombinant knowledge assimilation. This hybrid fuzzy-neural processing system forms what we describe as an Artificial Cognitive Neural Framework (ACNF) [57], which allows for artificially "conscious" software agents to carry "emotional memories," based on (REF}Dr. Levine's Autonomic Nervous System States [58]. These conscious software agents are autonomous agents that sense the environment and act on it, based on a combination of information memories (explicit spatio-temporal memories), emotional memories (implicit inference memories), and outside stimulus from the environment. These memories constitute into the pedigree of logged repositories of information content comprised of all related knowledge and context including the relationships which provide the input into the recursive processing. We will describe the constructs for basic emotions and short & long-term memories [59]. The short-term memories (non-recurrent associative memories) provide preconscious buffers as a workspace for internal activities while a transient episodic memory provides a content-addressable associative memory with a property consisting of a moderately fast decay rate [60].

In the ACNF, first the unconscious artificial neural perceptrons, each working toward a common goal, form a coalition. These coalitions will vie for access to the problem to be solved. The resources available to these coalitions depend on their combined nervous system state, which provides information on the criticality of their problem to be resolved. This nervous system state is defined by the Autonomic Nervous System States chart [57]. Based on this nervous system state, information are broadcast to all unconscious processes in order to recruit other artificial neural perceptrons that can contribute to the coalition's goals. The coalitions that understand the broadcast can then take action on the problem. What follows is a description of the overall ACNF architecture in the context of artificial neural emotions and artificial nervous system states.

23 RNA-ACNF Cognitive Coalition

Figure 1 illustrates a high-level view of the ACNF. This is similar to an artificial intelligence blackboard system, except that it is greatly extended to allow for system-wide action selection. The three main subsystems within the architecture are the Mediator, the Memory System, and the Cognitive System [61]. The Mediator gathers information and facilitates communication between agents. Hence, each decision handshake of a combined RNA-ACNF system is handled by the Mediator which takes information from perceptrons and from coalitions of perceptrons and updates the short-term, long-term and episodic memories or pedigree. The information available in memory (what the system has learned) is continually broadcast to the conscious perceptrons that form the cognitive center of the system (i.e., they are responsible for the cognitive functionality of perception, consciousness, emotions, processing, etc.).

The purpose of the ACNF is to:

1. Provide an architectural framework for "conscious" software agents.
2. To provide a "plug-in" domain for the domain-independent portions of the "consciousness" mechanism.
3. To provide an easily customizable framework for the domain-specific portions of the "consciousness" mechanism.
4. To provide the cognitive mechanisms for behaviors and emotions for "conscious" software agents.

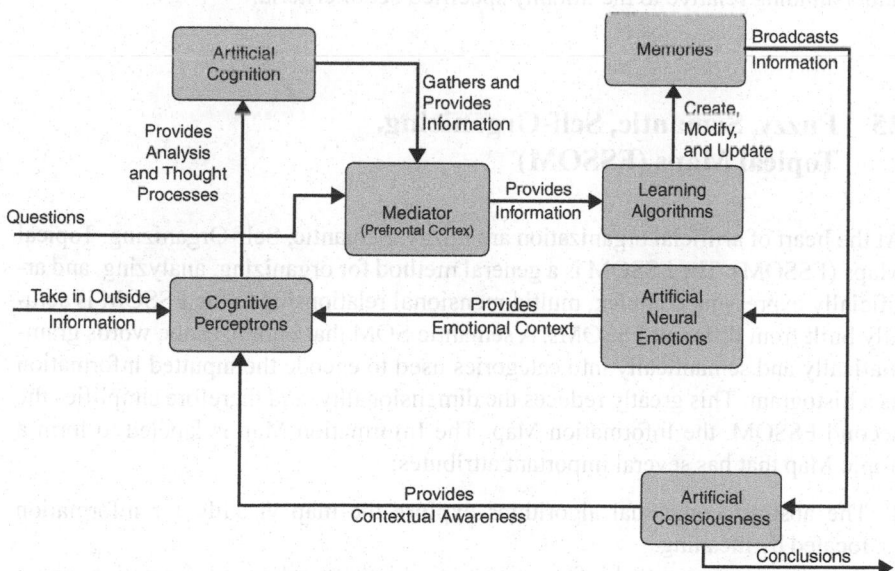

Figure 7. Artificial Cognitive Neural Framework (ACNF).

24   Artificial Neural Memories

The ACNF contains several different artificial memory systems (including emotional memories), each with specific purposes. Each of these memory systems are stored pedigree used in the recursive RNA process and are integrated during the processes of relationship formation between objects of knowledge and context [62].

1. Perceptual Memory – this memory enables identification, recognition, and characterization, including emotions.
2. Working Memory – contains preconscious buffers as a temporary workspace for internal activities.
3. Episodic Memory – this is a content-addressable associative memory with a rapid decay (very short-term memory).
4. Autobiographical Memory – long-term associative memory for facts and data.
5. Procedural Memory – long-term memory for learned skills.
6. Emotional Memory – both long-term (spatio-temporal) and implicit (inference) emotional memories.

When processing pedigree memory, RNA loosely categorizes the granularity of information content into knowledge and context based upon the criteria established by the cognitive human interaction input into the system. These loosely or fuzzy categories are only as fuzzy as the threshold of human understanding. Therefore, in order to artificially create this effect we use Intelligent Information Agents (I2A) to develop fuzzy organization over time, ultimately reaching a threshold of perceived understanding relative to the initially specified set of criteria.

25   Fuzzy, Semantic, Self-Organizing, Topical Maps (FSSOM)

At the heart of artificial organization are Fuzzy, Semantic, Self-Organizing, Topical Maps (FSSOM). The FSSOM is a general method for organizing, analyzing, and artificially expressing complex, multidimensional relationships. The FSSOM is actually built from different FSSOMs. A semantic SOM that organizes the words grammatically and semantically into categories used to encode the inputted information as a histogram. This greatly reduces the dimensionality, and therefore simplifies the second FSSOM, the Information Map. The Information Map is labeled to form a Topic Map that has several important attributes:

1. The abstract contextual algorithms explore the map visually for information located by meaning.
2. Searches use contextual information to find links to relevant information within other source documents.

Figure 8. The Fuzzy,
Semantic, Self-Organizing
Topical Map.

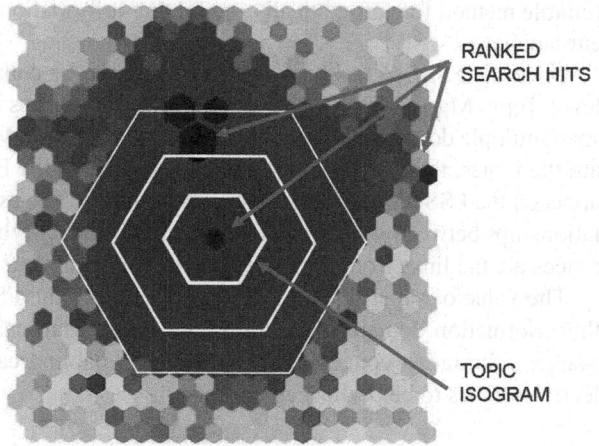

RANKED
SEARCH HITS

TOPIC
ISOGRAM

3. The Information Map is self-maintaining and automatically locates input on the grid, 'unsupervised'.
4. A semantic SOM and an information SOM are normalized representations of any physical information content (e.g. picture, text, object) used in the development of recombinant knowledge and context

Illustrated below is a FSSOM with information search hits superimposed. The larger hexagons denote information sources that best fit the search criterion. The isograms denote "closeness"; how close the hits are to particular information topics or criterion.

There are also other attributes to be explored that would provide significant benefit: as a natural language front end to relational data, like reseacrh abstracts; and, as a means to find information with common meaning located in a foreign language FSSOM through the use of common encoding [63]. This approach to mapping information by 'meaning' avoids problems common to classical natural language IR (information retrieval) methods. Classical IR requires extensive modeling of the innumerable forms of information representation. Such a modeling activity would always be on-going and expensive to maintain. In contrast, the FSOM promises a

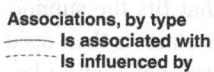

Associations, by type
——— Is associated with
------- Is influenced by

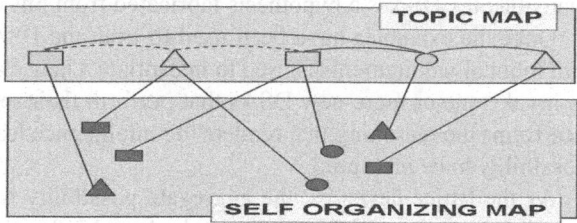

TOPIC MAP

SELF ORGANIZING MAP

Figure 9. Superimposing Hi Level Topical Maps on FSSOM.

reliable method for automatically organizing information for retrieval across different languages.

Once the FSSOM has been developed, it can be enhanced to include a higher-level Topic Map. This high-level Topic Map describes knowledge structures that span multiple documents. The key features of the Topic Map, illustrated in Figure 3, are the topics, their associations and occurrences in the FSSOM. The topics are the areas on the FSSOM that fall under a topic name. The associations describe the relationships between topics, such as 'biometric data' in 'bone fractures'. The occurrences are the links from the FSSOM into the documents used to form the FSSOM.

The value of superimposing a Topic Map onto the FSSOM is that it can define the information domain's ontology. The Topic Map enables end users to rapidly search information conceptually. Therefore, enabling candidate sophisticated dialectic searches to be performed upon them.

26 I2A Dialectic Search

The Dialectic Search [64] uses the Toulmin Argument Structure to find and relate information that develops a larger argument, or intelligence lead. The Dialectic Search Argument (DSA), illustrated in Figure 4, has four components:

1. Data: in support of the argument and rebutting the argument.
2. Warrant and Backing: explaining and validating the argument.
3. Claim: defining the argument itself.
4. Fuzzy Inference: relating the data to the claim.

The argument serves two distinct purposes. First, it provides an effective basis for mimicking human reason. Second, it provides a means to glean relevant information from the Topic Map and transform it into actionable intelligence (practical knowledge.) These two purposes work together to provide an intelligent system that captures the capability of the contextual sorting, to sort through diverse information and find clues to contextual closeness.

This approach is considered dialectic in that it does not depend on deductive or inductive logic, though these may be included as part of the warrant. Instead, the DSA depends on non-analytic inferences to find new possibilities based upon warrant examples. The DSA is dialectic because its reasoning is based upon what is plausible; the DSA is a hypothesis fabricated from bits of information.

Once the examples have been used to train the DSA, data that fits the support and rebuttal requirements is used to instantiate a new claim. This claim is then used to invoke one or more new DSAs that perform their searches. The developing lattice forms the reasoning that renders the intelligence lead plausible and enables the possibility to be measured.

As the lattice develops, the aggregate possibility is computed using the fuzzy membership values of the support and rebuttal information. Eventually, a DSA lattice is formed that relates information with its computed possibility. The computation, based on Renyi's entropy theory, uses joint information memberships to gen-

Figure 10. DSA Structure.

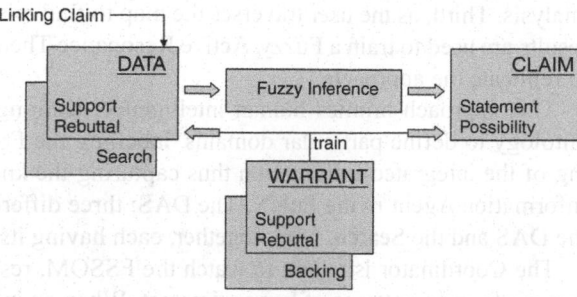

erate a robust measure of Possibility, a process that is not possible using Bayesian methods. Whereas the Topic Map builds and maintains itself, the Dialectic Argument Search (DAS) requires supervised training, meaning it must be seeded with knowledge from a domain expert. However, once seeded, it has the potential of evolving the Warrant to present new types of possible leads.

There is one other valuable attribute to using the FSSOM method. Because the vector that represents the information is a randomly constructed vector, it cannot be decoded to reformulate the source; the source must be reread. This is critical to protecting compartmentalized information. Using the FSSOM, the protected source can be included in the FSSOM and used to support/rebut an argument without revealing the detailed information. The approach to analyzing text information utilizing the FSSOM is threefold [65]. First the FSSOM is investigated to semantically organize the diverse information collected. Second, the map produced by the FSSOM is used to enhance the user's comprehension about the situations under

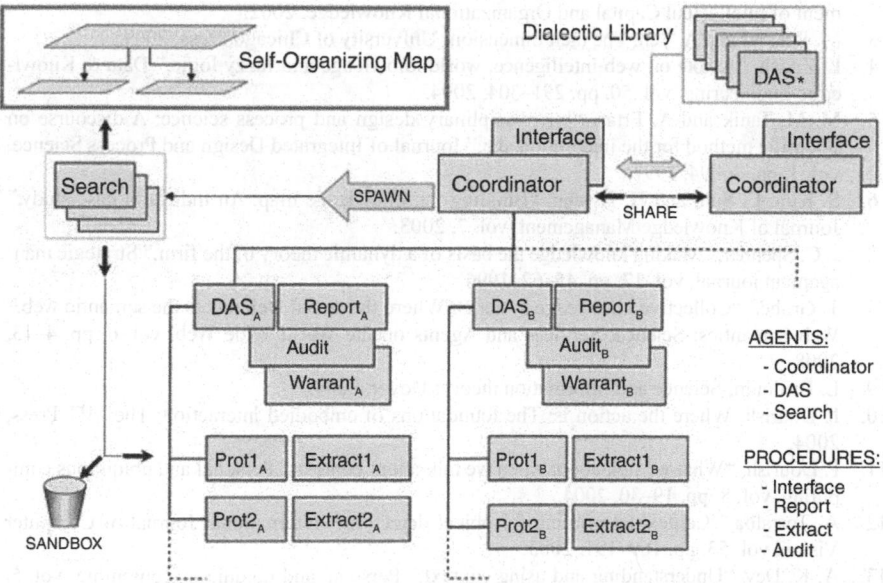

Figure 11. Intelligent DAS Software Agency.

analysis. Third, as the user traverses the map to find related and relevant events, the results are used to train a Fuzzy, Active Resonance Theory Neural Network (FuNN) to replicate the approach.

This approach mimics human intelligence, learning from documents using an ontology to define particular domains, labelling the FSSOM to capture the meaning of the integrated information thus capturing the knowledge of each Intelligent Information Agent in the FuNN. The DAS: three different agents, the Coordinator, the DAS and the Search, work together, each having its own learning objectives.

The Coordinator is taught to watch the FSSOM, responding to new hits (input) that conform to patterns of known interest. When an interesting hit occurs, the Coordinator selects one or more candidate DAS agents, and then spawns Search agents to find information relevant to each DAS. As time proceeds, the Coordinator learns which hit patterns are most likely to yield a promising lead, adapting to any changes in the FSSOM structure and sharing what it learns with other active Coordinators [66]. The Search agent takes the DAS prototype search vectors and, through the FSSOM, finds information that is relevant and related. The Search agent learns to adapt to different and changing source formats and could include parsing procedures required to extract detailed information. The figure below illustrates the DAS.

References

1. I. A. Nonaka and H. A. Takeuchi, The knowledge-creating company: How Japanese companies create the dynamics of innovation: Oxford university press, 1995.
2. I. Nonaka, "A dynamic theory of organizational knowledge creation," The Strategic Management of Intellectual Capital and Organizational Knowledge, 2002.
3. M. Polanyi and A. Sen, The tacit dimension: University of Chicago Press, 2009.
4. L. Zadeh, "A note on web intelligence, world knowledge and fuzzy logic," Data & Knowledge Engineering, vol. 50, pp. 291–304, 2004.
5. M. M. Tanik and A. Ertas, "Interdisciplinary design and process science: A discourse on scientific method for the integration age," Journal of Integrated Design and Process Science, vol. 1, pp. 76–94, 1997.
6. S. Kim, E. Suh, and H. Hwang, "Building the knowledge map: An industrial case study," Journal of Knowledge Management, vol. 7, 2003.
7. J. C. Spender, "Making knowledge the basis of a dynamic theory of the firm," Strategic management journal, vol. 17, pp. 45–62, 1996.
8. T. Gruber, "Collective knowledge systems: Where the social web meets the semantic web," Web Semantics: Science, Services and Agents on the World Wide Web, vol. 6, pp. 4–13, 2008.
9. L. Brillouin, Science and information theory: Dover, 2004.
10. P. Dourish, Where the action is: The foundations of embodied interaction: The MIT Press, 2004.
11. P. Dourish, "What we talk about when we talk about context," Personal and ubiquitous computing, vol. 8, pp. 19–30, 2004.
12. A. Torralba, "Contextual priming for object detection," International Journal of Computer Vision, vol. 53, pp. 169–191, 2003.
13. A. K. Dey, "Understanding and using context," Personal and ubiquitous computing, vol. 5, pp. 4–7, 2001. [14] J. Coutaz, J. Crowley, S. Dobson, and D. Garlan, "Context is key," Communications of the ACM, vol. 48, pp. 53, 2005.

14. J. Coutaz, J. Crowley, S. Dobson, and D. Garlan, "Context is key," Communications of the ACM, vol. 48, pp. 53, 2005.
15. C. Graham and J. Kjeldskov, "Indexical representations for context-aware mobile devices," 2003, pp. 3–6.
16. T. Winograd, "Architectures for context," Human-Computer Interaction, vol. 16, pp. 401–419, 2001.
17. J. I. Hong and J. A. Landay, "An infrastructure approach to context-aware computing," Human-Computer Interaction, vol. 16, pp. 287–303, 2001.
18. N. Howard and A. Qusaibaty, "Network-centric information policy," in The Second International Conference on Informatics and Systems, 2004.
19. D. Ejigu, M. Scuturici, and L. Brunie, "Hybrid approach to collaborative context-aware service platform for pervasive computing," Journal of computers, vol. 3, pp. 40, 2008.
20. A. Newell, J. C. Shaw, and H. A. Simon, "Preliminary description of general problem solving program-i (gps-i)," WP7, Carnegie Institute of Technology, Pittsburgh, PA, 1957.
21. H. A. Simon, "Modeling human mental processes," in Papers presented at the May 9–11, 1961, western joint IRE-AIEE-ACM computer conference. Los Angeles, California: ACM, 1961.
22. S. Sternberg, "High-speed scanning in human memory," Science, vol. 153, pp. 652–4, 1966.
23. J. Rowley and R. Hartley, Organizing knowledge: An introduction to managing access to information: Ashgate Pub Co, 2008.
24. G. Booch, Object-oriented analysis and design. New York: Addison-Wesley, 1996.
25. J. Rumbaugh, M. Blaha, W. Premerlani, F. Eddy, and W. Lorensen, Object-oriented modeling and design: Prentice-Hall, Inc. Upper Saddle River, NJ, USA, 1991.
26. D. Engelbart, "A conceptual framework for the augmentation of man's intellect," Computer-supported cooperative work: A book of readings, pp. 36–65, 1988.
27. M. Van Ittersum, F. Ewert, T. Heckelei, J. Wery, J. Alkan Olsson, E. Andersen, I. Bezlepkina, F. Brouwer, M. Donatelli, and G. Flichman, "Integrated assessment of agricultural systems–a component-based framework for the european union (seamless)," Agricultural Systems, vol. 96, pp. 150–165, 2008.
28. W. M. K. Trochim, "An introduction to concept mapping for planning and evaluation," CONCEPT MAPPING FOR EVALUATION AND PLANNING, vol. 12, pp. 1–16, 1989.
29. M. Davison, "Multidimensional scaling," New York, 1983.
30. B. Everitt, "Unresolved problems in cluster analysis," Biometrics, vol. 35, pp. 169–181, 1979.
31. N. Deo, Graph theory with applications to engineering and computer science: PHI Learning Pvt. Ltd., 2004.
32. M. Ali, "Metrics for requirements engineering," UMEA University, 2006.
33. J. Rumbaugh, I. Jacobson, and G. Booch, Unified modeling language reference manual, the: Pearson Higher Education, 2004.
34. D. Beckett and B. McBride, "Rdf/xml syntax specification (revised)," W3C Recommendation, vol. 10, 2004.
35. D. L. McGuinness and F. Van Harmelen, "Owl web ontology language overview," W3C Recommendation, vol. 10, pp. 2004–03, 2004.
36. S. Polyn and M. Kahana, "Memory search and the neural representation of context," Trends in Cognitive Sciences, vol. 12, pp. 24–30, 2008.
37. M. Howard and M. Kahana, "A distributed representation of temporal context," Journal of Mathematical Psychology, vol. 46, pp. 269–299, 2002.
38. N. Konstantinou, D. Spanos, and N. Mitrou, "Ontology and database mapping: A survey of current implementations and future directions," Journal of Web Engineering, vol. 7, pp. 001–024, 2008.
39. A. Zouaq, R. Nkambou, and C. Frasson, "An integrated approach for automatic aggregation of learning knowledge objects," Interdisciplinary Journal of Knowledge and Learning Objects, vol. 3, pp. 135–162, 2007.
40. J. Novak and A. Cañas, "The theory underlying concept maps and how to construct and use them," Florida Institute for Human and Machine Cognition Pensacola Fl,

www. ihmc. us. [*http://cmap. ihmc. us/Publications/ResearchPapers/T heoryCmaps/ TheoryUnderlyingConceptMaps. htm*], 2008.

41. S. S. Liao, J. W. He, and T. H. Tang, "A framework for context information management," Journal of Information Science, vol. 30, pp. 528–539, 2004.
42. A. K. Gupta and V. Govindarajan, "Knowledge flows within multinational corporations," Strategic management journal, vol. 21, pp. 473–496, 2000.
43. M. Pinto, "A grounded theory on abstracts quality: Weighting variables and attributes," Scientometrics, vol. 69, pp. 213–226, 2006.
44. N. Suh, "Application of axiomatic design to engineering collaboration and negotiation," in 4th International Conference on Axiomatic Design, Firenze, 2006.
45. J. Llinas, C. Bowman, G. Rogova, A. Steinberg, E. Waltz, and F. White, "Revisiting the jdl data fusion model ii," 2004.
46. J. Anderson, Cognitive psychology and its implications: John r. Anderson: Worth Pub, 2004.
47. D. Ausubel, J. Novak, and H. Hanesian, Educational psychology: A cognitive view: Holt Rinehart and Winston, 1978.
48. J. Habermas, "Knowledge and human interests: A general perspective," Continental philosophy of science, pp. 310, 2005.
49. A. Ertas, D. Tate, M. M. Tanik, and T. T. Maxwell, "Foundations for a transdisciplinary systems research based on design & process," The ATLAS Publishing, vol. TAM–Vol.2, pp. 1–37, 2006.
50. R. C. Hibbeler, Engineering mechanics: Dynamics: Prentice Hall, 2009.
51. J. R. Anderson, Cognitive psychology and its implications: John r. Anderson: Worth Pub, 2004.
52. N. P. Suh, Complexity: Theory and applications: Oxford University Press, USA, 2005.
53. A. S. Herbert, "Modeling human mental processes," in Papers presented at the May 9–11, 1961, western joint IRE-AIEE-ACM computer conference. Los Angeles, California: ACM, 1961.
54. J. Hartman and N. Tippery, "Systematic quantification of gene interactions by phenotypic array analysis," GENOME BIOLOGY, vol. 5, 2004.
55. S. Pei and I. Tam, "Effective color interpolation in ccd color filter arrays using signal correlation," IEEE Transactions on Circuits and Systems for Video Technology, vol. 13, pp. 503–513, 2003.
56. D. Stokols, "Toward a science of transdisciplinary action research," American journal of community psychology, vol. 38, pp. 63–77, 2006.
57. Crowder, J., "Artificial Emotions and Emotional," Proceedings of the International Conference on Artificial Intelligence, ICAI'10, Las Vegas, 2010.
58. Levine, P., "Walking the Tiger: Healing Trauma." North Atlantic Books, Berkeley, CA, 1997.
59. Eichenbaum H, "The cognitive neuroscience of memory." New York: Oxford University Press, 2002.
60. Crowder, J., Barth, T., and Rouch, R., "Learning Algorithms for Stochastically Driven Fuzzy, Genetic Neural Networks." NSA Technical Paper, 1999.
61. Crowder, J., Barth, T., and Rouch, R., "Neural Associative Memory for Storing Complex Memory Patterns." NSA Technical Paper, 1999.
62. Crowder, J. A., "Integrating an Expert System into a Neural Network with Genetic Programming for Process Planning." NSA Technical Paper, 2001.
63. Crowder, J., Barth, T., and Rouch, R., "Evolutionary Neural Infrastructure with Genetic Memory Algorithms: ENIGMA Theory Development." NSA Technical Paper, 1999.
64. Newell, A., "Unified Theories of Cognition." Cambridge MA: Harvard University Press 2003.
65. Crowder, J., "Anti-Terrorism Learning Advisory System: Operative Information Software Agents (OISA) for Intelligence Processing" Proceedings of the American Institute of Aeronautics and Astronautics Conference (Infotech@aerospace 2010) Atlanta, GA, 2010.
66. Crowder, J. A., "Machine Learning: Intuition (Concept) Learning in Hybrid Genetic/Fuzzy/ Neural Systems." NSA Technical Paper, 2003.

Chapter 3
DESIGN AND PROCESS DEVELOPMENT FOR SMART PHONE MEDICATION DOSING SUPPORT SYSTEM AND EDUCATIONAL PLATFORM IN HIV/ AIDS-TB PROGRAMS IN ZAMBIA

M. Willig, M. Sinkala, R. S. Sadasivam, M. Albert, L. Wilson, M. Mugavero, E. Msidi, S. Brande, J. Nikisi, J. Stringer, I. Tami, G. Henostroza, V. Gathibhande, J. Menke, M. Tanik, and S. Reid

1   Introduction

Once thought impossible, the dedication and efforts of countless international organizations and individuals have made HIV/AIDS therapy available in developing nations where it is estimated 90% of global cases are present. Despite the vastly improved access to antiretroviral therapy (ART), in 2007 World Health Organization (WHO) estimates indicated only 31% of those infected in low- and middle income countries had accessed HIV/AIDS therapy [1, 2, 3]. Reduced human resources are a key barrier to the provision of HIV care in resource limited settings. Contributing factors include a limited supply of new healthcare workers coming into the workforce, inadequate human resource management systems for recruitment, deployment and retention, attrition due to HIV/AIDS, limited career and professional opportunities, and increasing rates of international migration [1, 4, 5]. In response to these shortages, and to maximize available human resources, the World Health Organization (WHO) published guidelines related to task-shifting, the strategy of moving tasks from highly qualified health workers to workers with shorter duration training needs. Alongside task-shifting, the WHO stresses the need for efforts to increase the overall number of trained healthcare workers and to establish appropriate quality assurance mechanisms to evaluate and monitor clinical outcomes.

HIV care has become increasingly complicated due to a number of factors. The complexity of dosing antiretroviral therapy continues to expand as more agents are added to the treatment arsenal and older drugs are co-formulated into new combinations. Providers are prone to various dosing errors if they fail to account for multiple patient characteristics at the time of prescription [6, 7]. At the University

S. Brande (✉)
Birmingham, AL, USA
e-mail: sbrande@uab.edu

S. Suh et al. (eds.), *Biomedical Engineering*,
DOI 10.1007/978-1-4614-0116-2_3, © Springer Science+Business Media, LLC 2011

of Alabama at Birmingham HIV care clinic, we found dosing errors occurred in 6% of all prescriptions and in 31% of those in patients with renal dysfunction [8]. Dosing errors may lead to increased drug toxicity, which in turn may adversely affect adherence to therapy and hasten the development of drug resistant virus and regimen failure, which impacts HIV/AIDS morbidity and mortality. In the developing world, having to move from a first line regimen to a second line regimen increases the cost of ART tenfo50ld [9, 10, and 11]. At present, third line regimens do not exist; meaning the development of resistance to second line regimens leaves patients without therapeutic options.

An additional challenge is that patients with advanced HIV/AIDS often need concomitant treatment for co-existing opportunistic infections such as tuberculosis (TB). Agents used to treat TB often interact with drugs used to treat HIV, resulting in changes in serum levels that may contribute to toxicity (elevated drug levels) or treatment failure (lowered drug levels) [12, 13, 14, 15]. Faced with the challenge of appropriately dosing complex ART and TB regimens, providers would benefit from targeted informatics tools that would assist the prescription of appropriately dosed therapy. We are proposing to develop such a tool in collaboration with front line care providers in Zambia. We propose the utilization of mobile phone technology due to its penetration, ease of use and familiarity to users, and its potential to harbor software. Such a mobile phone system will not only allow for dosing decision support at the point of care, but it will also serve as a platform for the dissemination of continuing medical education, and enhance the skills of care providers longitudinally. Such a strategy holds the promise of helping patients by decreasing toxicity and maximizing effectiveness, as well as providing a practical tool that can monitor quality assurance. Monitoring patterns of prescription error can provide insight into areas where educational intervention is needed. What follows is a summary of a staged approach for the development of such software. We anticipate that such expert dosing software could play an important role in minimizing dosing errors, monitoring quality of care, and targeting educational interventions as task-shifting initiatives are increasingly implemented throughout the world.

2   Requirements

The goal is to create a flexible software tool that could be utilized at the front lines of HIV and TB care in resource limited settings. The tool must be:

2.1   User friendly

The tool should provide clear value to providers. The input of patient and planned treatment information should be simple, and result in readily interpretable and patient specific dosing recommendations. To accomplish this goal, patient specific characteristics (i.e. height, weight, pertinent laboratory values, etc.) will need to be entered before a planned treatment is formulated. The application will then calculate and present individualized treatment and dosing recommendations based on patient characteristics.

2.2   Flexible

It must allow for easy and accurate capture and storing of this information in a central database for later access. Though much data (i.e. drug dosing recommendations) will be portable across settings, the application must be flexible enough to allow for some individualization of treatment recommendations based on the needs and available drugs in specific settings.

2.3   Accessible

The application needs to be readily deployable in resource rich and resource limited settings. Due to their penetration and rapidly increasing computing power, we feel mobile phones provide an optimal platform for dissemination of this application. The application could be installed on mobile phones and provide recommendations even at times without ready access to phone networks. Educational modules could be automatically downloaded to provider phones when in a service area, to be accessed and completed later. Evaluations of these modules would be uploaded automatically, and scores sent back to providers once they return to service areas. In this fashion, one can envision each HIV treatment program establishing continuing education requirements and have a platform in place to readily disseminate such interventions. Likewise, in case it becomes necessary to communicate to providers new scientific findings that would have an impact on current practice, a network to send and track review of these data would be in place. This new network might also be used to disseminate information on new epidemics or other public health emergencies.

2.4   Prepared for additional linkage

At the central repository for information, individual patients will have a system wide identifier (primary key), around which the data structures will be organized. By structuring the data this way, the toolset may later be able to link to and display information captured throughout the healthcare system (laboratory, medication history/changes at other sites, dates/times/locations of previous visits, etc). Laying the foundation to be able to potentially organize and later share information from a central site will allow future healthcare providers to have reliable information on past care that patients have received. For example, such thorough capture of historical information will enable providers to make optimal treatment decisions based on comprehensive knowledge of past medication exposure. It has been reported that as many as 40-50% of patients are lost to follow up from where they began to receive HIV care in developing nations. Loss to follow up is associated with poor treatment outcomes and increased mortality. The capture of the proposed data through a national network of mobile phones could provide insight into patient histories across multiple sites of care. These data in turn could help guide the development of interventions to mitigate the risks of "losing" patients and ultimately improve HIV care outcomes. Laying the foundation to cross link to other databases from the very beginning will lay the groundwork for potentially expanding the functionality of this application in the future.

3 Functional Requirements

Initially, this application will provide point of care assistance in the dosing of antiretroviral and tuberculosis therapy. This will only represent the first stage of application development, with further phases and correspondingly expanded functionality to follow. Below we propose a "growth plan" for the application, the development of additional functionality seeking to respond to the most pressing needs of our partners.

3.1 Stage 1: Provision of educational materials

The mobile application will allow the display of documents such as national guidelines for the treatment of HIV and/or TB therapy. These will be updated and sent to providers periodically as soon as new iterations become available. Continuing medical education modules which test providers on their knowledge of important changes to treatment recommendations could be delivered as well. This will allow national programs ready access to a distance education network that will potentiate their ability to disseminate needed guidelines in a timely fashion to the field.

3.2 Stage 2: Decision support for adult dosing recommendations for HIV and/or TB therapy

Guided by comprehensive review of drug prescribing information, a set of rules governing the appropriate dosing of adult HIV and TB therapy will be completed.

These rules will include the detection of interactions between therapies for illnesses that could compromise drug levels to the detriment of those under care (high drug levels leading to toxicity, or low drug levels leading to loss of effectiveness). The clinical parameters that need to be taken into account in the dosing of each drug will be entered into a streamlined form. After entry, providers will be asked to select the planned therapeutic regimen. Upon completion, the application will generate dosing recommendations for the relevant drugs for the specific patient. In this manner, patient specific dosing will be calculated taking into account such factors as weight, CD4 value, and creatinine clearance, co-administration of HIV and TB therapy.

3.3   Stage 3: Decision support for pediatric dosing recommendations for HIV and/or TB therapy

Dosing TB and HIV therapy in children represents additional challenges. The addition of the patient characteristics needed to generate appropriate dosing recommendations for HIV and TB infected children will be included. Decision support taking into account the relevant clinical characteristics will be generated at the point of care.

3.4   Stage 4: Addition of drug interactions

We will extend the capabilities of the application to provide guidance on drug interactions covering all potential medications the patient could be using. The average life expectancy of HIV positive individuals is calculated to be approximately 70 years of age. As our HIV infected population ages and requires additional medication for other concomitant medical conditions (e.g., cardiac disease or diabetes), the ability to easily and reliably check for drug interactions with the existing HIV treatment regimen will increase in importance. We will explore standalone applications to add such functionality, and work with strategic partners or knowledge vendors to acquire databases that can be used for this purpose.

3.5   Stage 5: Linkage to existing databases and central information repository

Herein lies the potential to expand the utility of this application in the future. An example would be linkage to existing national laboratory databases that would allow access to laboratory data on all patients wherever providers have access to a mobile phone network. If one were to add linkage to patient HIV drug resistance reports, it will greatly aid in the selection of optimal new regimens. A consequence of the expansion of the use of mobile applications to receive feedback on appropriate dosing is that the recommended regimens could be stored in the central database along with stop dates. When another provider in a different geographic setting is starting subsequent therapy, a record of the initiation dates and agents utilized in the past would be available. Currently, a "history of prior agents" often depends exclusively on patient recollection. The ability to have access to well characterized past regimen data will present health care providers with comprehensive drug histories that will allow for the initiation of optimal subsequent therapy.

4   Looking Ahead

A mobile phone application, well adapted and appropriately implemented in daily clinical practice in the targeted settings, will enable providers to potentially avoid erroneous prescription of HIV and TB therapy. The establishment of a national network of providers using such an application will allow health ministries to disseminate guidelines and complete continuing medical education initiatives in rapid fashion. Use of tools, such as the application described, have the potential to allow for monitoring of HIV and TB outcomes for healthcare providers serving patients as part of task-shifting initiatives. These data will be necessary to evaluate the effectiveness of task-shifting initiatives, as well as allowing the detection of issues that will help to target continuing medical education for such providers. Building a system with the foundation in place to capture the data that will allow for the optimal selection of future therapy in resource limited settings will serve to provide reliable databases that will improve longitudinal patient care. This will be critical in HIV therapy where the current paradigm calls for lifelong ART, and subsequent treatment decisions are influenced by prior treatment and resistance history. Due to the expanding computing capacity, increased penetration of mobile phone networks and newer alternatives such as solar-powered cell phones, this technology allows for the deployment of software tools in remote settings where comprehensive informatics resources are lacking. Through carefully crafted software applications, we can bring many of the benefits of medical informatics to resource limited settings.

References

1. Munjanja, O. K., Kibuka, S., & Dovola, D., 2005, "The nursing workforce in sub-Saharan Africa", Geneva, Switzerland: International Council of Nurses, accessed February 3, 2008, <http://www.icn.ch/global/Issue7SSA..pdf>
2. PEPFAR, "President's Emergency Plan for AIDS relief", accessed July, 20 2009, <http://www.pepfar.gov>.

3. WHO, 2006, "Treat, train, retain. The AIDS and health workforce plan. Report on the consultation on AIDS and human resources for health", Geneva, Switzerland, World Health Organization, accessed July 20, 2009 <http://www.who.int/hiv/pub/meetingreports/ TTRmeetingreport2.pdf>.

4. WHO, 2008a, "Task shifting: Rational redistribution of tasks among health workforce teams: Global recommendations and guidelines", Geneva, Switzerland: World Health Organization, accessed July 20, 2009 at <http://www.who.int/healthsystems/TTR-TaskShifting.pdf>.

5. WHO, 2008b, "Towards universal access: scaling up priority HIV/AIDS interventions in the health sector: Progress Report", accessed 20 July 2009, <http://www.who.int/hiv/ mediacentre/2008progressreport/en/print.html>.

6. Purdy, B. D., Raymond, A. M. and Lesar, T. S., 2000, "Antiretroviral prescribing errors in hospitalized patients," *Ann Pharmacother,* Vol. 34, Iss.(7–8), pp. 833–8.

7. Rastegar, D. A., Knight, A. M. and Monolakis, J. S., 2006, "Antiretroviral medication errors among hospitalized patients with HIV infection," *Clin Infect Dis,* Vol. 43, Iss.(7), pp. 933–8.

8. Willig, J. H., Westfall, A. O., Allison, J., et al., 2007, "Nucleoside reverse-transcriptase inhibitor dosing errors in an outpatient HIV clinic in the electronic medical record era", *Clin Infect Dis,* Vol. 45, Iss.(5), pp. 658–61.

9. Sukasem, C., Churdboonchart, V., Chasombat, S., et al., 2007, "Prevalence of antiretroviral drug resistance in treated HIV-1 infected patients: under the initiative of access to the NNRTI-based regimen in Thailand", *J Chemother,* Vol. 19, Iss.(5), pp. 528–35.

10. Sungkanuparph, S., Manosuthi, W., Kiertiburanakul, S., et al., 2007, "Options for a second-line antiretroviral regimen for HIV type 1-infected patients whose initial regimen of a fixed-dose combination of stavudine, lamivudine, and nevirapine fails", *Clin Infect Dis,* Vol. 44, Iss.(3), pp. 447–52.

11. Marconi, V. C., Sunpath, H., Lu, Z., et al., 2008, "Prevalence of HIV-1 drug resistance after failure of a first highly active antiretroviral therapy regimen in KwaZulu Natal, South Africa," *Clin Infect Dis,* Vol. 46, Iss.(10), pp. 1589–97.

12. Onyebujoh, P. C., Ribeiro, I. and Whalen, C. C., 2007, "Treatment Options for HIV-Associated Tuberculosis," *J Infect Dis,* Vol. 196 Suppl 1, pp. S35–45.

13. Lalloo, U. G. and Pillay, S., 2008, "Managing tuberculosis and HIV in sub-Sahara Africa," *Curr HIV/AIDS Rep,* Vol. 5, Iss.(3), pp. 132–9.

14. Scano, F., Vitoria, M., Burman, W., et al., 2008, "Management of HIV-infected patients with MDR- and XDR-TB in resource-limited settings", *Int J Tuberc Lung Dis,* Vol. 12, Iss.(12), pp. 1370–5.

15. Manosuthi, W., Sungkanuparph, S., Tantanathip, P., et al., 2009, "A randomized trial comparing plasma drug concentrations and efficacies between 2 nonnucleoside reverse-transcriptase inhibitor-based regimens in HIV-infected patients receiving rifampicin: the N2R Study," *Clin Infect Dis,* Vol. 48, Iss.(12), pp. 1752–9.

Chapter 4
IMAGE-BASED METHODOLOGIES AND THEIR INTEGRATION IN A CYBER-PHYSICAL SYSTEM FOR MINIMALLY INVASIVE AND ROBOTIC CARDIAC SURGERIES

Erol Yeniaras and Nikolaos V. Tsekos

1 Introduction

A Cyber-physical-system (CPS) is a computational system that integrates computing modules with physically distributed dedicated elements such as robotic devices and actuators. The concept is closely related to robotics and real-time (RT) embedded systems. Principally, within a CPS, embedded software modules manage the physical processes by means of feedback loops and continuous computations. It is expected that new technological advancements will increase the prominence and usability of CPS by amending the link between the physical and computational elements [1]. One of the most promising applications of CPS is the use of robotics in minimally invasive surgeries (MIS). Nevertheless, such a system entails much higher reliability and robustness than the general purpose computing, as it is directly related to human health. For instance, the results of a crash in the control software of a robotic device during a surgery can be detrimental or even fatal. Therefore, a surgical CPS should be extremely rigorous and ready for unexpected situations as well as adaptable to any possible system failures, due to unpredictability of the area of operation (AoO).

MIS encompasses a specialized type of surgical methods that necessitate access to the targeted anatomy via small incision(s), rather than the more traditional "open surgeries". Minimal invasiveness offers certain benefits, such as substantially less operative trauma and post surgical complications, faster patient recovery and lower overall cost. As a result, MIS are becoming more popular while additional research is performed to expand their scope, especially in robotic applications. However, in contrast to its open-counterpart, in MIS the operator/surgeon requires a method for indirectly visualizing the surgical field. Although endoscopic visualization is a well accepted methodology for such guidance, it offers a limited perception of AoO. Re-

E. Yeniaras (✉)
Medical Robotics Lab, Department of Computer Science, University of Houston,
Houston, TX, USA
e-mail: eyeniaras@gmail.com

S. Suh et al. (eds.), *Biomedical Engineering*,
DOI 10.1007/978-1-4614-0116-2_4, © Springer Science+Business Media, LLC 2011

cently, RT image guidance in MIS has gained importance due to increasing demand for on-the-fly and accurate assessment of tissue motion before, during and after the surgical procedures as well as monitoring the tool(s) with three-dimensional perception AoO. Therefore, RT image guidance may alleviate the paradigm shift from endoscopic visualization toward the use of effective image modalities [2].

1.1 MRI Guidance

While computed tomography (CT), fluoroscopy, and ultrasonograpy are the most widely used imaging modalities for guiding interventional procedures, magnetic resonance imaging (MRI) has also emerged and been evolving as a modality for both pre-operative planning and intra-operative guidance. MRI is inherently robust and versatile in addition to its other unique to the modality features [2] such as:

- It offers a wide range of soft-tissue contrast mechanisms for assessing physiology and pathophysiology, such as heart wall motion myocardial perfusion, heart valve flow and coronary angiography,
- It is versatile allowing implementation of imaging protocols for visualizing interventional tools and robotic manipulators,
- It is a true three-dimensional (3D) and multislice modality with the capability for the electronic control of the imaging volumes or planes,
- Modern state-of-the-art scanners allow on-the-fly adjustment of the acquisition parameters directly from the robotic control core,
- It operates without ionizing radiation; vs. fluoroscopy or CT,
- The image quality is operator-independent; vs. ultrasound.

Besides, MR-compatible robotic systems for performing interventions with MR guidance are also evolving [3-5]. It should be noted that the development of such robotic manipulators is challenging due to the exceptionally strong magnetic fields, the rapidly switching magnetic field gradients, and the limited space inside the MR scanners [6]. In spite of those technical challenges, a number of MR-compatible robotic systems have been introduced [5]. The works of those groups have resulted in enabling technologies that may contribute to the goal of performing MR-guided complex MIS [5, 7-10]. Lately, the advent of RT MRI and its optimization emerged a wide range of interventional procedures, e.g., off-pump cardiac surgeries [11, 12]. In these studies, MRI was reported to provide excellent visualization of the AoO.

Herein, we describe a computational system for performing MRI guided interventions in the heart [13, 14]. The system provides the appropriate tools for pre-operative planning, intra-operative guidance and control of a robotic device. This prototype system includes dedicated software modules that operate synergistically for both controlling a semi-autonomous robot and adjusting on-the-fly the image acquisition parameters of the MRI scanner to better suit the particular conditions of the interventions as they evolve. In particular, the system includes five software modules for RT image processing, dynamic trajectory planning, three dimensional model construction using two dimensional contours, image acquisition and robot control [14]. The described system is the first prototype of the CPS, which it is referred herein as **M**ultimodal **I**mage-guided **RO**bot-assisted **S**urgeries (MIROS). The primary design directive and innovation of the MIROS system is the use of true and controllable sensing of the physical world with multi-contrast MRI to extract information needed for the control of a surgical robot, as well as generating an information-rich but intuitively comprehensive perception for the operator. In the following section we briefly describe different aspects and provide definitions about concepts and technologies that we used in designing the MIROS system [15].

2 System Architecture and Methodology

The MIROS system was designed and implemented to seamlessly integrate the operator with the robot and the MR scanner via multi-task central software, the "computational core". The computational core of the system receives on-the-fly (or off-line) MR images, processes them and provides the operator with the information to perform both the pre-operative planning and the RT intra-operative guidance phases of the procedure. Figure 1 shows the different modules and the information flow in the system, while Table 1 reviews the terminology and abbreviations of the different processes and components of the system.

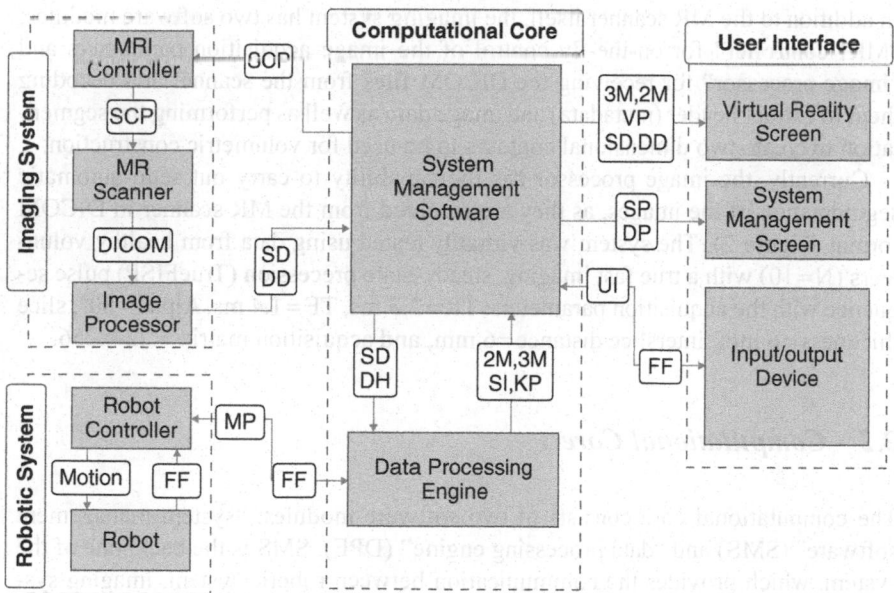

Figure 1. The architecture and information flow of MIROS.

Table 1. The nomenclature of the system is defined.

Abbreviation	Full name
DICOM	Digital Imaging and Communications in Medicine
SD	Segmented Data
DD	Decoded DICOM
DH	DICOM Header Information
3M	Three Dimensional Model
2M	Two Dimensional Model
VP	Visual Parameters
SI	Size Information
SP	System Parameters
KP	Kinematic Parameters
MP	Motion Parameters
DP	Diagnostic Parameters
FF	Force Feedback
UI	User Input
SCP	Scan Parameters

We used C++ as main programming language, OpenGL for visualization, CUDA for GPU programming and Microsoft Visual Studio 2008 as the integrated development environment.

2.1 Imaging system

In addition to the MR scanner itself, the imaging system has two software modules; "MRI controller" for on-the-fly control of the image acquisition parameters and "image processor" for receiving the DICOM files from the scanner and decoding them to extract header (metadata) and image data as well as performing the segmentation to create two dimensional contours to be used for volumetric construction.

Currently, the image processor has the capability to carry out semi-automatic segmentation of the images, as they are received from the MR scanner in DICOM format (Figure 2). The system was virtually tested using data from healthy volunteers (N=10) with a true fast imaging, steady-state precession (TrueFISP) pulse sequence with the acquisition parameters: TR=2.3 ms, TE=1.4 ms, Alpha=80°, slice thickness=6 mm, interslice distance=6 mm, and acquisition matrix = 224x256.

2.2 Computational Core

The computational core consists of two software modules: "system management software" (SMS) and "data processing engine" (DPE). SMS is the backbone of the system, which provides the communication between robotic system, imaging system, user interface and DPE. As shown in Figure 1, all the necessary information is collected and distributed by SMS.

Figure 2. Four selected long axis images depict the segmentations of the inner boundaries of the left ventricle in different heart phases.

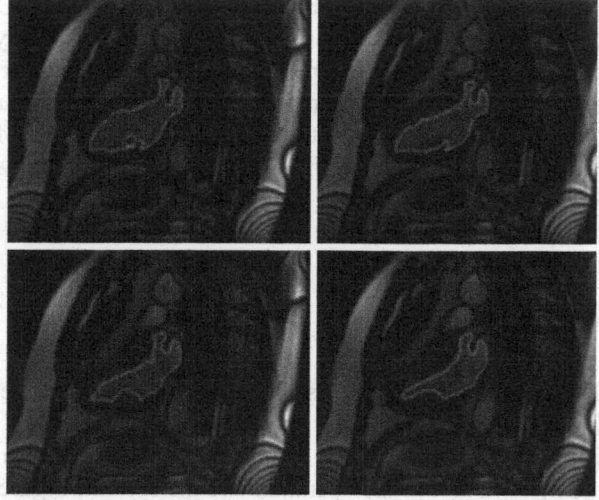

DPE's main job is to get SD and DH to build a safe trajectory and update it continuously for the robotic catheter. This module shouldered computationally the heaviest part, since it must process the data in RT, i.e., it should finish processing the data for one time frame before the next arrives.

In order to achieve the necessary fast image processing, we used a graphics processing unit (GPU) (NVIDIA® Tesla™ C1060; 240 Streaming Processor Cores; and total dedicated memory of 4 GDDR3) for DPE. GPU is a specialized multicore processor designed for getting high performance in graphics rendering. Besides, due to its parallel structure and excellent floating point performance, a GPU is far more effective than a conventional CPU for an application such as MIROS. Recently, GPUs are used as general purpose parallel processors that come with readily available Software Development Kits. CUDA (Compute Unified Device Architecture), the parallel computing architecture introduced by NVIDIA®, is the computing engine of the GPU. It enables the developers to reach an incomparable performance via high level programming languages such as C/C++, and standard APIs like OpenCL and the Microsoft .NET Framework.

In MIROS, the data is parallelized in time frame level for every MRI slice, copying from main memory to the memory of a graphics processing unit (GPU), by accompanying CUDA with C++. Then the GPU gets this data and CPU's instruction for parallel execution in different cores.

2.3   Robotic system

Robotic system includes robot controller software and the robotic manipulator itself. The necessary actuation commands are fed the robot controller in the form of

MP. These parameters can be automatically created by DPE, as well as defined by the operator. In other words, the robot has the capability to deploy autonomously based on the trajectory determined by DPE, however operator can override the commands given by the system to prevent any possible harm. On the other hand, if operator gives a command that can move the robot out of its safe trajectory (due to the unexpected motion of the surgical field) the system will correct his motion and keep the robot in course.

2.4 User Interface

The user interface provides the interaction between the system and the operator. It comprises five input devices for not only controlling the robot but also managing the system parameters, and two 22-inch Widescreen Flat Panel (Dell™ E2210) monitors to screen the operation and the system. Figure 3 shows the current user interface setting.

To control the motion of the virtual robot, we used Logitech® Flight System G940 which has a force feedback joystick, a dual throttle and rudder pedals with toe brakes. For the rest of the user input we used Logitech® Cordless Desktop® which includes a cordless ergonomic split keyboard and cordless high-performance laser mouse. One of the monitors, called virtual reality screen (VRS), is dedicated to virtual follow-up of the robot and the patient; while the other one, "system management screen", is assigned to show the system parameters and diagnostic information.

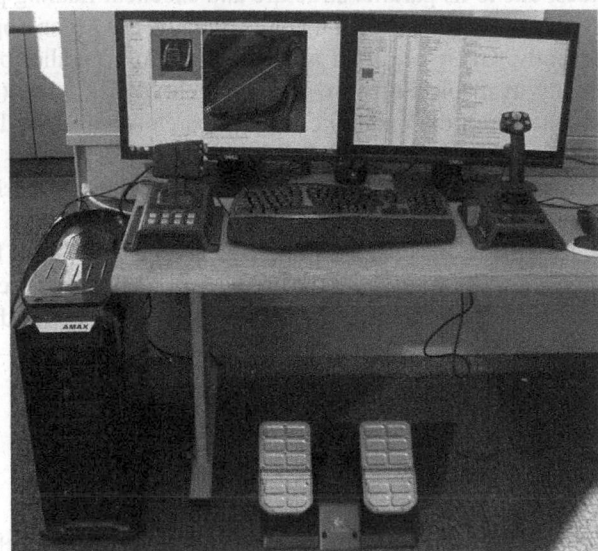

Figure 3. The photo shows the current setting of the user interface of MIROS.

3 Investigated Operation Scenario

MIROS was designed as a modular system that can be adopted for performing different types of procedures. Its specific software modules were built around a backbone in a way that modifications can be easily implemented. To better describe the structure and functionality of the MIROS, a sample practice scenario will be used that requires certain steps:

1. The operator selects the appropriate imaging planes and the acquisition parameters for the specific needs of the intervention.
2. The SMS gets UI from Input/output Device and sends it to the MRI Controller in the form of SCP.
3. The MRI controller makes the necessary adjustments on the scanner based on SCP.
4. The scanner collects the images and sends them to the Image Processor in DICOM format.
5. Image Processor decodes the DICOM files in order to extract the image and header data (DD). Then it performs the necessary image segmentation to create desired two dimensional contours (SD) and sends SD and DD to SMS.
6. The SMS sends SD and DH to DPE for processing, while it sends SD and DD to VRS for the operator to visually evaluate the images.
7. The DPE creates 2-dimensional (2M) and 3-dimensional (3M) models of the AoO, calculates the dimensions of the path (SI) and performs the kinematics calculations to generate KP for the robot.
8. Then DPE sends 2 M, 3 M, SI and KP to SMS for distribution to the User Interface and the Robotic System.
9. At this step, the operator has full control over the system. In other words, he/she can see the AoO in detail and give commands to the Robot Controller as well as the MRI Controller.
10. The SMS gets UI on-the-fly and send it to the RC in the form of MP via DPE.
11. The RC provides the desired actuation for the robot.
12. The SMS updates the VRS continuously as the procedure evolves.
13. The Robot gets force feedback from the surgical field and transfers it to the operator via RC, DPE and SMS.

4 Conclusion

In this work we investigated the feasibility of image guidance in minimally invasive cardiac surgeries and described the prototype version of a CPS for performing MR-guided cardiac interventions in the beating heart. Although at its early stages, the system has the required data processing and control modules with the appropriate architecture to perform the desired task. Our current works-in-progress include: (1)

Implementing an automatic image segmentation algorithm [16], (2) Accessing to a clinical scanner via a dedicated TCP/IP connection for receiving data in real time and controlling the scanner on-the-fly [17], (3) In vitro testing of the robotic manipulator on a phantom [18] and (4) Adapting a robust tissue tracking algorithm for motion compensation to account for the beating of the heart [19].

Based on the design challenges and the preliminary results of the prototype system, we conclude that surgical CPS development entails coherent integration of the robot, MR scanner, the patient, the operator and a robust methodology to provide an intuitive perception of the AoO.

Acknowledgements This work was supported by NSF CPS-0932272. All opinions, findings, conclusions or recommendations expressed in this work are those of the authors and do not necessarily reflect the views of our sponsors. The authors wish to thank to Prof. Zhigang Deng, Nikhil Navkar, Nick von Sternberg, Yousef Hedayati and Johann Lamaury for their continuous input and contributions to this project.

References

1. E.A. Lee, "Cyber Physical Systems: Design Challenges," Book Cyber Physical Systems: Design Challenges, Series Cyber Physical Systems: Design Challenges, ed., IEEE Computer Society, 2008.
2. F.A. Jolesz, "Future perspectives for intraoperative MRI," Neurosurg Clin N Am, vol. 16, no. 1, 2005, pp. 201–213.
3. K. Cleary, A. Melzer, V. Watson, G. Kronreif, and D. Stoianovici, "Interventional robotic systems: applications and technology state-of-the-art," Minim Invasive Ther Allied Technol, vol. 15, no. 2, 2006, pp. 101–113.
4. M. Moche, R. Trampel, T. Kahn, and H. Busse, "Navigation concepts for MR image-guided interventions," J Magn Reson Imaging, vol. 27, no. 2, 2008, pp. 276–291.
5. N.V. Tsekos, A. Khanicheh, E. Christoforou, and C. Mavroidis, "Magnetic resonance-compatible robotic and mechatronics systems for image-guided interventions and rehabilitation: a review study," Annu Rev Biomed Eng, vol. 9, 2007, pp. 351–387.
6. C. Keroglou, N.V. Tsekos, I. Seimenis, E. Eracleous, C.G. Christodoulou, C. Pitris, and E.G. Christoforou, "Design of MR-compatible robotic devices: magnetic and geometric compatibility aspects," 9th International Conference on Information Technology and Applications in Biomedicine, 2009, ITAB 2009.
7. E. Christoforou, E. Akbudak, A. Ozcan, M. Karanikolas, and N.V. Tsekos, "Performance of interventions with manipulator-driven real-time MR guidance: implementation and initial in vitro tests," Magn Reson Imaging, vol. 25, no. 1, 2007, pp. 69–77.
8. A. Ozcan, E. Christoforou, D. Brown, and N. Tsekos, "Fast and efficient radiological interventions via a graphical user interface commanded magnetic resonance compatible robotic device," Conf Proc IEEE Eng Med Biol Soc, vol. 1, 2006, pp. 1762–1767.
9. L.H. Cohn, "Future directions in cardiac surgery," Am Heart Hosp J, vol. 4, no. 3, 2006, pp. 174–178.
10. M.M. Reijnen, C.J. Zeebregts, and W.J. Meijerink, "Future of operating rooms," Surg Technol Int, vol. 14, 2005, pp. 21–27.
11. G. Li, D. Citrin, K. Camphausen, B. Mueller, C. Burman, B. Mychalczak, R.W. Miller, and Y. Song, "Advances in 4D medical imaging and 4D radiation therapy," Technol Cancer Res Treat, vol. 7, no. 1, 2008, pp. 67–81.

12. E.R. McVeigh, M.A. Guttman, R.J. Lederman, M. Li, O. Kocaturk, T. Hunt, S. Kozlov, and K.A. Horvath, "Real-time interactive MRI-guided cardiac surgery: aortic valve replacement using a direct apical approach," Magn Reson Med, vol. 56, no. 5, 2006, pp. 958–964.

13. E. Yeniaras, J. Lamaury, Z. Deng, and N.V. Tsekos, "Towards A New Cyber-Physical System for MRI-Guided and Robot-Assisted Cardiac Procedures," Proceedings of the 10th IEEE International Conference on Information Technology and Applications in Biomedicine, ITAB, 2010.

14. E. Yeniaras, Z. Deng, M. Davies, M.A. Syed, and N.V. Tsekos, "A Novel Virtual Reality Environment for Preoperative Planning and Simulation of Image Guided Intracardiac Surgeries with Robotic Manipulators," Medicine Meets Virtual Reality, vol. 18, 2011.

15. E. Yeniaras, N. Navkar, M.A. Syed, and N.V. Tsekos, "A Computational System for Performing Robot-assisted Cardiac Surgeries with MRI Guidance," Proceedings of the 15th International Transformative Systems Conference, SDPS.

16. M. Fradkin, C. Ciofolo, B. Mory, G. Hautvast, and M. Breeuwer, "Comprehensive segmentation of cine cardiac MR images," Med Image Comput Comput Assist Interv, vol. 11, and no. Pt 1, 2008, pp. 178–185.

17. N.V. Tsekos, A. Ozcan, and E. Christoforou, "A prototype manipulator for magnetic resonance-guided interventions inside standard cylindrical magnetic resonance imaging scanners," J Biomech Eng, vol. 127, no. 6, 2005, pp. 972–980.

18. N.V. Sternberg, Y. Hedayati, E. Yeniaras, C. E., and N.V. Tsekos, "Design of an actuated phantom to mimic the motion of cardiac landmarks for the study of image-guided intracardiac interventions," Proceedings of the 10th IEEE International Conference on Robotics and Biomimetrics, vol. 2010, 2010.

19. Y. Zhou, E. Yeniaras, P. Tsiamyrtzis, N. Tsekos, and I. Pavlidis, "Collaborative Tracking for MRI-Guided Robotic Intervention on the Beating Heart," Proceedings of the 13th International Conference on Medical Image Computing and Computer Assisted Intervention (MICCAI), 2010.

Chapter 5
IMPLEMENTING A TRANSDISCIPLINARY EDUCATION AND RESEARCH PROGRAM

U. John Tanik

1 Introduction

The objective of this chapter is to introduce the implementation of a graduate education program integrated with research which can be identified as Transdisciplinary Healthcare Engineering. Before the introduction of the proposed program, we will review healthcare engineering as proposed by the National Academy of Engineering (NAE) and Institute of Medicine (IOM). This review will provide the basis and rationale for the proposed transdisciplinary healthcare engineering program introduced in the second section of the paper. Our transdisciplinary approach to advancing healthcare engineering is to create a new integrated graduate educational and research paradigm in which students are trained to conduct research seamlessly across disciplinary boundaries. This new graduate program will be designed to draw students from all disciplines and geography leveraging distance education technology in order to provide general education and training on how to apply their current and new knowledge towards a common goal in healthcare engineering. The outcomes of this healthcare engineering education will produce awareness of healthcare engineering issues, applications, and job opportunities in a collaborative transdisciplinary environment based on the guiding expertise of faculty from Engineering, Arts and Sciences, Medicine, Business, Law, and Education.

2 National Healthcare Crisis Driving Healthcare Engineering Demand

Healthcare is expected to comprise nearly 20% of the Gross Domestic Product of the United States within 10 years, making it the largest industry in this country [1]. Yet, despite such a huge investment and budget set aside to treat our citizens,

U. J. Tanik (✉)
Fort Wayne, IN, USA
e-mail: jtanik@uab.edu

S. Suh et al. (eds.), *Biomedical Engineering*,
DOI 10.1007/978-1-4614-0116-2_5, © Springer Science+Business Media, LLC 2011

the number of deaths attributed to *preventable* medical errors reaches as high as 100,000 persons *each year* [2]. This figure represents those patients who have received treatment through the current medical system and does not attempt to count the unnecessary injuries occurring during treatment that numbers in the millions. These findings and other critical factors about the unacceptable state of the current health system were reported by a joint effort between the National Academy of Engineering and the Institute of Medicine in 2005. Simply put, the current health system needs an improved system to keep pace with technological advances. In fact, according to Purdue University Healthcare Engineering Program, the problem is so great that an "estimated 30 – 40% of healthcare expenditures go to overuse, underuse, misuse, duplication, system failures, unnecessary repetition, poor communication, and inefficiency. Furthermore, the industry is massively under-invested in information technology, with fewer than 15% of patient records available electronically." [1]

2.1 Transdisciplinary Approach to Healthcare Engineering

In order to address this national health crisis, a new field of study called *healthcare engineering* is emerging that primarily applies systems engineering solutions that can benefit from a transdisciplinary approach. One recently established representative model to develop programs to educate and train professionals in this field can be found at Purdue. In January 2005, the Regenstrief Center for Healthcare Engineering (RCHE) began operating with $3 million in initial funding [3]. The transdisciplinary character to healthcare engineering can be seen in this internationally-acclaimed effort at Purdue, considering that dozens of faculty members from a spectrum of disciplines are vigorously collaborating on various joint projects in fields of Engineering, Management, Pharmacy, Nursing, Health Science, Communications, Consumer and Family, Veterinary Medicine, Computer Science, and Technology. However, in order to tackle national problems, success in one university is not enough. A consortium of universities need to work collaboratively to achieve the success found at Purdue and UAB to make a greater impact on a national scale. We propose that progressive universities that can adapt to the needs of the nation should apply their resources and strengths in engineering and medicine to develop such programs. A good starting point of applying the lessons learned at Purdue would be collaboration between IPFW and UAB by leveraging their mutual strengths and resources in engineering and medicine.

2.2 Transdisciplinary Education and Research

The traditional disciplinary efforts of the 20th century do not seem to meet all our needs in the 21st century as evidenced by commonly found cost-overruns and waste in all facets of industry. The growing interconnections and dependencies among academic disciplines and the explosion of knowledge have transcended discipline-specific academic fields, resulting in different approaches to solving complex problems, involving disciplinary paradigms such as Multidisciplinarity, Interdisciplinarity, Crossdisciplinarity, and Trandisciplinarity. Specifically, *Multidisciplinarity* joins together two or more disciplines without integration; *Interdisciplinarity* integrates two or more disciplines without dissolving disciplinary boundaries; *Crossdisciplinarity* crosses disciplinary boundaries to explain one subject in terms of another. *Trandisciplinarity* joins, integrates, and/or crosses two or more disciplines by dissolving disciplinary boundaries to form a cohesive whole (Table 1) [5].

We propose that educational programs must explicitly step beyond the discipline–based approach to learning and instead focus on knowledge domains that cut across traditional academic fields using *the transdisciplinary approach to teaching and research.* Transdisciplinarity provides an essential communications bridge between disparate disciplinary projects. The need for this transdisciplinary approach grows as the silos of knowledge in each discipline grows exponentially over time in the Information Age.

To improve disciplinary thinking, Transdisciplinarity appears as a promising approach to integration which can be woven into existing disciplines. In so doing, the isolated silos can be melded together for better understanding and management. With this approach, process issues become an integral part of the design problem, a premise for founding the Society for Design and Process Science [4]. This international society is positioned to form foundational pillars for design and process science. Furthermore, SDPS international will be able to manage the development of the SDPS student chapters in the United States in order to support and spearhead educational programs in universities that will advance NISTEM goals at IPFW and UAB at the outset. Understanding such design and process techniques will take time, but it will be worth the effort. Many startup ventures may also benefit from the advances made in corporate design theory developed in these student chapters.

Axiomatic design theory can be used as a starting point to teach the two principles of design based on the information axiom and the independence axiom. In the case of Axiomatic Design, the most appropriate design choice can be made during the integration process by optimally taking into account a myriad of disciplinary

Table 1. Brief Descriptions of Various Integrated Disciplinary Approaches.

Approach	Short Description
Multidisciplinarity	*Joins* together disciplines without integration
Interdisciplinarity	*Integrates* disciplines without dissolving disciplinary boundaries
Crossdisciplinarity	*Crosses* disciplinary boundaries to explain one subject in terms of another
Trandisciplinarity	*Joins, integrates, and/or crosses* disciplines by dissolving disciplinary boundaries

solutions to fulfill any given functional requirement. Around the world, universities are working to refine their vision of education and research as new technologies emerge, creating distance education opportunities that remove many geographic barriers to learning. These technologies can be readily implemented in healthcare engineering as well.

In the years prior to 1995, a group of visionary leaders convened under the auspices of the United States, in Austin, Texas, to develop the first transdisciplinary graduate education program. Since these years of expansion, the quality and precision of thinking for this group has reached new heights of success. However, a transformational paradigm must be introduced to take some aspects of university education to levels needed to meet the needs of industry demands. We believe that the intellectual enrichment process introduced by SDPS can reinvigorate some major university course delivery systems. This will not occur overnight. In fact, the long, but fruitful road ahead depends on the efforts of bright students in the university sector who rely on the rich educational landscape and historical legacy provided by the US higher education system. This success can be achieved by introducing universally accepted standards of professionalism and integrity, while ensuring that the primary tenets of design and process science are introduced in an orderly fashion. A Transdisciplinary Master of Engineering and Ph.D. Programs in Design, Process and Systems were developed for employees at a major defense corporation based on the tenets of transdisciplinary thinking [5]. The program has been very well received by Raytheon and by the students. Our intention is to apply the lessons learned from this first transdisciplinary experience and apply it to healthcare engineering. Therefore, first we will review the various topics addressed by healthcare engineering below and in the last section we will discuss the implementation.

3   Healthcare Engineering Topics

As the expected need for services increases for the baby boomer generation, the national hospital and clinic construction efforts are attempting to keep pace with an aging population. As the growth in this sector continues, the 2008 Hospital Building Report published detailed findings in its article "No End in Sight, Facilities still in major construction mode, but will patients come?" In the case of facility construction, health care engineering efforts have doubled since 2005, reaching $41 billion in the fourth quarter of 2007 [6]. This means more students need to be educated in the field of healthcare engineering, which we believe can benefit from a transdisciplinary education approach. In addition to playing a role in facilities design, healthcare engineering covers other topics in Operations Management, Clinical Environmental Management, Medical Imaging, Pharmaceuticals, Clinical Decision Support, Distance Health, Patient Informatics, Assisted Mobility, and Perception-Based Engineering [3].

4   Healthcare Engineering Recommendations advancing Education and Research

A landmark study conducted by the National Academy of Engineering and the Institute of Medicine [2] reports on the medical crisis facing the United States with a series of recommendations. Among these high level recommendations, many needs are articulated for healthcare engineering education and training. We provide a survey (Table 2) of some of these recommendations as a perspective on how they may apply to our notion of Healthcare Engineering education and research.

In this study "Building a Better Delivery System: A New Engineering/Health Care Partnership," the Committee on Engineering and the Health Care System from the Institute of Medicine (IOM) and National Academy of Engineering (NAE) states that a divide exists between disciplines that needs to be bridged. We believe that this gap can be overcome by implementing a transdisciplinary approach to healthcare engineering education by addressing this report.

> In a joint effort between the National Academy of Engineering and the Institute of Medicine, this book attempts to bridge the knowledge/awareness divide separating health care professionals from their potential partners in systems engineering and related disciplines. The goal of this partnership is to transform the U.S. health care sector from an under-performing conglomerate of independent entities (individual practitioners, small group practices, clinics, hospitals, pharmacies, community health centers et. al.) into a high performance "system" in which every participating unit recognizes its dependence and influence on every other unit. By providing both a framework and action plan for a systems approach to health care delivery based on a partnership between engineers and health care professionals, Building a Better Delivery System describes opportunities and challenges to harness the power of systems-engineering tools, information technologies and complementary knowledge in social sciences, cognitive sciences and business/management to advance the U.S. health care system.

A partial list of recommendations from the NAE and IOM with applications to transdisciplinary implementation is provided in Table 2.

5   Proposed Transdisciplinary Healthcare Engineering Program

Transdisciplinary Healthcare Science & Engineering Program can bring a systems-analysis approach for improving the processes of healthcare delivery. The application of engineering, management, and scientific principles can be taught systematically by training professionals in order to reshape the healthcare delivery landscape. Furthermore, education and research can now be accomplished online with effectiveness as well as efficiency for greater geographic reach. For instance, Virtual Health Systems development in which "process replaces products" is expected to become a significant force to keep pace with technology convergence and the global service industry [10]. The Society for Design and Science can lead efforts with

Table 2. Relevant Recommendations from NAE and IOM for Implementation.

Recommendations with Direct Impact on Education and Research	Implementation
Recommendation 5-1a. The federal government, in partnership with the private sector, universities, federal laboratories, and state governments, should establish multidisciplinary centers at institutions of higher learning throughout the country capable of bringing together researchers, practitioners, educators, and students from appropriate fields of engineering, health sciences, management, social and behavioral sciences, and other disciplines to address the quality and productivity challenges facing the nation's health care delivery system. To ensure that the centers have a nationwide impact, they should be geographically distributed. The committee estimates that 30 to 50 centers would be necessary to achieve these goals.	SDPS and collaborating universities, e.g. UAB, and Purdue can support the establishment of transdisciplinary centers.
Recommendation 5-1b. These multidisciplinary research centers should have a three-fold mission: (1) to conduct basic and applied research on the systems challenges to health care delivery and on the development and use of systems engineering tools, information/communications technologies, and complementary knowledge from other fields to address them; (2) to demonstrate and disseminate the use of these tools, technologies, and knowledge throughout the health care delivery system (technology transfer); and (3) *to educate and train a large cadre of current and future health care, engineering, and management professionals and researchers in the science, practices, and challenges of systems engineering for health care delivery.*	SDPS and collaborating universities, e.g. UAB and Purdue can implement Point 2, which is directly relevant to education.
Recommendation 5-2. Because funding for the multidisciplinary centers will come from a variety of federal agencies, a lead agency should be identified to bring together representatives of public- and private-sector stakeholders to ensure that funding for the centers is stable and adequate and to develop a strategy for overcoming regulatory, reimbursement, related, and other barriers to the widespread application of systems engineering and information/communications technologies in health care delivery.	SPDS and collaborating universities, e.g. UAB and Purdue can establish the pilot delivery mechanisms.
Recommendation 5-3. Health care providers and educators should ensure that current and future health care professionals have a basic understanding of how systems-engineering tools and information/communications technologies work and their potential benefits. Educators of health professionals should develop curricular materials and programs to train graduate students and practicing professionals in systems approaches to health care delivery and the use of systems tools and information- communications technologies. Accrediting organizations, such as the Liaison Committee on Medical Education and Accreditation Council for Graduate Medical Education, could also require that medical schools and teaching hospitals provide training in the use of systems tools and information/communications technologies. Specialty boards could include training as a requirement for recertification.	SPDS and collaborating universities such as UAB and Purdue can introduce transdisciplinary education through the identified core courses in Table 3.

Table 2. (continued)

Recommendations with Direct Impact on Education and Research	Implementation
Recommendation 5-4. Introducing health care issues into the engineering curriculum will require the cooperation of a broad spectrum of engineering educators. Deans of engineering schools and professional societies should take steps to ensure that the relevance of, and opportunities for, engineering to improve health care are integrated into engineering education at the undergraduate, graduate, and continuing education levels. Engineering educators should involve representatives of the health care delivery sector in the development of cases studies and other instructional materials and career tracks for engineers in the health care sector.	SPDS, and collaborating universities such as UAB and Purdue are well-suited to cover these collaborative aspects.
Recommendation 5-5. The typical MBA curriculum requires that students have fundamental skills in the principal functions of an organization—accounting, finance, economics, marketing, operations, information systems, organizational behavior, and strategy. Examples from health care should be used to illustrate fundamentals in each of these areas. Researchers in operations are encouraged to explore applications of systems tools for health care delivery. Quantitative techniques, such as financial engineering, data mining, and game theory, could significantly improve the financial, marketing, and strategic functions of health care organizations, and incorporating examples from health care into the core MBA curriculum would increase the visibility of health care as a career opportunity. Business and related schools should also be encouraged to develop elective courses and executive education courses focused on various aspects of health care delivery. Finally, students should be provided with information about careers in the health care industry.	SPDS and collaborating universities such as UAB and Purdue can cover the business and computing synergy essential for advancing the transdisciplinary paradigm.
Recommendation 5-6. Federal mission agencies and private-sector foundations should support the establishment of fellowship programs to educate and train present and future leaders and scholars in health care, engineering, and management in health systems engineering and management. New fellowship programs should build on existing programs, such as the Veterans Administration National Quality Scholars Program (which supports the development of physician/scholars in health care quality improvement), and the Robert Wood Johnson Foundation Health Policy Research and Clinical Scholars Programs (which targets newly minted M.D.s and social science Ph.D.s, to ensure their involvement in health policy research). The new programs should include all relevant fields of engineering and the full spectrum of health professionals.	SPDS and collaborating universities such as UAB, and Purdue can initiate Fellowships and Training Programs.

other systems organizations, such as the International Society for the Systems Sciences (ISSS) and International Council for Systems Engineering(INCOSE), to advance goals in education and research for healthcare engineering that leverage new media technologies for distance education [12]. Furthermore, a grounding in system of systems (SoS) thinking may be needed to handle the complexities of systems integration [13]. In this context, a representative set of core courses is proposed in Table 3.

5.1   Proposed Implementation

The paradigm of Transdisciplinary Science Engineering (TSE) introduces a significant potential shift in science and engineering. The trend is towards a more synthetic approach, external to the disciplines, yet guided through successful integrated solutions to complex problems. Contrary to interdisciplinary or multidisciplinary thinking, the transdisciplinary approach seeks to find a synthesis of knowledge through a unifying idea or concept. It attempts to find a framework for bringing together disciplinary knowledge that enhances both the discovery and application processes.

Large–scale, complex problems include not only the design of engineering systems with numerous components and subsystems which interact in multiple and intricate ways; they also involve the design, redesign and interaction of social, political, managerial, commercial, religious, biological, and medical systems. Furthermore, these systems are likely to be dynamic and adaptive in nature. Solutions to such large–scale, complex problems require many activities which cross disciplinary boundaries.

In his SDPS 2000 talk Nobel Laureate Herbert Simon stated ... *Today, complexity is a word that is much in fashion. We have learned very well that many of the systems that we are trying to deal with in our contemporary science and engineering are very complex indeed. They are so complex that it is not obvious that the powerful tricks and procedures that served us for four centuries or more in the development of modern science and engineering will enable us to understand and deal with them. We are learning that we need a science of complex systems, and we are beginning to construct it...* [8]: Transdisiplinarity science and Engineering: integrating science and engineering principles [11]. Thus, we seek to expand Simon's transdisciplinary approach by tackling complex systems in healthcare engineering.

6   Conclusion

In the last decade we have witnessed a surge of technology innovations enabling humans to transcend time and place, accompanied with shifting clinical systems initiatives and patient needs, which have changed the way we perceive healthcare delivery today. We expect our classical healthcare, delivered by professionals in primary care and hospitals, to be extended by having personalized self-care, home-care, or any other customized healthcare service, which can be delivered at any time and any place. Transdisciplinarity addresses these evolving patient expectations and global system transformations through the integration of healthcare environments, supported by communication and information technology, which provide seamless access to healthcare services and ultimately improve the delivery of healthcare. Leading transdisciplinary organizations such as SDPS can also advance numerous educational and research initiatives recommended by governments around the world. In addition to proactively upgrading and preserving national health, lasting economic benefits can be achieved by overhauling the ailing healthcare systems in a way to keep pace with aging population and life-style related diseases. Therefore, introducing the transdisciplinary approach to healthcare engineering research and education can make an impact on national health and economics.

References

1. Purdue College of Engineering Homesite: https://engineering.purdue.edu/Engr/Research/Initiatives/HE/ [accessed April 2008]
2. Reid, Proctor P., Compton, W. Dale, Grossman, Jerome H., and Fanjiang, Gary, Committee on Engineering and the Health Care System; *Building a Better Delivery System: A New Engineering/Health Care Partnership,* Institute of Medicine and National Academy of Engineering 2005 Report, The National Academies Press: www.nap.edu [accessed April 2008]
3. The Regenstrief Center for Healthcare Engineering at Purdue University http://www.purdue.edu/discoverypark/rche/ [Accessed April 2008]
4. Society for Design and Process Science home site: www.sdpsnet.org [Accessed April 2008]
5. Ertas, A., Maxwell T., Rainey, Vicki P., and Tanik, M. M., Transformation of Higher Education: The Transdisciplinary Approach in Engineering: IEEE Transactions On Education, Vol. 46, No. 2, May 2003.
6. Carpenter, D., Hoppszallern, S., Hospital Building Report "No End in Sight, Facilities still in major construction mode, but will patients come?" February 2008, Health Facilities Management Magazine www.hfmmagazine.com [Accessed April 2008]
7. Tanik, U. J., December 2006. *Artificial Intelligence Design Framework for Optical Backplane Engineering.* Dissertation. University of Alabama at Birmingham.
8. 5th International Conference on Integrated Design and Process Technology (SDPS/IDPT), June 4th-8th 2000, Dallas, Texas, USA.
9. The Task Force on Access to Health Care in Texas, composed of a non-partisan group sponsored by all ten of the major academic health institutions in Texas. Statement issued in April 17, 2006 the "Challenges of the Uninsured and Underinsured" at a press conference and at the James A. Baker III Institute for Public Policy at Rice University. http://www.coderedtexas.org/files/Report_Preface.pdf [Accessed April 2008]
10. General Systems Bulletin Volume XXX,2001 The International Society for the Systems Sciences http www.isss.org/isss/ [Accessed April 2008]
11. Simon, H. A.,1981, *The Sciences of the Artificial,* MIT Press, Cambridge, MA.
12. INCOSE,2006, "The International Council on Systems Engineering," http://www.incose.org/ [Accessed April 2008]
13. Ring, J. and A. M. Madni,2005, "Key Challenges and Opportunities in 'System of Systems' Engineering," *IEEE International Conference on Systems, Man, and Cybernetics,*Vol. 1, pp. 973–978.

14. GUARDS,1997, "Approach, Methodology and Tools for Validation by Analytical Modelling, "Technicatome/PDCC Second Part of D302: Functional Specification and Preliminary Design of GUARDS Validation Environment", by E. Jenn and M. Nelli, 1997,*ESPRIT Project*, 20716 GUARDS Report, February.

15. HIDE,1999, "High Level Integrated Design," http://www3.informatik.uni-erlangen.de/Publications/Articles/dalcin_words99.pdf [Accessed April 2008]

16. Gruber, T. R.,1993, *Toward principles for the design of ontologies used for knowledge sharing*, Report KSL-93-04, Stanford University.

17. IHMC,2006, "Florida Institute for Machine Cognition," http://www.coginst.uwf.edu/newsletters/IHMCnewslettervol3iss1.pdf [Accessed April 2008]

Chapter 6
JIGDFS: A SECURE DISTRIBUTED FILE SYSTEM FOR MEDICAL IMAGE ARCHIVING

Jiang Bian and Remzi Seker

1 Introduction

The growth in use of medical imaging resulted in great challenges, such as handling, storing, retrieving and transmitting biomedical images. The Health Information Technology for Economic and Clinical Health (HITECH) Act went into effect in February 2010 [1]. The HITECH Act has drawn great attention to healthcare IT. One of the major concerns in healthcare IT is the development of highly secure medical imaging archiving systems. Although, the Digital Imaging and Communications in Medicine (DICOM) standard [2] enables the integration of modalities, servers, workstations and printers into a picture archiving and communication system (PACS), it does not specify how to handle the underlying storage system efficiently and securely. Further, a commercial PACS system often does not actively address the security concerns of healthcare professionals. Meanwhile, due to the lack of expert knowledge in the computer and information security fields, the physicians and medical imaging technicians are often reluctant to embrace new medical imaging technologies. In some cases, healthcare professionals are not able to fully appreciate security issues. Without proper education, it is hardly possible to increase the medical professionals' level of confidence, over the security features of a medical imaging system. On the other hand, when developing such a system, the software engineers need to have a deep understanding of not only the importance of security in such an environment, but also possess a certain level of knowledge about the medical practice environment. Therefore, cross-disciplinary collaboration is inevitable.

The large number of studies performed at the Nuclear Medicine Department of University of Arkansas for Medical Sciences (UAMS) generates a decent amount PET/CT images daily. There are two SIEMENS PET/CT modalities with full workload (i.e. on average, 15 patients/studies each day). Each study generates three sets of images: a set of Computed Tomography (CT), a set of raw Positron Emission

R. Seker (✉)
Liitle Rock, AR, USA
e-mail: rxseker@ualr.edu

S. Suh et al. (eds.), *Biomedical Engineering,*
DOI 10.1007/978-1-4614-0116-2_6, © Springer Science+Business Media, LLC 2011

Tomography (PET), and a set of corrected PET, which results in on average 1,800 image slices and requires 350 megabytes of storage space for each study. Each year, the Nuclear Medicine Department alone demands three terabytes of storage space. Considering other departments that rely on medical imaging technologies to make clinical decisions, the amount of data generated for the overall campus is enormous. Commercial PACSs are widely used on campus, however because of a lack of collaboration between software developers and medical professionals, those systems are far from satisfactory in terms of user requirements and experiences. Many PACS implementations increase the storage capacity simply by adding more disk space. In most cases, a PACS relies on the Redundant Array of Independent Disks (RAID) technology to improve storage reliability. Most commercial PACSs rely on one commodity computer hardware to handle all storage, as well as backups. However, there certainly is a limit in terms of the overall size of storage one PACS server can have. Law requires that the medical images, which are part of patients' medical records, must be stored safely by the healthcare provider for a long period of time. The retention policy of health information varies for every state. For example, in Arkansas, all medical records shall be retained using acceptable methods for ten years after the last discharge. In some other states, like Massachusetts, the hospitals shall keep records of the patients for 30 years following discharge or last contact of patient. In general, any diagnostic image is recommended to be retained for at least 5 years [3]. Storing these long-term imaging archives securely and efficiently is certainly a challenge.

Most PACSs favor tape drives for backup as well as for archiving. Although magnetic tape is relatively cheap and has a high storage capacity, it certainly has disadvantages such as slow and sequential access. Moreover, the security issues of using a tape drive certainly constitute a bigger concern. Although DICOM standard does have a supplemental specification for security, its goal is to protect DICOM files during media interchanges rather than for long-term storage. Originally, DICOM suggested using Transport Layer Security (TLS) [4] standard for secure communications between the client and server. Later, in a security supplement (SUP-51), DICOM addresses the security issues at the level of DICOM media storage by using the Cryptographic Message Syntax (CMS) [5]. However, since CMS is derived from PKCS #7 [6] and has a certificate-based key management system, it requires a priori knowledge of all recipients and a public key infrastructure. This approach is often referred as too hard to deploy and implement in practice. For example, if there is a new facility to which a patient is transferred, it may be difficult to pre-determine the new facility's public key, so that a re-encryption can take place. DICOM CP-895 [7] mentions that it is possible to use a "password-based encryption for CMS" as specified in RFC 3565 [8]. However, the complexity associated with constructing a secure DICOM object makes its full implementation difficult in practice.

Through collaborative research with the medical experts and imaging technicians, the authors of this chapter believe that the PACS system should be separated from its storage system. Also the security and reliability of the underlying storage system should not be a concern for the PACS system. Therefore, we propose a secure distributed file system, JigDFS that can meet the storage requirements of medical images. As a file system, JigDFS provides fault tolerance that is competitive to traditional mirroring schema while using less storage. JigDFS uses an Information Dispersal Algorithm (IDA) [9] to add error correction capability that increases availability of the files in the system as well as the possibility of recovery when a certain number of node failures occur. The data slices making up a file are encrypted on each child node and then sliced further and encrypted again to be propagated down the tree. In JigDFS itself, the files and file slices are not digitally signed, and the recursive IDA and layered encryption enhance privacy and provide certain level of plausible deniability. However, to be HIPAA compliant, a JigDFS-based PACS system should implement entity authentication, access controls as well as audit trails apart from the JigDFS storage system. In another words, the sole goal of JigDFS is to provide a reliable and secure storage system that ensures data integrity, secrecy and recovery. An actual PACS implementation over JigDFS is straightforward.

The rest of the chapter is organized into five sections. We will discuss the potential threats and design goals of a secure distributed file system especially when used as a medical imaging archive in the next section. In the background section, we present our background study about the DICOM standard, the Information Dispersal Algorithm as well as the hashed-key chain structure. In implementation section, we will introduce the overall design of JigDFS and some of the implementation details, especially, how files are indexed, stored and retrieved in JigDFS. Fault tolerance, data integrity, security analysis and system performance will also be discussed. An architecture overview of building a PACS using JigDFS as its backend long-term storage service will also be presented. Finally, we conclude with the importance of developing a JigDFS-like system for medical image archiving systems.

2   Potential Threats and Design Goals

Since JigDFS aims to serve as an underlying storage system where the files are stored securely and reliably, our design is instructed by potential threats that exist in many of the present systems and assumptions that help find a cost-effective solution, which is specific for archiving large volume of medical images.

2.1   Potential Threats

Medical images in the DICOM format contain protected health information (PHI), such as name, birth date, admission date, etc. Under the HIPAA regulations, any information about health status or provision of health care that can be linked to a specific individual needs to be protected in the Hospital Information Systems (HISs). Strong encryption helps protect information, since the adversary cannot decrypt the data without the key used for encryption. However, the security of modern encryption algorithms relies heavily on mathematical assumptions, such as computational infeasibility to factor n as pq for large p and q in a reasonable time frame. This might be a challenge since an archiving system is presumed to store its contents for a long time. The rapid development in computing technologies raises the available computational power dramatically. For example, in the future, quantum computing may increase the chance of breaking current PKI systems within a reasonable time.

It is also possible that a powerful adversary may be monitoring some of the network traffic, if he/she has control over the routers/switches to which users are connected. There may also be one or more internal adversaries who have access to some of the nodes, which can compromise the information stored on or visible to these nodes. Ideally, any leaked information should not reveal the secret of a whole document/file.

Malware is a common problem in today's computer world. A well-educated adversary may have the ability to take full control of one or more nodes by exploiting software bugs, such as buffer overflows, to launch shell-code on user computers. Even in such situations, where one or more JigDFS nodes are compromised, JigDFS should still protect user assets within its security boundary.

Many of these potential threats have not been fully addressed in any of the existing systems similar to JigDFS, which makes JigDFS different and unique. In the security analysis section, we will discuss in detail the approaches we employed to tackle each of the security issues mentioned above.

2.2 Design Goals

JigDFS is assumed to be deployed where hardware issues and operating system errors take place frequently on the nodes. Especially, to be a long-term storage system for archiving medical images, the file system needs to be fault-tolerant and the files should be recovered easily in the presence of node failures.

Medical image files such as positron emission tomography (PET), computed tomography (CT) and magnetic resonance imaging (MRI) scans are relatively large in size. For example, at the nuclear medicine department of UAMS, each PET/CT scan of a patient generates about 1,800 image slices and over 350 MB in size. Currently, these images are stored on a DICOM compatible Enterprise Image Repository (EIR) [10] using simple mirroring strategy to ensure data availability. The annual data storage consumption is over 3 TB. By using IDA, we can achieve the same level of data redundancy compared to simple mirroring while using less actual hard drive storage. Additionally, JigDFS is designed to be deployed on inexpensive commodity hardware. Comparing to costly RAID technologies, the JigDFS solution for archiving data is much more reasonable and easier to expand.

Most of the medical images, especially archived data, are rarely modified once written. Therefore, the write operation traffic is rather small. However, the system should have relatively higher traffic in sequential reads, notably for files like videos, sound files, high-resolution images, etc. Random reads and file modifications are supported. However, these operations do not have to be as efficient as read operation. The system should be more concerned with throughput than low-latency data access. Our initial benchmarks show that the read/write throughputs of JigDFS are acceptable as a user-space file system, while the read speed is reasonably higher than write, which conforms to the requirements of an archival system.

The system should efficiently support a large number of concurrent reads with desirable throughputs. Although two or more nodes uploading exactly the same file simultaneously can happen rarely, this should not cause any problem or corruption of data, but rather result in storing redundant data. Atomicity and synchronization are not concerns for storing files in JigDFS and are handled on each node by the underlying operating system. Since, JigDFS is a fully decentralized P2P system, the system scales relatively better than client/server architecture. In another words, the file operations throughputs are not affected by adding more nodes.

To meet these assumptions while tackling the security concerns, the main design decisions we made are 1) the use of IDA not only ensures the availability of files and provides fault tolerance against node failures, but also enhances the overall security; 2) the use of the hashed-key chain algorithm accommodating key management which provides strong encryption and some level of plausible deniability. In the next section, we will present our considerations for these design decisions in detail.

3   Background

3.1   The DICOM Standard

DICOM [2] is undoubtedly known as the most popular standard for medical image storage and communication. The DICOM standard was started by the American College of Radiology (ACR) and the National Electrical Manufacturers Association (NEMA) jointly in 1983. It standardizes the communication of medical images and related clinical information independent of the various device manufacturers. Also, the DICOM standard promotes the development and expansion of Picture Archiving and Communication Systems (PACSs) that help with the distribution and viewing of medical images. Part 10 of the standard describes a file format for storing and distributing images. Unlike most other Electronic Medical Record (EMR) standards, DICOM images use a binary encoding with hierarchical data sets. A DICOM file is constituted of a DICOM header and the image itself. The DICOM header stores patient/study information and some other meta-data such as study instance id, image dimensions, etc.

The DICOM standard describes a set of services to help with the transmission of images/data over a network in addition to the file format. A list of DICOM services includes:

- Store service is used to send image/data objects to a PACS or workstation.
- Storage Commitment is used to confirm that an object has been permanently stored.
- Query/Retrieve provides a workstation a way to find and retrieve certain image objects.
- Modality Worklist and Modality Performed Procedure Step give modalities the ability to obtain a list of scheduled examinations, as well as the patient information.
- Printing service is used to send DICOM images to a DICOM capable printer. DICOM standard also helps to ensure consistency between various display devices.
- DICOM Files is a file format standard to help store medical images and related clinical data on removable storage devices.

3.2   DCM4CHE DICOM Toolkit and DCM4CHEE DICOM Clinical Data Manager System

Gunter Zeilinger wrote the popular JDicom utility suite using Java DICOM Toolkit (JDT) in 2000. Overtime, this utility evolved and expanded into the current DCM4CHE DICOM toolkit and DCM4CHEE DICOM Archive system [11]. DCM4CHE is a high performance and open source implementation of the DICOM standard. It includes a set of libraries to manipulate DICOM objects or facilitate the communications among DICOM-enabled devices and systems. DCM4CHE is developed using platform independent Java technologies and is envisioned to be used on various devices running different operating systems.

DCM4CHEE is built on the success of DCM4CHE that is intended to provide a set of open-source applications to provide a number of clinical services. Mostly, DCM4CHEE is considered to be used as a DICOM image manager or a PACS, to provide services for a DICOM viewer such as OsiriX [12], K-PACS [13], ClearCanvas [14], etc. DCM4CHEE provides a set of modules to the radiologist as well as the system administrators. The set of modules includes:

- A robust Web-base User Interface to ease administrative tasks.
- DICOM interfaces to provide image archiving services for modalities.
- HL7 interfaces to ease the integration with existing HISs and RISs.
- Web Access to DICOM Objects (WADO) and IHE Retrieve Information for Display (RID) interfaces to enable access the images from the Web.
- An Audit Record Repository (ARR) for security purposes.
- Media Creation (CDW) service to export images onto a removable storage device.
- Both XDS/XDS-I as a Document Repository and an Imaging Document Source.
- Xero component that provides a thin-client for clinical access to patient records1 and studies.

3.3   Information Dispersal Algorithm

Files in the JigDFS are split into n segments, where the number n is set based on the configuration of the IDA. While recovering, however, only m segments are needed to reconstruct the complete file, where m is smaller than n. The IDA was first introduced by Michael O. Rabin [9] to design a fault-tolerant, secure and transmission efficient information storage system. However, the secret splitting scheme was first proposed by Shamir [15] to construct robust key management systems that can function securely and reliably even when a few of the file segments are destroyed accidentally. The IDA is actually a special use of *erasure codes* a.k.a. *forward error correction* (FEC) codes. The most well-known erasure code is the one used in RAID level 5, known as the parity driver. In this system, there are at least 3 disk drives, the first two store different data but the third drive stores the XOR value of the data on the other two drives. Under this setting, given any two of the drives, one can recover all of the data being stored. The basic idea of erasure codes is to add redundant data (error correction codes), in addition to the original data before transmission, and this extra information allows the receiver or reader to detect and correct data errors without the need to ask the sender for a resubmission. Other storage specific systems including PASIS [16] [17], Mnemosyne [18] [19], GridSharing [20] as well as Tahoe [21] and Cleversafe [22] have used or suggested the use of IDA for fault tolerant and secure data storage.

Irving Reed and Gus Solomon introduced a class of error correcting codes called Reed-Solomon (R-S) codes [23] in 1960, which are mostly used in CD/DVD storage devices to recover the data after scratch. R-S codes have been suggested to be slow compared to some of its descendants, such as LDPC-codes [24], Turbo-codes [25], LT-codes [26], Raptorcodes [27], etc. Although, both LDPC-codes and Turbo-codes are theoretically very fast, near Shannon limit [28], LDPCshave been neglected by researchers because of its complexity (especially software implementation), while Turbo-codes are avoided because of patent issues. Raptor-codes are closely related to LT-codes, since both of them are fountain codes. A fountain code produces a potentially limitless stream of output symbols for a given set of m input symbols. The decoder can recover the original m symbols from any set of k output symbols with high probability. Such codes are best suitable for data transmission over networks, where the receiver can cut off the communication as soon as it collects enough segments to reconstruct the original data. However, in JigDFS, each child node has only one unique segment of the dispersed file to minimize a node's knowledge about the original file. In such a case, a fountain code appears to be too hard to control and might also increase the probability of a security breach (i.e. ensuring that each child node will only get one segment of the original dispersed file).

The advantage of using an IDA algorithm over a simple hard drive mirroring schema is significant. There are two major factors we need to consider when designing a redundancy schema for a reliable file system with high fault tolerance. First, the *Recovery Rate* (R), which indicates the possibility of recovering data when one or more nodes fail. Second, the *Storage Overhead* (O) of the distributed file system, which gives the amount of overall extra hard drive space needed when maintaining a certain recovery rate. A well designed distributed system should provide a reasonable recovery rate in case of node failures while storing redundant data as less as possible. For example, let us assume that a storage node has a fail rate of p (i.e. practically, even commodity hardware is pretty reliable, and the hardware failure rate is lower than 10%). For a mirroring redundant schema, where a file is simply duplicated and stored on a two different node, the chance of recovery is $R = 1 - p^2$ (i.e. you will lose the data when both nodes fail). And the storage overhead of such system is 100%, while, an m/n IDA schema only have an overhead of $(n - m)/n$. Here, n is the total number file slices including the redundant information. As a definition, you will need only m segments to recover the entire file, where m is less than n, so that $(n - m)/n$ is less than 100%). Also, the recovery rate of an m/n ($R = 1 - p^{n-m+1}$) IDA schema increases dramatically when the difference between n and m becomes larger. Under the assumption that commodity hardware failure rate is less than 10%, a 3/7 IDA schema will provide 99.99% recovery rate with a relatively low (57%) storage overhead, while a mirroring system have 100% overhead with slightly less recovery rate (99.00%). Through analysis, we suggest to use ($m=3$, $n=7$) in JigDFS for the IDA.

For the sake of implementation simplicity, we have chosen to use an optimized R-S code known as Cauchy Reed-Solomon [29], which has been previously used in [22] and suggested in [30].

3.4   Data Secrecy via IDA

Various secure distributed file/storage systems such as Coda [31] [32], OceanStore [33], Farsite [34], etc. use encryption to implement data secrecy. Encryption algorithms normally work well and are considered reasonably safe as long as the key remains secret. However, most of the encryption algorithms are designed based on mathematical assumptions that are hard to solve computationally. With the rapid development of computing technologies and new cryptanalysis methods' emergence, the encryption algorithms become less safe. For example, when the Data Encryption Standard (DES) [35] became a standard, it was considered a strong encryption with a 56-bit key space. A DES encrypted message was first broken in the DES challenge project in 1997. One year later, in 1998, a DES cracker (Deep Crack) could break a DES key in 56 hours [36]. Nowadays, DES is insecure for many applications.

An encryption-based secure system can provide reasonably good short-term protection. This may not be the case for long-term data storage. The use of an IDA schema in a storage system cannot only provide fault tolerance and improve system reliability, but also strengthens the secrecy of the data. In an m/n IDA schema, a file is dispersed into n segments and the IDA ensures that combing k segments ($k < m <= n$) reveals no information about the file.

Like Tahoe [21], CleverSafe [22], and Mnemosyne [18], JigDFS uses encryption followed by IDA to provide stronger secrecy. In JigDFS, the dispersed file segments are distributed onto different nodes. In other words, no node will store more than one segment of the same file. Doing so minimizes each node's knowledge about the entire file, so that if one or more nodes (less than k) are compromised, the secrecy of the entire file will not be at risk.

3.5   One-way Hashed-key Chain Algorithm

The Role-based Hashed-key (RBHK) chain algorithm was proposed to develop a Role-based Secure Group Communication (RBSGC) framework [37], which aims to implement the Bell-LaPadula security model in a group communication environment. A key generation algorithm is introduced to enforce the "no read up" and "no write down" policies. In a nutshell, a user with high clearance will get key x, while his/her immediate child will be given $h(x)$, where $h()$ is a cryptographic one-way hash function. Under this setting, "no read up" is possible. The files encrypted with x are hidden from lower level users, since it is hardly possible to compute x from $h(x)$.

Such algorithms have been also used in securing Wireless Sensor Networks (WSNs). In [38] and [39], a one-way hashed-key chain is used to ensure the authenticity of the packets road casted from the base station. First, the base station uses a one-way function h () to generate a sequence of keys k_0, k_1, KN, such that $k_i = h(k_{i+1})$. k_0 and hash function h () are pre-distributed to every node. In the first broadcast round, the base station uses k_1 to sign its packet, and the child nodes can verify the signature by comparing $h(k_1)$ with a known k_0. Since h () is a one-way hash function, an adversary cannot compute k_{i+1} from k_i. The authenticity is ensured, since the base station is the only one who knows k_{i+1} at the ith round of communication. Moreover, even if a node is compromised by an attacker, k_i is useless to the next round of communication. However, based on the number of communication rounds needed, this method may require the base station to compute a long chain of keys for pre-distribution, which will dramatically increase the setup time.

In JigDFS, the one-way hashed-key chain algorithm is used to generate the keys for layered encryption, which also "secretly" links file segments to their parents and the original file. A parent node will first encrypt the file with key x and then slice the output according to an IDA. Each file segment will be sent to a child node along with a new key $h(x)$. The child nodes will then encrypt their file slices, divide and distribute them further until reaches the demanded file security level. As h () is a one-way function, it is not feasible for a child node to compute x from $h(x)$ and decrypt the received file slice. Also, since the actual data segment is stored separately from its encryption key, it minimizes the node's knowledge about the segment. Even if a node is compromised, the adversary gains no useful information from the encrypted data segment.

4 The Jigsaw Distributed File System

While sharing many of the same goals, such as performance, scalability, reliability and availability, as previous distributed file systems such as the Google file system [40], Frangipani [41] [42] etc, the design of JigDFS is mainly driven by many security concerns in protecting user privacy. Similar to other secure distributed file systems, the main concerns of the JigDFS are data secrecy, recoverability, and data integrity.

In JigDFS, a file F is first encrypted using a symmetric encryption algorithm and then the encrypted file is split into n segments using an Information Dispersal Algorithm (IDA) [9]. Each data segment is then delivered onto different child nodes. The process is recursive, thus, the child node will encrypt and split the received data segment further based on the desired File Security Level (FSL). Upon retrieving request, the file segments will be routed to the retriever and because of the property of the IDA, only m intact segments are needed to reconstruct the original encrypted file, where $m \leq n$ (i.e. m/n values are configurable and a typical configuration is $m=3$, $n=7$). The user will then need to provide the correct decryption key to access the file. The use of IDA not only provides good fault tolerance, but also enhances the level of security [9] [15]. Any $m - 1$ segments give no information about the original file F.

In JigDFS, there is not a single node acting as a central server. JigDFS cluster is more like a decentralized peer-to-peer network. In the current prototype, nodes in JigDFS are all connected to one ethernet switch, which builds a small local area network (LAN). So, message broadcasting is supported. The node discovery protocol in JigDFS is simply implemented through broadcasting a JigDFS "ALIVE" command to announce its existence. The "ALIVE" command is broadcast both in the initialization process of a JigDFS node and periodically as a node's "heartbeat". Each node also maintains a list of other known nodes in a cache file for easy access. Although, there is no hard restriction to prevent running two JigDFS service instance on one node as long as the node's system resources permit, it's not recommended, since two segments for one file may end up on the same node. If hardware/ system failures occur on such a node, it will result in losing two segments of one file, which rapidly reduces the redundancy introduced by the IDA.

There a few basic aspects of implementing a distributed file system, which are file naming/indexing, node identification and discovery, and file operations. The following sections describe the implementation of these aspects in JigDFS.

4.1 Naming

All of the data entities in JigDFS (i.e. original files, intermediate file segments, and final data segments) are identified by SHA256 hash values. The file segments are stored in a JigDFS data folder on each node and named using their JigDFS identifiers. There is no folder structure defined within the JigDFS data folder. Since SHA256 is a cryptographic one-way hash function that has strong collision resistance, two different segments should result in two different SHA256 identifiers. Therefore, file name duplication is avoided in the JigDFS data folder on each node. It is possible that the same data segment is received by one node more than once, since the data segments are moved around with the JigDFS cluster to not only balance the nodes' load but also to gain better user privacy. On a JigDFS client, there is no special routine to check whether this segment has been stored before or not. The JigDFS will simply replace the existing one if it receives a segment with the same identifier. It is safe to do so, since the file segments should be the same if they have the same SHA256 identifier.

Using hash value to name files and file segments is not only for its uniqueness. It also helps to avoid revealing information of the segments through names. Moreover, it is an important attribute in ensuring data integrity. In all data transformation and transmission procedures, the identifiers also serve as the checksum of the data segments. If a data segment does not match the recorded checksum, the JigDFS client will simply abandon this segment and request a new one from a different node. Recall that, in an m/n IDA schema, any m data blocks can reconstruct the original data. A corrupted file segment should not affect the integrity of the data. However, the corrupted segment will be recorded and the JigDFS cluster will try to remove all the corrupted data segments and consolidate all file segments when the system is idle or under low work load. Segment consolidation will be discussed in a later section.

4.2   Node Identification and Discovery

Each node in the JigDFS cluster keeps a list of other nodes and the list is updated constantly by the nodes discovery protocol. Currently, the nodes discovery protocol in JigDFS is achieved by broadcasting and monitoring the nodes' heartbeat messages. If there are not enough known nodes, a node can also broadcast a HEARTBEAT-REQUEST command in the cluster to renew its local node list. All live nodes would respond by broadcasting its heartbeat message immediately.

The nodes in the cluster are also identified uniquely through each node's MAC address, and a node identifier (NID) is generated by hashing the node's MAC address. In the current proof of concept JigDFS implementation, node discovery is merely based on the broadcast mechanism supported by the LAN. However, it is possible to implement a JigDFS cluster in a non-broadcast network environment. There are various ways of constructing a P2P network. For example, one server can be used as a master that keeps track of all nodes in the JigDFS cluster. All other nodes will report to the server for joining or leaving the cluster. During the segment placement process, the node will first contact the master to get a list of known clients. In the real world, a true P2P file system would use other methods to locate nodes and achieve "broadcasting" behavior, for example, the use of a Distributed

Hash Table (DHT). The detailed implementation of the node discovery protocol in such environment is rather straightforward.

4.3 File Operations

Although, JigDFS does not conform to any API standard such as the POSIX, most of the common file operations are supported and can be interfaced with the third party user applications. In the current version of JigDFS, the implemented file operations are: create/write, read, and delete. File update operations are achieved through deleting than inserting the updated file.

4.3.1 File Write

In order to use the JigDFS storage services, each node should have a JigDFS client installed and running in daemon mode. The JigDFS clients run in user space as a user applica- tion rather than running in kernel space and serving as a kernel service. To insert a new file into a JigDFS cluster, as shown in Figure 1, the users upload their files to the JigDFS client on one of the JigDFS nodes. Also, a password (p) and the desired File Security Level (FSL) need to be supplied along with the file.

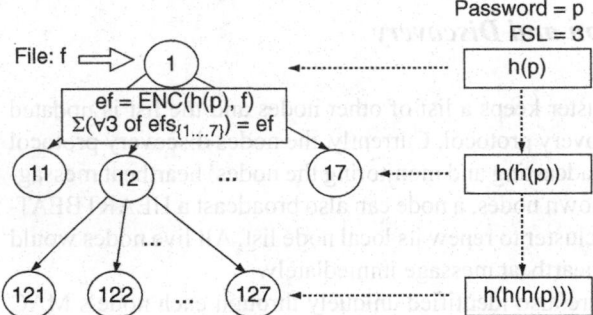

Figure 1. Writing a file into JigDFS. In this example, the JigDFS is configured to use a 3/7 IDA schema. The user supplies a password p and defines the desired security level FSL=3. On the node (Node 1) where the file is uploaded, the file will be first encrypted using a symmetric encryption algorithm *ENC* () with key h (p) and produce the encrypted file *ef* (i.e $ef = ENC$ (h (p), f)). Then the encrypted file is split using a 3/7 IDA schema into 7 unique segments ($efs_{\{1...7\}}$), and any 3 of them can be used to restore the original *ef*. The file segments (*efs*) are then distributed to 7 differ- ent child nodes (Node 11 to Node 17) along with the next level encryption key $h(h(p))$. The same data transformation process will be repeated on each child node that received a file segment until it reaches the required security level. Remember that the intermediate nodes, in this case Node 1 and its children Node 11 to Node 17, do not store the actual file segments they processed.

The password's hash value ($h(p)$ using SHA256 in current implementation) will be computed and used by the server as the master/root encryption key of the hashed-key chain algorithm. In the current JigDFS implementation, the user files will be first encrypted using AES [43] in 256 bits key size mode, while the block size is 128 bits. Then, the encrypted file will be split into 7 segments using a 3/7 IDA schema. The client will find 7 different child nodes within the JigDFS cluster and send one segment to each along with the required security level (FSL) and the next level encryption key computed according to the hashed-key chain algorithm ($h(h(p))$). The selected child nodes will repeat the same procedure unless the desired security level is reached. As shown in Figure 1, for the 3/7 IDA schema and $FSL=3$, a hierarchy tree structure is constructed by the nodes for each file. The file is split into segments and propagated down the file tree. The nodes in between the bottom of the file tree and the root only store the metadata of the segments they received, but not the actual intermediate file segments.

Before retrieving a file, there must be a mechanism in a file system to index and locate files. Keeping a globally synchronized file-indexing table is a difficult undertaking, since the resources are distributed. In JigDFS, each node keeps a file-indexing list to track the segments it has processed. There are four basic fields in this file-indexing table, which are:

- **_H (OF):_** The hash value of the original file using SHA-256.
- **_H (FS):_** The hash value of the processed file slice using SHA-256.
- **_PNID:_** The identifier of the parent node, which has the key used to encrypt the file slice stored on this node. Obviously, there is no parent node for the root node. However, the root node will randomly pick one to fill this field; otherwise, an adversary can easily identify the root node by reading the file indexing table.

Moreover, each node uses a Key Table (KT) to store the encryption keys. The key table is securely stored on each node locally. It is safe to do so because the keys stored on one node do not correspond to any of the file data stored on the same node. The fields in this table are:

- **_H (OF):_** The same value as the one in the file indexing table. This is the link between the keys and the file segments. Note that $H(OF)$ points to the original file rather than a specific file segment.
- **_K:_** The encrypted key received from the parent node.

The key table on each node stores the key it received from its parent (i.e. again, it is not the key used to encrypt the received file segment), which is also used to encrypt the file slices it will produce and send to its children nodes. Since there is no parent node for the root node, if the root node is identifiable and it stores $h(p)$, the compromise of this node can risk revealing all files originated from it. Therefore, the root node will fake an entry in the KT. Since the root has the highest level encryption key, $h(p)$, from which all subsequent keys are derived. Theoretically, the root node can forge itself to be any child node down the tree. This also helps to improve user's deniability.

4.3.2 File Read

To read a file from JigDFS, the process is reversed. The user can send a read command including the file identifier from any JigDFS node, and the message will be "broadcasted" to all other nodes within the network (i.e. For the ease of developing a prototype, we used a cluster where broadcasting a message is possible. In the real world, a true P2P file system would use other methods to locate nodes and achieve "broadcasting" behavior, for example, use a Distributed Hash Table (DHT)). The nodes, which have a segment of this file, will then report to the requester, and do handshakes according to the JigDFS communication protocol before sending the actual data segment. There are two ways of retrieving data segments for the requester.

Bazaar Read

When a node receives the *REQ* message, it will first check its local file indexing table, to see whether it has a segment of the requested file or knows where to find a slice. The file segments then climb up from the bottom of the file tree back to the original uploader node. The data decoding is performed on each node where the segments go through. The file segments are first combined using the IDA decoder, and then decrypted with the key recorded in the key tables when the file was created. Then the decrypted segment is sent back to the node's parent along with the *RNID*, so that eventually the file can be transferred to the requester node. The same data decoding process will be repeated on the parent nodes and after all the encrypted file will be reconstructed on the original uploader node and then transmitted to the requester node referenced by *RNID*.

As an example, Figure 2 shows the process of reading a file from a JigDFS cluster using the bazaar read.

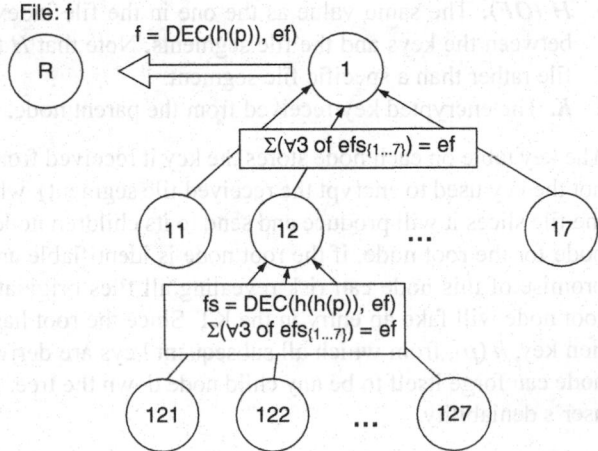

Figure 2. Bazaar Read: The file segments propagate from the bottom of the file tree through the same nodes used in the original file creation. Eventually, the file is transferred from the original uploader node to the requester node.

It is worthwhile to remember that we only need 3 segments to reconstruct the original data in a $3/7$ IDA schema. For example, as shown in the Figure 2, when node N_{127} receives the *REQ* message for file F, it checks against its local file indexing table and finds that it does hold a segment of the file. From the indexing table, it can find out both the segment identifier and the parent node identifier (*PNID*). Since it is at the bottom of its branch, there is no decryption process needed. The data stored was encrypted with the parent node's key. It reads the segment efs_{127} from its data folder and sends the slice directly to its parent pointed by *PNID*, which should be node N_{12}. Another two segments need to be collected on N_{12} before the data transformation starts, which are segments efs_{121} and efs_{122} in this case. Again, the three segments are combined to efs_{12} using the $3/7$ IDA decoder and then decrypted with key $h\ (h\ (p))$ held in N_{12}'s local key table for this file, which should be the original encryption key. Each node uses checksums to detect corruption of stored data. Since we use the segments' SHA256 hash value as their identifiers, the integrity of a segment can be ensured by re-computing its SHA256 hash values and verifying it against the segment's identifier. If they do not match, data corruption is detected. Note that since the encryptions are layered, the decryption on this node only removes the outermost encryption layer. Then N_{12} sends the transformed data segment to its parent N_1, which is the original uploader. The same process occurs again on N_{11} as well as on N_{17}, and finally the original uploader node N_1 collects enough segments to reconstruct the original file. However, as mentioned above, during file saving, N_1 does not keep the original encryption key $h\ (p)$ for security reasons. Thus, only IDA decoding is performed on N_1, and the reconstructed file is sent back directly to the requester. If the user has the right p, there should be no problem in decrypting the file and reading its content.

Cathedral Read

As shown in Figure 3, as compared to the first approach, the segments are directly sent from the nodes at the bottom of the file tree directly to the requester instead of sending to their parent nodes. The data transformation processes that were conducted on the parent nodes in the bazaar read are now recursively performed on the requester node. As all the encryption keys are derived from the user's password, there should not be a problem in regenerating all the keys on the requester node. However, the requester has neither access to the intermediate nodes' local file-indexing tables nor their key tables. There is not an easy way for the requester to pin point the right number of hash layers needed to generate the correct keys for specific segments. Also, without the intermediate segments' identifiers, there is no way to perform checksum checks after intermediate data transformation processes. For example, since the requester node does not know the identifier of the segment (efs_{11}) that was further split into seven smaller slices, $efs11_{(1...7)}$, when reconstructing efs_{11} from three of those segments, $efs11_4$, $efs11_5$ and $efs11_7$, using the IDA decoder, there will be no way to verify the integrity of the restored efs_{11} segment.

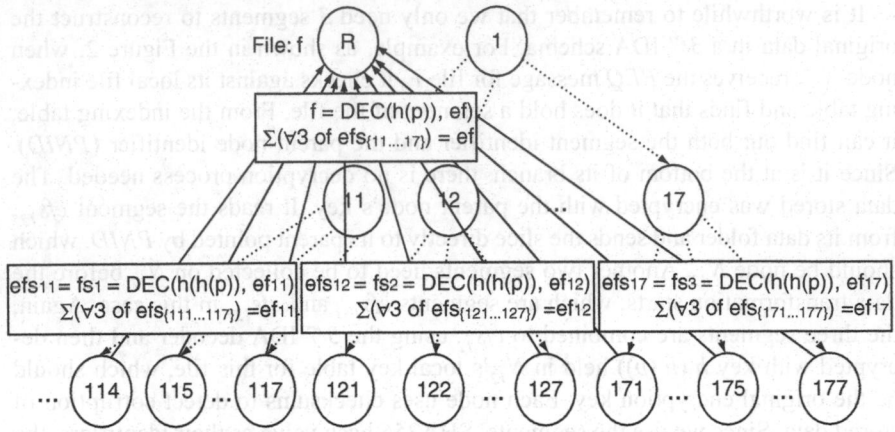

Figure 3. Cathedral Read: The requester node collects the smallest file segments directly from the nodes at the bottom of the file tree.

Hence, the intermediate identifiers need to be attached to each segment when the file is inserted into the cluster. Using the same example shown in Figure 3, when efs_{11} is divided into $efs11_{\{1\ldots7\}}$, the identifier for efs_{11} will be attached to $efs11_{\{1\ldots7\}}$ as well. The internal data structure on the hard drive for each segment on $N_{11\{1\ldots7\}}$ will then be:

$$[efs11_i,\ h\,(efs11_i),\ h\,(efs_{11})],\ (1 <= i <= 7)$$

where $efs11_i$ is the actual segment data, $h\,(efs11_i)$ is the identifier of $efs11_i$ segment, and $h\,(efs_{11})$ is the identifier of $efs11_i$'s parent segment efs_{11}.

Moreover, when N_{114} sends $efs11_4$ segment to the requester, it should also send the key $(h\,(h\,(h\,(p))))$ it holds in its local key table as well. This helps the requester node to compute the right decryption quickly. To that end, the data sent from N_{114} to the requester R will be:

$$\{DATA,\ [efs11_4,\ h\,(efs11_4),\ h\,(efs_{11})],\ h\,(h\,(h\,(p)))\}$$

After R collects all three segments ($efs11_4$, $efs11_5$, $efs11_7$), it first combines them into efs_{r11} using the IDA decoder and compute its checksum $h\,(efs_{r11})$. If the value $h\,(efs_{r11})$ matches $h\,(efs_{11})$, then the reconstructed data (efs_{r11}) is intact and $efs_{r11} = efs_{11}$. However, in order to decrypt efs_{11}, R needs to compute the right decryption key from the user password p. It is obvious that the encryption key (k) of efs_{11} is the parent of $h\,(h\,(h\,(p)))$. Although, since $h\,()$ is a one-way function, it is not possible to compute k from $h\,(k)$. However, from p, a simple loop can easily narrow down the number of hash layers needed to meet $h\,(k) = h\,(h\,(h\,(p)))$ and extract the key k.

Both read approaches have advantages and disadvantages. The bazaar read involves more nodes, and the intermediate nodes help to share the data transformation workload that is computationally intensive. However, it takes the risk that the read path could be blocked if the intermediate nodes fail. In terms of the cathedral read, all the workload of data transformation bears on the file requester, which may reduce the read performance significantly. On the other hand, the cathedral read does not rely on the help of the intermediate nodes in reconstructing the files. In the real JigDFS implementation, we make the most of a mixed avenue using both approaches. To gain maximum read performance, read operations are usually conducted using the bazaar read approach, since several data transformation processes for different segments can take place on different intermediate nodes simultaneously. However, if one of the intermediate nodes fails, the cathedral read will take place instead, where the data segments are directly sent to the requester for reconstruction. However, not all segments need to be routed to the requester node directly. Only the segments whose read paths are blocked by a failed parent node will jump start at the requester node.

4.3.3   File Deletion

Deleting a file in JigDFS is a straightforward process. All file segments associated with the same file in the cluster need to be removed and securely deleted from the nodes that formed the file tree. However, there exists a problem of confirming the identity of the user who issued the delete command and whether the user is authorized to do so. Therefore, when a user sends a request to delete a file from the JigDFS cluster, the user needs to provide the password p and the file security level he/she used when the file is created. The following is the delete request message:

$$\{DELETE, h(f)\}$$

When a node which has a piece of the file receives this message, it will send a "challenge" message encrypted with the key it holds in its key table for this file in order to verify the authorization of the requester. Since, all the keys are derived from the password p using the hashed-key chain algorithm; it should be no problem for the file owner to resolve the challenge messages. When confirmed, the segment will then be removed from that node. And eventually, all file segments will be removed from the JigDFS cluster.

There remains a potential security problem. In the threat model, we mentioned that it is possible for an intermediate node of a file tree to be compromised. The attacker can then use the information stored in the local key tables to issue DELETE commands to delete arbitrary files listed in the local file indexing tables. One potential solution is to implement a voting system for nodes on the file tree to decide collaboratively whether the issued DELETE command is authenticated or not. For a specific file tree, the key stored in the compromised intermediate node cannot fulfill the "challenge" message sent from nodes above it.

4.3.4   File Deletion

Since the files are divided into small segments and the content of the file segments appear to be scrambled random data on each node, when a file is updated, all segments generated from the file will be different from the previous one. Even, the file SHA256 identifier changes, the file will be considered a different file rather than an updated version of the previous one. Therefore, file update in JigDFS is rather trivial and implemented through deleting the old file, and then inserting the new file. For a large file, this process will be not only time-consuming but also inefficient. However, as discussed before, in the case of inserting large files, the file needs to go through a preprocessing layer. In these cases, the large file will be broken down into fixed-sized smaller fragments and produce an XML document, which contains the SHA256 identifiers of all fragments as well as the original file. So, when a large file is updated, the JigDFS client will only do an update on the changed fragments.

4.4   Fault Tolerances and Data Integrity

JigDFS is envisioned to be used in a large-scale cluster environment, and a typical such cluster often has hundreds of nodes. Disk failures and other hardware and software issues that cause data corruption or data loss on both read and write paths are inevitable. With the help of the IDA, in an m/n IDA redundant schema, we can recover from data corruption as long as m nodes survive. Moreover, since the IDA is applied recursively in the JigDFS cluster for each file, the chance of losing data is even lower, and the recovery rate will be higher than 99.99% in a 3/7 IDA schema. As mentioned previously, reading a file from the JigDFS cluster can involve not only the nodes at the bottom of the tree which hold the file segments but also the intermediate nodes that have processed the file during file creation. However, it is possible to do offline reconstructions using the cathedral read, in the case of broken network connectivity.

All files and segments, including the intermediate segments, are identified by their SHA256 hash values as described above. The identifiers are also used as checksums to ensure the integrity of segments and detect data corruptions. Each node scans and preforms data integrity checks on all inactive segments it stores locally during its idle periods. This allows us to detect corruption in segments that are rarely read. Once the corruption is detected in a segment, the node will notify its parent for that specific segment. The parent node will collect related segments from other child nodes and try to reconstruct an uncorrupted copy of the original parent segment. If the parent segment is recovered, the file will then go through the data transformation process again, and be divided and distributed onto n new child nodes. The corrupted segment and related segments are then deleted from the cluster. Although checksumming using SHA256 has performance penalty on both reads and writes, considering its benefits of not only ensuring data integrity and helping detect data corruptions but also serving as a unique identifier, it is worth implementing.

4.5   Security Analysis

At the beginning of this chapter, we presented a list of potential threats and security concerns that JigDFS is trying to address. Here, we highlight the approaches we used to tackle each of these security issues.

All files including file slices are encrypted using strong encryption. The encryption algorithm is pluggable, so that users can use any encryption algorithm of their choice as long as it is symmetric. The key size is also changeable. One can easily swap SHA-256 with SHA-512 to generate the hashed-key chain, so that the key size can be increased from 256 bits to 512 bits. Currently, in the proof of concept JigDFS implementation, the default encryption algorithm is AES [43] with PKCS1 padding with a 256-bit key size and 128-bit block size. Strong encryption helps to protect not only the user assets stored on each node, but also the necessary communication between nodes. Therefore, even an adversary captures a JigDFS communication packet or takes control of one or more nodes; he/she would not be able to read the information without a correct decryption key.

The files are sliced into small segments and spread across the network. One slice of the file being compromised will not reveal the information stored in the original file. Actually, according to IDA, the adversary would not be able to reconstruct the original file, if he/she only has access to less than m data slices. Moreover, the IDA is applied recursively, so that more nodes are involved in constructing a file tree. It is hardly possible for an adversary to invade enough nodes to gather the mandatory number of file slices, besides the fact that each file slice is protected with strong encryption as well. Furthermore, if any node along the tree is compromised and both the key and data held by that node is acquired, the key would not be the one used to encrypt that data. According to the hashed-key chain algorithm, the key and the data are separated, where only the parent node has the key to decrypt the data stored on its child nodes. Therefore, the data that an attacker acquires from a compromised node appears to be nothing more than a random block of bytes. As mentioned above, it is true that the keys stored on a compromised node along with their subsequent keys are also compromised. However, because of IDA, the adversary needs to first collect enough segments from its children nodes to reconstruct the original data slice, otherwise, the keys are useless. Also, notably, each file has a different file tree structure, which means each encryption key corresponds to a different set of children nodes. Moreover, in JigDFS, the parent node has no knowledge of its children nodes after the encoding process. Thus, it is hardly possible for an adversary to find the right set of children nodes that can resemble the right data segment. Further, a user can also submit encrypted data to the root node, in order to get around the problem of root node being compromised. Submission of an encrypted file to the root node is rather trivial.

In JigDFS, file slices can be randomly moved from node to node, hence the name Jigsaw. The actual implementation is much like corruption detection described in the previous section. By doing so, we increase the complexity of tracing the flow of file segments.

4.6   Security Analysis

In this section we present a few micro-benchmarks to illustrate the usability of JigDFS. All the simulations are conducted and measured on a single machine running 64 bit Ubuntu 9.04. The machine is configured with a 2.66 GHz Quad Core Intel processor, 8 GB of ram, and a 750 GB 7200 rpm disk. The Java JVM is version 1.6.0_15 running in 64 bit mode.

The read/write throughputs measurements of the complete process involve all the nodes on the entire file tree. The number of nodes entangled can be massive and depends on the required file security level. There is no easy way to synchronize the times of all machines in the cluster with microsecond precision. Therefore, the overall read/write throughputs presented here are formularized values rather than gathered for the real cluster. However, the measurements of the data transformation on each node are collected through simulations. On average, for a 3/7 IDA redundancy schema, the encoding throughput is 6.40 MB/s while the decoding throughput is 9.23 MB/s. Note that, the encoding process on each node includes a AES encryption, a 3/7 Cauchy Reed-Solomon IDA encoding, as well as performing SHA256 hashes over file and file segments.

We assume that the nodes are interconnected using gigabyte Ethernet cards. The theoretical transmission speed between two nodes is 1000 Mbps, which are 125 MB/s. Second, the throughputs heavily depend on the depth of the file tree. Using the previous example, where the depth of the file tree is 3.

To measure the write throughput, we only need to take into account the longest path on the file tree, since on the same level, the data transformation processes are performed in parallel on each node. Therefore, the write throughputs for a 3 level file tree and a 3/7 IDA is 3.89MB/s.

We have two different read approaches in JigDFS, bazaar read and cathedral read. The detailed throughput calculation formulas are trivial. However, the final result is 5.23 MB/s for bazaar read and 4.29 MB/s for cathedral read. Since JigDFS uses both approaches to read a file, the overall read throughput should be somewhere between 4.29 MB/s and 5.23 MB/s on average.

5   Building a PACS on JigDFS

PACS consists of various functional elements that facilitate the management of medical images, such as image acquisition, archival and retrieval, communication, image processing, distribution, etc. More often, PACS also integrates the image management system with other Radiology Information Systems (RISs) and Hospital Information Systems (HISs). Its system infrastructure contains hardware components such as interfaces to imaging modality, storage devices, workstations in the reading room and display system that are integrated by network and software systems for image management, storage management, patient management, etc. PACS

Figure 4. Architecture overview of a PACS built on JigDFS.

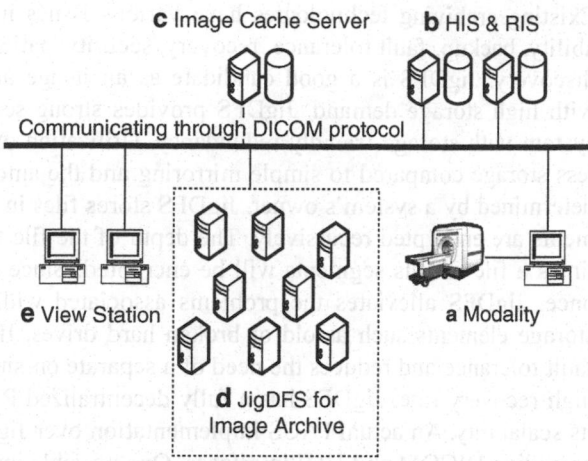

offers an efficient way of reading, analyzing and archiving medical images, documenting study results as well as communicating study results to the referring physicians or researchers. Figure 4 shows a typical PACS configuration using JigDFS as its archival storage system. Here are the five basic components/functionalities that a PACS contains:

1. Image Acquisition: Digital image acquisition requires interfacing the PACS with the imaging modalities such as CT, MRI, etc. The modality interfaces require that the devices to be manufactured comply with the DICOM standard.
2. Patient Data Interface: PACS interfaces with the Hospital Information Systems (HISs) and Radiology Information Systems (RISs) through Health Level 7 (HL7) messages [44] [45]. HL7 message is the standard of electronic data exchange in the clinical domain, and the most widely implemented standard in the healthcare informatics field.
3. Image Cache Server: Most PACS employ a hierarchical storage strategy, in which new images are stored on an image cache server and older images are archived on an off-line tape library for long-term storage.
4. Image Archival System: A long-term archival system is indispensable in PACSs, since medical images are required by law to be stored for a long period of time. The selection of archiving technology mainly depends upon archiving media. The common archiving media are magnetic discs, a redundant array of inexpensive discs (RAID), optical disc (i.e. CD/DVD), or magnetic tape. Most PACSs favor digital linear tape (DLT) due to its affordability, high capacity and reliability. However, DLT has a very slow read speed and only supports sequential read. Moreover, it takes a lot of physical storage room and it is hard to manage when the data volume is huge.
5. Image Display and Interpretation: Viewing stations are connected to PACS via the DICOM communication protocol.

Existing archiving technologies have various issues including availability, scalability, backup, fault tolerance, recovery, security, patient privacy, and knowledge discovery. JigDFS is a good candidate as an image archival system for PACSs with high storage demand. JigDFS provides strong security as a distributed file system with storage-friendly redundancy. Utilization of IDA enables JigDFS use less storage compared to simple mirroring and the amount of redundancy can be determined by a system's owner. JigDFS stores files in segments and the file segments are encrypted recursively. The depth of the file tree determines how many times a file and its segments will be encrypted. Since a file is encrypted at least once, JigDFS alleviates the problems associated with the decommissioning of storage elements such as old or broken hard drives. IDA improves the system's fault tolerance and reduces the need of a separate on-site backup system due to its high recovery rate. JigDFS has a fully decentralized P2P structure, which boosts its scalability. An actual PACS implementation over JigDFS is straightforward by providing DICOM and HL7 interfaces. One possible implementation is to provide underline storage service for DCM4CHEE [11], an open source clinical image and object management system.

6 Conclusion

We presented a JigDFS-based PACS store, which will store medical images. JigDFS is a good candidate as a file system for a PACS system with high storage demand. JigDFS provides security as a distributed file system with storage-friendly redundancy. Utilization of IDA enables JigDFS use less storage compared to simple mirroring and the amount of redundancy can be determined by a system's owner.

JigDFS stores files in segments and the file segments are encrypted. Depth of the file tree determines how many times a file segment will be encrypted. Since a file segment is encrypted at least once, JigDFS alleviates the problems associated with decommissioning of storage elements such as old or broken hard drives.

As JigDFS doesn't support audit trails due to its very design principles, the PACS system must implement an auditing schema. Implementing an auditing system side by side with JigDFS is not a difficult undertaking and is rather straightforward.

References

1. Congress of USA. (2009, September) U.S. Government Printing Office. [Online]. http://fdsys.gpo.gov/fdsys/pkg/BILLS-111hr1ENR/pdf/BILLS-111hr1ENR.pdf
2. NEMA. (2009, June) National Electrical Manufacturers Association, DICOM. [Online]. ftp://medical.nema.org/medical/dicom/2008/
3. AHIMA. (2002, June) American Health Information Management Association. [Online]. http://library.ahima.org/xpedio/groups/public/documents/ahima/bok1_012545.hcsp?dDocName=bok1_012545

4. Tim Dierks and Christopher Allen. (1999, January) Rfc 2246: The tls protocol. [Online]. http://www.ietf.org/rfc/rfc2246.txt

5. Russell Housley. (1999, June) Cryptographic Message Syntax. [Online]. http://www.ietf.org/rfc/rfc2630.txt

6. Burt Kaliski. (1998, March) PKCS #7: Cryptographic Message Syntax Version 1.5. [Online]. http://www.ietf.org/rfc/rfc2315.txt

7. NEMA. (2009) The DICOM standard. [Online]. ftp://medical.nema.org/medical/dicom/final/cp895ft.pdf

8. Jim Schaad. (2003, July) Use of the Advanced Encryption Standard (AES) Encryption Algorithm in Cryptographic Message Syntax (CMS). [Online]. http://tools.ietf.org/rfc/rfc3565.txt

9. Michael O. Rabin, "Efficient dispersal of information for security, load balancing, and fault tolerance," *J. ACM*, pp. 335–348, 1989.

10. Jiang Bian, Umit Topaloglu, and Cheryl Lane, "EIR: Enterprise imaging repository, an alternative imaging archiving and communication system," in *EMBC '09: Proceedings of the 31st Annual International IEEE EMBS Conference*, Minneapolis, Minnesota, 2009.

11. Gunter Zeilinger. (2009) Open Source Clinical Image and Object Management. [Online]. http://www.dcm4che.org/

12. The OsiriX Foundation. (2009) OsiriX Imaging Software, Advanced Open-Source PACS Workstation DICOM Viewer. [Online]. http://www.osirix-viewer.com/

13. Andreas Knopke. (2009) K-PACS DICOM Viewing Software. [Online]. http://www.k-pacs.net/

14. ClearCanvas Inc. (2009) clearcanvas. [Online]. http://www.clearcanvas.ca/dnn/

15. Adi Shamir, "How to share a secret," *Commun. ACM*, vol. 22, no. 11, pp. 612–613, 1979.

16. J. J. Wylie et al., "Survivable information storage systems," *Computer*, vol. 33, no. 8, pp. 61–68, Aug 2000.

17. G. R. Ganger et al., "Survivable storage systems," in *DARPA Information Survivability Conference & Exposition II, 2001. DISCEX '01. Proceedings*, 2001, pp. 184–195 vol.2.

18. Steven Hand and Timothy Roscoe, "Mnemosyne: Peer-to-Peer Steganographic Storage," in *IPTPS '01: Revised Papers from the First International Workshop on Peer-to-Peer Systems*, London, UK, 2002, pp. 130–140.

19. D. Hayashi, T. Miyamoto, S. Doi, and S. Kumagai, "Design and implementation of autonomous distributed secret sharing storage system," in *APCC 2003. The 9th Asia-Pacific Conference on Communications, 2003.*, 2003, pp. 57–60 Vol.1.

20. Arun Subbish and Douglas M. Blough, "An approach for fault tolerant and secure data storage in collaborative work environments," in *StorageSS '05: Proceedings of the 2005 ACM workshop on Storage security and survivability*, New York, NY, 2005, pp. 84–93.

21. Zooko Wilcox-O'Hearn and Brian Warner, "Tahoe: the least-authority filesystem," in *StorageSS '08: Proceedings of the 4th ACM international workshop on Storage security and survivability*, New York, NY, USA, 2008, pp. 21–26.

22. Clesafe.verorg. (2007, June) Cleversafe.org. [Online]. http://www.cleversafe.org/documentation/

23. Irving S Reed and Gustave Solomon, "Polynomial Codes Over Certain Finite Fields," *Journal of the Society for Industrial and Applied Mathematics*, pp. 300–304, 1960.

24. Robert G. Gallager, "Low Density Parity-Check Codes," Cambridge, MA, USA, 1963.

25. C. Berrou, A. Glavieus, and P. Thitimajshima, "Near Shannon limit error-correcting coding and decoding: Turbo-codes," *Communications 1993 ICC 93 Geneva Technical Program Conference Record IEEE International Conference on (1993)*, pp. 1064–1070, 1993.

26. Michael Luby, "LT Codes," in *FOCS '02: Proceedings of the 43rd Symposium on Foundations of Computer Science*, Washington, DC, USA, 2002, p. 271.

27. Amin Shokrollahi, "Raptor codes," *IEEE/ACM Trans. Netw.*, pp. 2551–2567, 2006.

28. C. E. Shannon, "A mathematical theory of communication," *SIGMOBILE Mob. Comput. Commun. Rev.*, pp. 3–55, 2001.

29. Johannes Blömer et al., "An XOR-Based Erasure-Resilient Coding Scheme," International Computer Science Institute, 1995.

30. James S. Plank and Lihao Xu, "Optimizing Cauchy Reed-Solomon Codes for Fault-Tolerant Network Storage Applications," in *NCA '06: Proceedings of the Fifth IEEE International Symposium on Network Computing and Applications*, Washington, DC, USA, 2006, pp. 173–189.
31. Mahadev Satyanarayanan et al., "Coda: A Highly Available File System for a Distributed Workstation Environment," *IEEE Trans. Comput.*, pp. 475–459, 1990.
32. M. Satyanarayanan, "The evolution of coda," *ACM Trans. Comput. Syst.*, pp. 85–124, 2002.
33. John Kubiatowicz et al., "OceanStore: an architecture for global-scale persistent storage," *SIGARCH Comput. Archit. News*, pp. 190–201, 2000.
34. William J. Bolosky, John R. Douceur, and Jon Howell, "The Farsite project: a retrospective," *SIGOPS Oper. Syst. Rev.*, vol. 41, no. 2, pp. 17–26, 2007.
35. NIST. (1979) National Institute of Standards and Technology. [Online]. http://csrc.nist.gov/publications/fips/fips46-3/fips46-3.pdf
36. Electronic Frontier Foundation, *Cracking DES: Secrets of Encryption Research, Wiretap Politics and Chip Design*. Sebastopol, CA, USA: O'Reilly & Associates, Inc., 1998.
37. Jiang Bian, Umit Topaloglu, Remzi Seker, Coskun Bayrak, and Chia-Chu Chiang, "A role-based secure group communication framework," in *Proceedings of the third International Conference on System of Systems Engineering*, Monterey, California, United States, 2008.
38. Jing Deng, Richard Han, and Shivakant Mishra, "INSENS: Intrusion-tolerant routing for wireless sensor networks," *Computer Communications: Dependable Wireless Sensor Networks*, vol. 29, no. 2, pp. 216–230, January 2006.
39. Jing Deng, Richard Han, and Shivakant Mishra, "A performance evaluation of intrusion-tolerant routing in wireless sensor networks," in *IPSN'03: Proceedings of the 2nd international conference on Information processing in sensor networks*, Palo Alto, CA, 2003, pp. 349–364.
40. Sanjay Ghemawat, Howard Gobioff, and Shun-Tak Leung, "The google file system," *SIGOPS Oper. Syst. Rev.*, vol. 37, no. 5, pp. 29–43, 2003.
41. Chandramohan A. Thekkath, Timothy Mann, and Edward K. Lee, "Frangipani: a scalable distributed file system," SIGOPS Oper. Syst. Rev., vol. 32, no. 5, pp. 224–237, 1997.
42. Frank B. Schmuck and Roger L. Haskin, "GPFS: A Shared-Disk File System for Large Computing Clusters," in FAST '02: Proceedings of the Conference on File and Storage Technologies, Berkeley, CA, USA, 2002, pp. 231–244.
43. J. Daemen and V. Rijmen, "AES Proposal: Rijndael," 1999.
44. Health Level Seven International. (2007) Health level 7 (HL7) V2 messages standard. [Online]. http://www.hl7.org/implement/standards/v2messages.cfm
45. George W. Beeler, "HL7 version 3–an object-oriented methodology for collaborative standards development," International Journal of Medical Informatics, vol. 48, pp. 151–161, 1998.

Chapter 7
SEMANTIC MANAGEMENT OF THE SUBMISSION PROCESS FOR MEDICINAL PRODUCTS AUTHORIZATION

Rajae Saaidi, Pavandeep Kataria, and Radmila Juric

1 Introduction

The pharmaceutical corporations are confronting big challenges when applying for Marketing Authorization (MA) licenses in extremely restricted environments, which are built around a variety of regulations, legislations and regulatory requirements. For novel pharmaceutical products, companies are required to submit thousands of pages of documents as part of the regulatory review process, which very often restricts companies' and governments' initiatives to speed up 'time-to-market' for medicines and minimize costs of the whole process. Furthermore, pharmaceutical companies are supposed to submit their Marketing Authorization Applications (MAA) according to their national specifications. Thus the formatting and organization of MA submission documents are dictated by the regulatory requirements of each country. Hence, pharmaceutical companies had to reformat and adjust the contents of MAA if they were applying for MA licenses across the world.

The International Conference on Harmonization (ICH) harmonizes MA procedures in the three ICH regions: Europe, Japan, and USA. Consequently they created the Common Technical Document (CTD) that represents a common format and structure of the contents required for MAA. The CTD has five modules: Module 1 contains regional administrative information, Module 2 supplies quality, non-clinical and clinical summaries, Module 3 gives chemical, pharmaceutical and biological data, Module 4 provides non-clinical reports and Module 5 produces clinical study reports. CTD was supposed to offer many benefits compared to old fashioned MA submission procedures: it was expected to minimize contents preparation time and costs, and to be reusable across the regions [1]. However, the pharmaceutical companies and the licensing authorities have found that the huge amounts of paper-based documents are unmanageable, and need large physical space for transfer and storage [2]. Therefore the ICH suggested electronic submissions through the electronic Common Technical Document (eCTD), which is considered to be an efficient way

R. Juric (✉)
London, UK
e-mail: R.Juric@westminster.ac.uk

S. Suh et al. (eds.), *Biomedical Engineering,*
DOI 10.1007/978-1-4614-0116-2_7, © Springer Science+Business Media, LLC 2011

of transmitting documents and metadata from applicants to the licensing authorities [3]. eCTD is composed of PDF documents organized in folders and referenced from a hierarchical data file in XML format, which is also called XML backbone of the MAA. eCTD has been a mandatory submission format in the US since 2008. It is recommended in the EU with expectations to be obligatory in 2010 [4].

It has been claimed by [5 and 6] that eCTD submissions decrease the amount of paperwork and shipping expenses, help with keeping track of the document amendments over time, simplify archiving and distribution and minimizing 'time-to-filling submissions' and thus accelerating 'time-to-market' for medicines. However, there are problems with eCTD submissions [7 and 8]. eCTD's XML backbone has a rigid hierarchical structure, which becomes especially problematic during the coding of many of its elements which refer to the numerous submission files within the eCTD. Pharmaceutical companies find it difficult to hand code a valid XML document. Incorrect or missing navigation aids (Bookmarks, Hyperlinks), inaccurate or outdated table of contents of eCTDs, and inadequate documents contents (i.e. they are not created according to the granularity of the CTD provided by the guidelines) may be the reasons for outright rejection of eCTD submissions [7 and 9].

Current software support for eCTD submissions provides solutions for the automatic creation of XML backbone [5], helps to build the folder structure [10] and validates the eCTD submissions against the technical validation criteria provided by the licensing authorities. They are able to check the file format, missing folders, file size, and the existence of valid DTD for XML files. They have also provided Microsoft word templates to guide applicants to properly utilize styles, formats, bookmarking and similar, to produce 'regulatory compliant formatting' [11]. However, none of these vendors provides a solution to check the validity of the content of the eCTD submissions. In other words, they do not ensure that these documents contain the right and relevant information in the correct place.

In this paper we offer an ontological solution for solving the problem of storing the correct content within the correct location of the eCTD submission. We describe the domain of MAA, CTD, its transfer towards the eCTD and the role of ICH in the process of harmonizing MA. We analyze the current eCTD submissions and highlight the problems related to the XML/PDF formats and limitations of software which supports them. We also explain how we model the ontology and use its constraints and reasoning rules in order to ensure that the correct content of an eCTD document is stored at the correct place within it.

2 The Domain

2.1 Marketing Authorisation Procedures

Within the European Union (EU) we have national, centralized and mutual MA or recognition procedures. In the national procedures, the pharmaceutical company applies for MA to the national competent authority of the member state, in which the pharmaceutical product will be placed. However, if the pharmaceutical company applies for MA in more than one country in the EU then the choice should be made between the Centralized Procedure (CP) and Mutual Recognition Procedure (MRP) [12]. The CP is administered by the European Medicines Agency (EMEA) in London. In 1995 the EMEA became the first regulatory body to grant a product licence that was valid across the EU [13]. The initial purpose of the agency was to make the MA process for new pharmaceuticals faster than before. However, they focused later on improving patient's and professional's information on the accurate use of pharmaceutical products for human and veterinary use, by providing scientific advice on questions regarding their quality, safety, and efficacy [14]. The centralised procedure is composed of a single application which, when accepted, gives MA for all markets within the EU. Most of the MA for generic medicines is granted through the MRP, which means that the assessment of the MAA of one Member State should be mutually recognized by other member states. The MRP is currently considered the most widely utilized way of acceptance for medicinal products in the EU.

Pharmaceutical companies looking for using MA in the US have to submit to the US Food and Drug Administration's Centre for Drug Evaluation and Research (CDER) [15] the proof which demonstrates that the pharmaceutical product is safe and efficient for its intended use. The centre doesn't actually test drugs itself, although it does lead limited research in the areas of drug quality, safety, and effectiveness standards. Once a pharmaceutical is accepted for sale in the US the FDA's mission of protecting consumer's health goes on. It monitors the use of marketed medicines for unanticipated health risks [16].

2.2   Common Technical Document (CTD)

The ICH of Technical Requirements for Registration of Pharmaceuticals for Human Use (ICH) assembles together the regulatory authorities of Europe, Japan and the United States and experts from the pharmaceutical industry. They have been dealing with the communication of scientific and technical aspects of MA with emphasis on "Safety, Quality, Efficacy" to reflect the three standards which represent the foundation for accepting and authorizing novel pharmaceutical products. Their main goal has been to harmonize regulatory requirements for medicinal products in order to rationalize the MA process for novel medicines, and stay away from needless duplication of effort by pharmaceutical corporations and licensing authorities [14].

Therefore they have reported that the major differences are in the technical content of the sections of the reporting data, which has resulted in duplications and repetitions in the MA process. The ICH have looked at ways of minimizing time-consuming reworking of data for each region and concentrated on providing a harmonized set of requirements, for both the content and the format in which the content is presented in the MAA. It is the harmonization of the presentation of the content that has led to the definition of the CTD which was first approved in 2000 [2].

The 'CTD' is structured into five modules that consist of the scientific data and summaries: Module1 (m1) is a regional module which contains the prescribing and administrative data and documents particular to every regional MA authority. For instance, in EU these documents have to consist of the application form, the proposed summary of product characteristics, labeling and package leaflet etc.

Module2 (m2) consists of 'the quality overall summary (QOS), the nonclinical overview and the clinical overview followed by nonclinical written summaries and the clinical summary. The non-clinical overview presents an interpretation of the data, the clinical relevance of the findings, cross-linking with the quality aspects of the medicinal product and the implications of the nonclinical findings for the safe use of the medicine. The clinical summary produces an evaluation of the clinical development and clinical efficacy pertaining to the safety data. Also, it has to evaluate the quality of the design and performance of all studies. The nonclinical summary has to generate an abstract of the pharmacological, pharmacokinetic and toxicology studies in the following order: *in vitro, in vivo*, species, route and then duration. The clinical summary presents a comprehensive, factual summation of the clinical information in Module 5 of the CTD and any post marketing data for products that have been marketed in other regions.

Module 3 (m3) is related to quality standard and gives an examination of the chemical, pharmaceutical and biological' data. Modules 4 and 5 (m4) (m5) consist of 'the nonclinical and clinical study reports respectively'. 'A table of contents' has to be included for every 'module that lists all the nonclinical/clinical study reports' and provides the position of 'each study report in the CTD'.

2.3   The transition to electronic CTD (eCTD)

eCTD was developed by the ICH in 2003 to allow an electronic submission for the marketing authorization applications [4]. Confronted with rising pace of innovation, complicated MA regulatory conformity, extensive development lifecycle of MAA and enhanced novel product development cost, pharmaceutical companies are forced to implement newer techniques in order to minimize research development cost through alternative measures and reduce time-to-market for medicines by optimizing the MA processes. Thus, the eCTD is considered the appropriate solution that helps pharmaceutical companies to achieve these objectives [6]. The transition to eCTD claims to be beneficial for both the regulatory agencies and the pharmaceuticals. They experience great decrease in MA documentation cost, and sending expenses. They also have simpler archiving and distribution, thus the lifecycle management of MAA is much easier: applicants have the ability to track modifications and updates over time. However, the eCTD constitutes a critical challenge, as pharmaceutical companies have to adjust their business processes and implement a planned strategy towards managing the transition towards the eCTD.

Small and Mid-size companies are the most affected by the transition towards the 'eCTD' [16] because they do not have enough in-house expertise or money for increased expenses to retain skilled and knowledgeable staff. In contrary, the eCTD is a preferred way of managing MAA for licensing agencies.

3 The Current Situation

3.1 XML Backbone for eCTD

The eCTD is an electronic version of CTD, thus the organization and the detailed elements described in the CTD are the source for the structure and content of eCTD However, as a submission format, eCTD includes further technical elements which permits the lifecycle of each single document within the MAA, if there will be a need to add more information or amend the existing information within the eCTD. The eCTD submission is made up of the following three elements:

1. The Directory structure: contains a structure of folders and files (File and folder names are uniquely defined. Each folder and file name in the eCTD has to be expressive, reasonable and short.
2. XML eCTD backbone: contains elements (called leaf elements) that point to the documents belonging to the eCTD application. It is built on an XML Document Type Definition (DTD), which describes the structure of the application, from modules 2 to 5.
3. Content files: are PDF files which contain all relevant data/information for a particular MA procedure.

The XML backbone is designed to:

- deal with the meta-data for the whole application and every single file included in the MAA, and
- form a complete table of contents and using the leaf elements, thus generate the navigation links to every separate document.

XML meta-data related to the submission of the eCTD contains the sending and receiving companies, manufacturer, publisher, ID, type of the MAA, and connected information. The examples for meta-data related to a document within the eCTD are: versioning information, language, and descriptive information such as document names [17].

The present ICH regulations utilize the XML only for structuring information within the e-CTD. Therefore, the regular formats that have to be incorporated within an eCTD submissions are PDF for narrative files and XML for structured documents.

4 The Problem

4.1   Potential Problems with XML and PDF documents

Problems with XML and PDF files are listed below and compiled from [7, 9 and 18]:

- applicants don't provide accurate elements and leaf titles in the XML document, as these have to be informative to allow the MA assessors to deduce the documents' contents
- applicant do not reference all the documents in the Xml backbone, i.e. unreferenced documents are a consequence of absent or mislocated directory references in xlink:href, therefore they cannot be located by MA assessors;
- applicants usually send non-standard eCTD, which means that the XML document does not respect the standard DTD and style-sheet provided by the ICH;
- applicants do not provide accurate and updated table of contents and do not include navigation aids such as bookmarks and hyperlinks;
- applicants submit an inaccurate granularity of the PDF documents, thus the documents contents are not compliant with and organized according to the structure of the CTD.

4.2   Problems with Software Support for eCTD Submission

There are currently 19 software products provided by the 8 'eCTD software vendors' [19], which assist with the authoring of the submission documents and the technical validation of the eCTD against the ICH validation criteria. They provide a variety of functionalities for creating an eCTD, which helps pharmaceutical companies to create an MAA with the correct structure. In EMEA, Extedo/IABG Life Sciences's product 'EURS is Yours' is being used. It informs 'whether an eCTD-based submission complies with the official interpretation of the eCTD format and offers access to submissions allowing the user to search, view or print any needed regulatory file [20]. EURS is Yours is particularly intended for the validation, acceptance, import, review and maintenance requirements of the EMEA and related national competent authorities (NCAs). It can be incorporated within an information system, keeping full functionality [20].

The US FDA is using 'GlobalSubmit VALIDATE 2009, which also provides error analysis and structure quality checks, to ensure the compliance with the current ICH and worldwide regulatory specifications [21]. 'iRegulatory' provides customized eCTD/CTD templates, which have been developed using 'Microsoft word 2003'. They contain the eCTD/CTD headings and offer a set of styles that can be utilized to guarantee consistent formatting and support automatic tables of contents, bookmarks, and cross-referencing'

In spite of offering significant help to pharmaceutical companies with MAA, these software solutions do not check the validity of the contents within the MAA submission documents, which were supposed to use the eCTD format. In other words, the most common error is that an applicant stores the wrong "contents" in the wrong PDF file, or the wrong "contents" is placed under the wrong section within the PDF document itself. For instance, information about the drug substance is placed under the section within a drug product document. On top of that, there are cases of:

- improper software vendor's hand coding of XML, due to its complex hierarchical structure. They have to code too many elements which refer to the numerous files for the eCTD submission, and
- software vendor's misinterpretation of the XML document type definition (DTD).

There are many software providers that produce customized 'Microsoft Word templates' allowing the applicants to focus more on content rather than formatting. These templates can guide the author to correctly use styles, formats, header/footer information, and bookmarking. The templates also support automatic tables of contents, section numbering, and cross-referencing, which allow improved document quality through consistent and regulatory compliant formatting. However, it's still the responsibility of the applicant to make sure that documents prepared through these templates meet relevant ICH, regional and national guidance with regard to structure, format, and content'. To summarize:

1. managing the creation of more than 200 individual PDF documents for eCTD submission is not an easy task. Making sure that the right information is put in the right place is an ordeal [22];
2. importing the final submission documents in the accurate place within the eCTD is difficult [11].

In fact, the ICH provides simply an 'an empty folder template' as an example of an eCTD submission folder structure. It shows all the possible Module 2-5 folders which are organized according to the 'eCTD specifications' and that can be populated with the relevant submission content (i.e. 'pdf files'). Therefore pharmaceutical companies, usually involve 'writers to assist in determining the correct placement of the individual documents based on content' [2]. However, due to the huge number of files this may take time, and lead towards placing the wrong file in the wrong location within the eCTD.

5 Related Works

We could not find any academic work which addresses the deficiency of the current eCTD submissions. There are also no examples where ontological modeling concepts and reasoning upon them help with the document contents within the eCTD.

However, in our earlier works on MAA we proposed the solution for an automation of MA procedures, through the generic framework based on service architectures and component technologies, where MA submissions and evaluations are conducted through a set of 'rules' applicable to a variety of MAA. The framework leaves enough 'space' for creating any format of 'rules', and it should be able to accommodate specific rules for handling the correct choice of PDF documents, at the correct position within any eCTD submission. In [23] the framework facilitates the automation of MA procedures and enables a dynamic generation of applications and supports evaluation procedures across regulatory requirements. In order to carry out the deployment of service oriented software components within the framework, we have designed and implemented a database [24], with the ultimate goal to support such an MA automation. We focused on the issue of the interoperability of such MA procedures across the world, therefore the database is reusable across a family of MA procedures. A complete J2EE software application built upon such a database, which also accommodates a set of rules for creating an MAA is available at [25 and 26]. This solution does not include the eCTD as it does not rely exclusively on the ICH requirements.

6   The Proposal

6.1   Ontology for Storing Semantics of the Submission Documents of the MAA

We propose a solution which will guarantee a correct eCTD in terms of both content and structure. To illustrate the proposal, we use Module 2 of an eCTD. It is considered to be of greater importance than other modules, as it contains a summary of the data related to all aspects of the discovery and development processes, used to characterize the physiochemical properties, safety, efficacy, and quality of the medicine.

Our proposal consists of an ontology named PDF_ONTOLOGY, with its ontological classes and their hierarchies, a set of constraints imposed on them and a set of SWRL rules, run upon them, which places a certain element (PDF file) of the eCTD navigational structure at the correct place. In this section we describe the PDF_ONTOLOGY. Constraints and reasoning are given in the MODEL IMPLEMENTATION section. PDF_Ontology stores semantics of the guidance-compliant contents of all PDF files for Module 2 of the eCTD. The main super classes of the PDF_Ontology are: Navigation_Structure, Module2, Content, and Section, as shown in Figure 1.

Navigation_Structure superclass mirrors the PDF documents in Module 2 of the eCTD, which is defined according to their correct contents. Therefore, Navigation_ Structure class consists of 21 subclasses named according to the standard file names recommended by the ICH for eCTD. It is important to keep these names, because

Figure 1. PDF_Ontology.

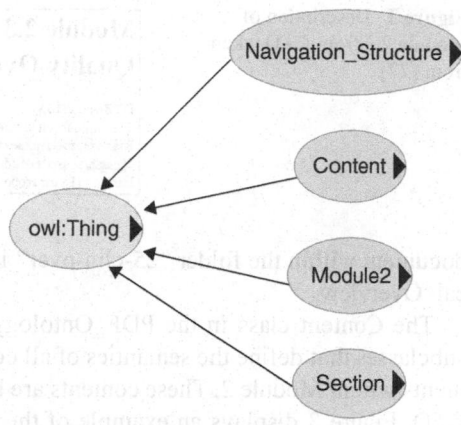

they help to eliminate misunderstandings when placing PDf document in the eCTD. It also enhances communication between the applicants and regulatory authority.

Figure 2 gives the subclasses of the Navigation_Structure superclass in the PDf_ Ontology (left hand side of Figure 2) and the suggested structure of an eCTD by ICH [27] (right hand side of Figure 2). The right hand side in Figure 2 is directly copied from the Module 2 of the eCTD. The Navigation_Structure superclass mimics all the PDF documents included in Module 2 of the eCTD. For example, "drug-substance" PDF file within the folder "23-qos" is modelled through the subclass Drug_Substance of the Navigation_Struture class. The "clinical-overview" PDF

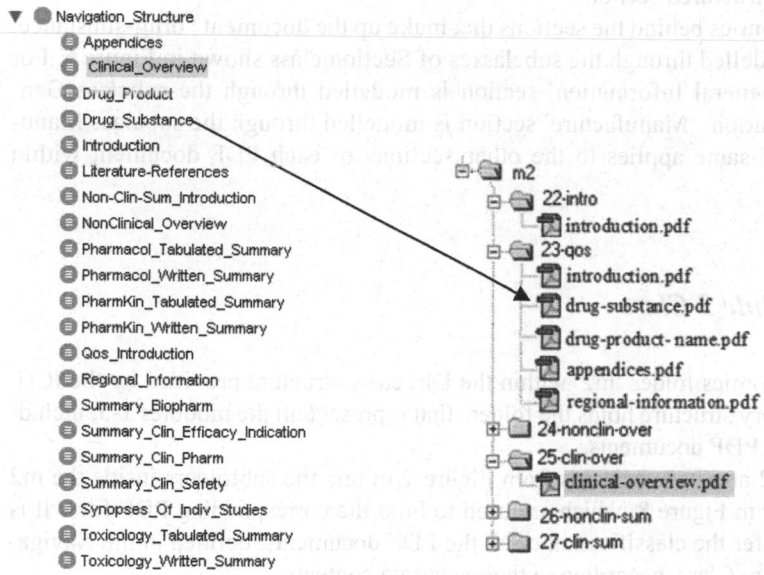

Figure 2. Example of Module 2 in the PDF_Ontology.

Figure 3. Description of
Module 2.3 from CTD taken
from [27].

Module 2.3
Quality Overall Summary
INTRODUCTION
The introduction should include proprietary name, non-proprietary name, European Pharmacopoeia name or common name of the drug substance, company name, dosage form(s), strength(s), route of administration according to the current version of the Standard Terms of the European Pharmacopoeia and proposed indication(s)

document within the folder "25-clin-over" is modelled through the subclass Clinical_Overview.

The Content class in the PDF_Ontology (Figure 1) has a number of disjoint subclasses that define the semantics of all contents that make up all the PDF documents within Module 2. These contents are based on the guidelines provided in the CTD. Figure 3 displays an example of the PDF contents that make up the document "introduction.pdf" of the "23-qos" folder within module 2 (m2 right side of Figure 2).

The semantics behind all the information shaded as grey in Figure 3, and which are included in the document "introduction.pdf", are modelled through the subclasses of the Content Class, as shown in Figure 4. For instance, the semantic of proprietary name of the drug substance is modeled through the ontological subclass Proprietary_Name_Of_Drug_Substance.

The Section class in the PDF_Ontology (Figure 1) contains subclasses to mirror all the various sections of PDF documents. These sections are also described in the CTD specification. For example, Figure 5 displays a fragment from the "drug-substance.pdf" document, within the folder "23-qos" of Module 2 (m2), which shows some of its structured sections.

The semantics behind the sections that make up the document "drug-substance.pdf" are modelled through the subclasses of Section class shown in Figure 6. For instance, 'General Information' section is modelled through the subclass General_Information, 'Manufacture' section is modelled through the subclass Manufacture. The same applies to the other sections of each PDF document within Module 2.

6.2 Module 2 Class

This class mimics folder 'm2' within the Directory structure provided by the ICH. This Directory structure holds the folders that represent all the modules 1-5, including the final PDF documents.

Module 2 and its subclasses from Figure 7 mimic the subfolders inside the m2 folder given in Figure 8, which are used to hold the corresponding PDF files. It is needed to infer the classification of all the PDF documents, defined in the Navigation_Structure Class, according to their accurate contents.

Figure 4. A fragment of PDF_Ontology with the class Content and its subclasses that model semantics of the PDF documents contents described in Module 2.

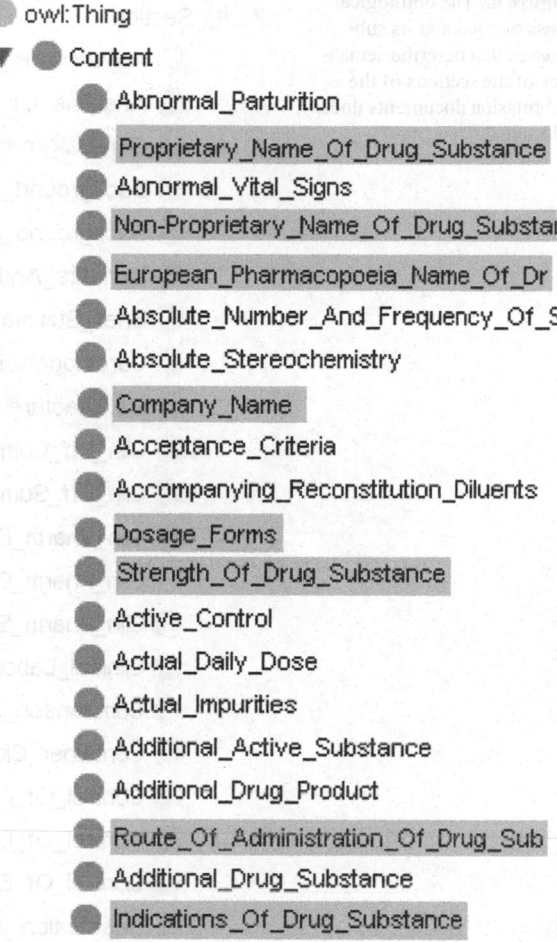

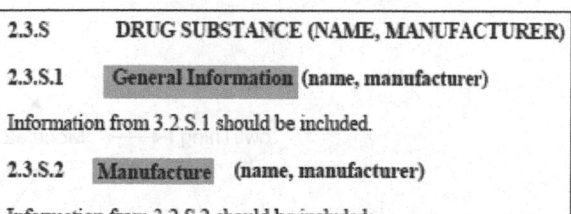

Figure 5. The structure of Drug Substance from CTD taken from [27].

Figure 6. The ontological class Section and its subclasses that describe semantics of the sections of the submission documents under Module 2.

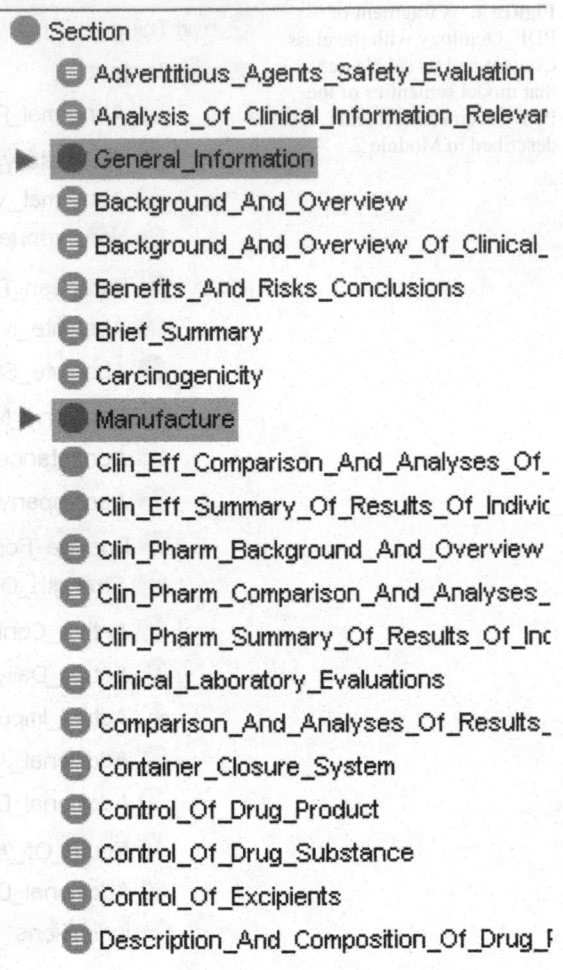

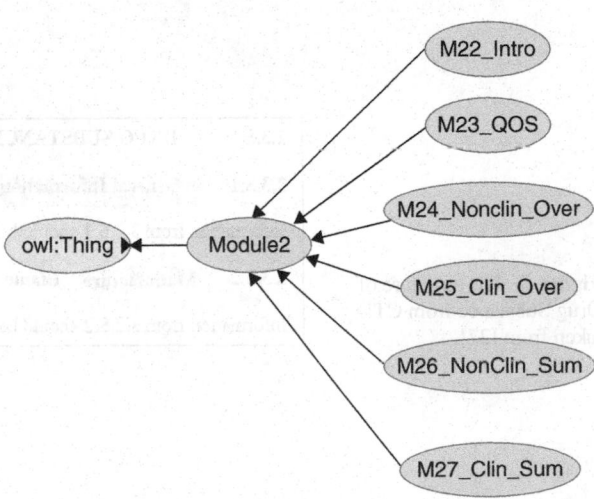

Figure 7. Module 2 subclasses.

Figure 8. Folder structure of
Module2 (m2) in the eCTD.

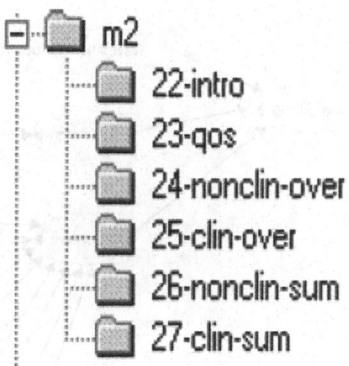

6.3   Model Implementation

To build the ontological model, we use 'OWL and its modelling constructs' [28] because it introduces axioms based on description logic (DL) paradigm [29]. The PDF_Ontology contains 1307 concepts as it stores semantics of the eCTD Module 2 PDF document contents. These concepts are implemented through the Protégé 2000 Ontology Editor, version 3.4 [30], The Pellet Reasoner [31] is used to check the consistency of the model's properties, classes, and individuals. The SWRL rules are created in order to reason upon the ontological concepts and to infer the location of each PDF document at the correct position in the eCTD Module 2. These rules were implemented using the SWRL Tab plug-in in Protégé' and executed by the Jess Rule engine.

6.4   Object Properties, Restrictions and Necessary and Sufficient Conditions

We use OWL object properties to describe the relationships between a particular PDF document and its contents and sections. The object properties used in the PDF_Ontology are (i) "has_Section" which relates individuals from the class Navigation_Structure ('domain') to individuals from the class Section ('range') and (ii) "has_Content" which relates individuals from the class Navigation_Structure to individuals from the class Content.

Having created these properties, OWL has allowed us to use them for specifying restrictions in order to define the classes of the Navigational Structure. We have used the following restrictions to describe and define the subclasses under the Navigation_Structure Class:

- *existential restrictions* (∃) to specify that a PDF document must have the specified section or content,

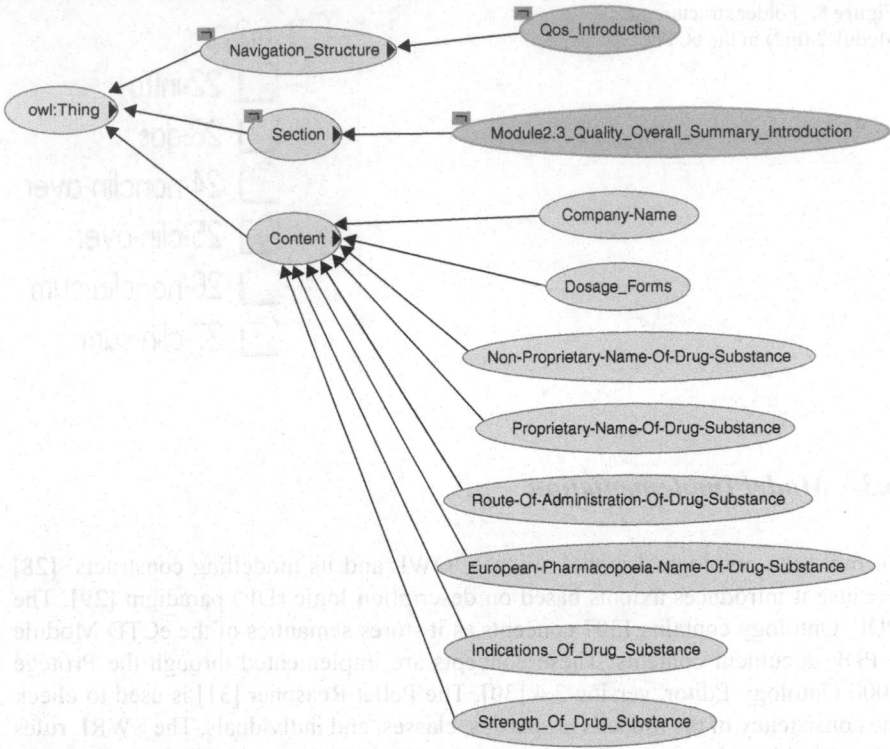

Figure 9. Class hierarchy with semantics of the contents of QOS Introduction document.

- *universal restrictions* (∀) to describe that a document must only have the specified sections, and
- *cardinality restrictions* to determine that a document has exactly the specified number of sections.

To illustrate the above, we use the example of a PDF document which stores "QOS Introduction" part of the eCTD. The QOS Introduction Document is modeled through the subclass "Qos_Introduction" of the Navigation_Structure class. It includes one section ('Module2.3 Quality Overall Summary Introduction') and contents ('proprietary name, non-proprietary name, European Pharmacopoeia name of the drug substance, company name, dosage forms, strength, route of administration, and proposed indications'), as described in the CTD. Figure 9 shows the class hierarchy that mirrors the contents and the sections of QOS Introduction document, where we can apply restrictions outlined in the bullets above.

Figure 10 displays the appropriate OWL necessary and sufficient conditions used to define the subclass Qos_Introduction, of Navigation_Structure class, in OWL DL. We give one example of a existential restriction: *the first eight 'existential restrictions' use the property "has_Content" to 'denote necessary relationships between individuals' from the class "Qos_Introduction" and the classes through*

Figure 10. Neccesary and sufficient conditions.

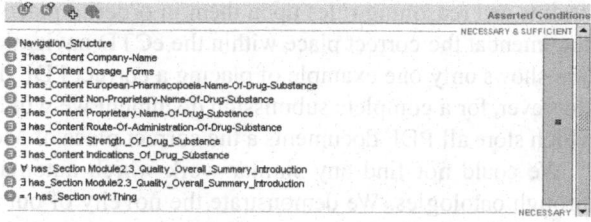

Figure 11. A SWRL rule which infers classification.

```
Navigation_Structure(?x)
∧ Module2.3_Quality_Overall_Summary_Introduction(?y)
∧ has_Section(?x, ?y)
∧ has_Content(?x, company_name)
∧ has_Content(?x, dosage_forms)
∧ has_Content(?x, european_pharmacopoeia_name_of_drug_substance)
∧ has_Content(?x, non_proprietary_name_of_drug_substance)
∧ has_Content(?x, proprietary_name_of_drug_substance)
∧ has_Content(?x, route_of_administration_of_drug_substance)
∧ has_Content(?x, strength_of_drug_substance)
∧ has_Content(?x, indications_of_drug_substance)
→ M23_QOS(?x)
```

which the contents are modeled, i.e. the document "QOS Introduction" must have all the specified contents: 'company name, dosage forms, EU pharmacopoeia name of the drug substance, etc.

These necessary and sufficient conditions ensure that the PDF document named "Qos_Introduction" of the navigation structure, has the correct set of sections and contents, i.e. if any individual of the Navigation_Structure class meets all these conditions, then it must be a member of the class Qos_Introduction. We use the same approach in order to establish the correct content and sections of other PDF documents.

6.5   Reasoning with SWRL Rules

We illustrate one SWRL rule in Figure 11. Its purpose is to ensure that the correct content of the PDF document named "Qos_Introduction" is moved into the correct subclasses of Module2 class (i.e. ontological class) of the Navigation_Structure class, which in turn mirrors the content of the "M23_Qos" folder within Module 2 of the eCTD.

Therefore, the SWRL in rule Figure 11 moves ontological individuals from the "Qos_Introduction" class into the subclass M23_QOS of the Module2 class in the Navigation_Structure class.

7   Conclusion

We have proposed an ontological environment which contains the semantic of the eCTD, and the MA procedures, specified by the ICH, in order to eliminate current errors and limitation of their MAA submissions. We use ontological concepts, constraints and reasoning rules upon them in order to place a correct content of a PDF document at the correct place within the eCTD navigational structure. Our illustration shows only one example of placing a correct PDF document within the eCTD. However, for a complete submission document we will have to run 87 SWRL rules which store all PDF documents a the correct place of Module 2 in the eCTD.

We could not find any similar work which automates the MAA submissions through ontologies. We demonstrate the novelty of our work through the example of (a) well defined and correct structure to the submission documents through their stored contents and sections and (b) SWRL rules which infer the placement of the navigation structure documents into Module 2 of CTD, according to their adequate contents.

Our future work should improve the licensing process of MAA, i.e. we should build an extension of the ontological model to perform the evaluation of the scientific data corresponding to the stored contents within the submitted eCTD. This would be a complete automation of the regulatory procedures for the MAA, because it would support the assessment of the quality, safety, and efficacy of a new pharmaceutical product.

References

1. Datafarm Inc, "eCTD: Will You Be Ready for 2009", epc, Data Management and IT Solutions, available at http://www.datafarminc.com/PDF/CS-EPC.pdf, 2007.
2. T. Felgate, "The evolution of the eCTD", available at http://www.appliedregulatory.com/articles/FelgateTPharmafocusMarch2009.pdf, 2009.
3. G.E. Overend, "Introduction to eCTD and first principles. TOPRA", The Organisation for Professionals in Regulatory Affairs, available at http://www.topra.org/files/focus1_0.pdf, 2008.
4. A. Neuer, "Learning to embrace the eCTD", Bio IT World, available at http://www.bio-it-world.com/issues/2009/may-jun/feature.html, 2009.
5. iRegulatory Ltd, "Building eCTDs", available at http://www.ectd-resources.com/readarticle.php?article_id=7, 2008.
6. A. Goel and M.K. Sundararajan, "Managing Electronic submissions through eCTD with strategic partners", eyeforpharma Briefings, Guest Feature, available at http://www.eyeforpharma.com/briefing/ectdfinal.pdf, 2006.
7. J. Pickett, "XML a frequent problem with eCTD submissions; speaker says refuse-to-files are painful", BioResearch Compliance Report Electronic clinical trials, available at http://www.entrepreneur.com/tradejournals/article/175407584.html, 2007.
8. C. Mathis, "A History of eSubmission" epc, Data Management and IT Solutions, available at http://www.reg123.com/Documents/LORENZ_EPC0609.pdf, 2009.
9. D. Duggan, "Implications of eCTD errors", FDA/CDER/OBPS, available at http://www.fda.gov/downloads/Drugs/DevelopmentApprovalProcess/HowDrugsareDevelopedandApproved/ApprovalApplications/AbbreviatedNewDrugApplicationANDAGenerics/UCM166278.pdf, 2008.
10. B.M Noel, "Managing a Major eCTD Filing", Clinical Trials, Issue 11, http://www.pharmafocusasia.com/clinical_trials/managing.htm, 2009.
11. P. Boe, "Templates: Taking the first step towards eCTD submissions in Europe", available at http://www.emwa.org/JournalArticles/JA_V15_I1_Boe1.pdf, 2006.

12. EGA, "European Generic Medicine Association: Authorisation", available at http://www.egagenerics.com/gen-authorisation.htm, 2004.
13. B. Avison, "Pharma Market Authorization Strategies: A guide to launching drugs quickly and efficiently in Europe", Business Insights, available at http://www.globalbusinessinsights.com/content/rbhc0093t.pdf, 2003.
14. P. Evers, "Pharmaceutical Regulatory Affairs Outlook 2002", available at https://www.globalbusinessinsights.com/content/rbhc0072m.pdf, 2002.
15. FDA, "Development & Approval Process (DRUGS)", available at http://www.fda.gov/Drugs/DevelopmentApprovalProcess/default.htm, 2009, 2009.
16. P. James and G. Archbold, "The secret life of a dossier", available at http://www.publicservice.co.uk/article.asp?publication=European%20Union&id=275&content_name=Health&article=8395, 2007.
17. J. Ramsden, "Common Technical Document", Pharmabiz, available at http://www.pharmabiz.com/article/detnews.asp?articleid=11405&sectionid=46, 2002.
18. G. Ventura, "Optimizing Your eCTD Submission: How to Achieve an Efficient and Timely Review", available at http://www.forumsci.co.il/Landau/routine%20updates/drugs/Optimizing%20Your%20eCTD%20Submission.doc, 2008.
19. Laszlo Letter, "eCTD Software Vendors", [online], Laslo, http://laszloletter.typepad.com/the_laszlo_letter/2007/02/ectd_software_v.html, 2007.
20. EXTEDO GmbH, "EURS is Yours", EXTEDO, available at http://www.extedo.com/products/submission-reviewing-validation/eurs-is-yours/, 2009.
21. Global Submit, "eCTD validation for submission assurance and quality", available at http://globalsubmit.com/home/Products/VALIDATE/tabid/243/Default.aspx, 2009.
22. Brilliant Leap Blog, "The eCTD/ Document Management Connection- What say the vendors", Document Content Management, available at http://www.brilliantleap.com/blog/2008/09/the-ectd-document-management-connection–what-say-the-vendors.html, 2008.
23. R. Juric and J. Juric, "Applying Component Based Modelling in the Process of Evaluation of Medicinal Products", Proceedings of the 11th International Conference on Integrated Design and Process Technology (IDPT), 2002.
24. L. Slevin, R. Shojanoori, and R. Juric, "Developing a database for automating regulatory affairs in the pharmaceutical industry", Journal of Integrated Design and Process Science, Vol. 9, Issue 4, pp.1–11, 2005.
25. R. Juric, L. Slevin, R. Shojanoori, and S. Williams, "Software Support in Automation of Medicinal Product Evaluations", Proceedings of the International Council on Medical and Care Computerics event, (ICMCC) 2005.
26. R. Juric, and S. Williams, "Experiences of Creating COTS Components when Automating Medicinal Product Evaluation", Proccedings of the 16th International Conference on Software Engineering and Knowledge Engineering (SEKE), 2005.
27. European Commission, "Volume 2B Notice to Applicants Medicinal products for human use", available at http://ec.europa.eu/enterprise/pharmaceuticals/eudralex/vol-2/b/update_200805/ctd_05-2008.pdf, 2008.
28. R. Stevens, M. E. Aranguren, K. Wolstencroft, U. Sattler, N. Drummond, M. Horridge, and A. Rector, "Using OWL to model biological knowledge", International Journal of Human-Computer Studies, Vol. 65, Issue 7, pp. 583–594, 2007.
29. J.S. Brunner, L. Ma, C. Wang, L. Zhang, D.C. Wolfson, Y. Pan, and K. Srinivas, "Explorations in the Use of Semantic Web Technologies for Product Information Management", Proceedings of the World Wide Web Conference, 2007.
30. Protégé, "Welcome to protégé", available at http://protege.stanford.edu/, 2009.
31. E. Sirin, B. Parsia, B.C. Grau, A. Kalyanpur, and Y. Katz, "Pellet: A Practical OWL-DL reasoned", Journal of Web Semantics, Vol. 5, Issue 2, pp. 51–53 2004.

Chapter 8
SHARING HEALTHCARE DATA BY MANIPULATING ONTOLOGICAL INDIVIDUALS

Pavandeep Kataria and Radmila Juric

1 Introduction

The purpose of this chapter is twofold:

A) We primarily want to disseminate our experiences of using the power of onto-logical engineering when dealing with interoperability and the problem of data sharing in heterogeneous computational environments in healthcare.

B) At the same time we illustrate the complexity of the manipulation of ontological concepts in order to achieve A) above, by giving a detailed description of the steps involved in the process of delivering data sharing.

To demonstrate A) and B) we create an ontological environment through ontologi-cal editing tools and reasoning engines attached to them, and we use OWL/SWRL languages which support the manipulation of ontological concepts stored in OWL/SWRL enabled ontologies. To illustrate the feasibility of ontological reasoning for the purpose of data sharing across heterogeneous healthcare environments, we use our own process which consists of the following steps:

1. Translation of databases into ontologies. The translation requires the introduc-tion of datatype properties representing semantics stored in database schemas (this applies to column names, attributes and data types within the database schemas).
2. Adding new ontological classes in order to manipulate datatype properties. The manipulation of datatype properties requires the existence of ontological classes which accommodate the results of our manipulation. These classes also define which conditions must be met by classes' ontological individuals.
3. Running of SWRL rules to perform the manipulation of ontnological individu-als. SWRL rules determine which datatype properties are to be grouped together in order to achieve data sharing.

R. Juric (✉)
London, UK
e-mail: R.Juric@westminster.ac.uk

S. Suh et al. (eds.), *Biomedical Engineering*,
DOI 10.1007/978-1-4614-0116-2_8, © Springer Science+Business Media, LLC 2011

4. Retrieval of the output of the manipulation. Once the results of SWRL rules are classified against a reasoning engine, such as JESS, datatype properties defined in the manipulation from 3 are automatically converted into ontological individuals, which can then be retrieved through an application as 'shared data / information' across heterogeneous databases.

2 The Scenario

We give a scenario of data retrievals across three heterogeneous environments: General Practitioner (GP), Hospital and Clinic, in order to create a medical summary for a particular patient. Database definition statements for patient records in each environment are given in Figure 1. The medical summary may consist of patient demographic and clinical data which are scattered across these three environments. If we assume that a GP wants to obtain such a medical summary for a particular patient X, then Figure 2 shows which data is retrieved from which environment. GP would like to obtain clinical data for patient X created by consultants in the hospital and clinic databases: *Hospital_DB* and *Clinic_DB*, but would be happy to

CREATE TABLE **sdps_gp.PATIENT(** PATIENT_ID INTEGER (6), FIRST_NAME VARCHAR (10), LAST_NAME VARCHAR (20), SEX CHAR(1)CHECK (sex IN ('M' , 'F')), DOB DATE, ADDRESS VARCHAR (100), NEXT_OF_KIN CHAR (30), NIN CHAR (7), EMAIL VARCHAR (25), TEL VARCHAR (14), PRIMARY KEY (PATIENT_ID));	CREATE TABLE **sdps_hospital.PATIENT(** PATIENT_ID VARCHAR (6), NAME VARCHAR (20), SEX CHAR(1) CHECK (Gender IN ('M' , 'F')), DOB DATE, HOSPITAL_MEDICAL_SUMMARY VARCHAR (300), MAJOR_ILLNESS VARCHAR (100), CHRONIC_DISEASE VARCHAR (100), PRIMARY KEY (PATIENT_ID));
CREATE TABLE **sdps_clinic.PATIENT(** PATIENT_ID VARCHAR (6), FIRST_NAME VARCHAR (10), LAST_NAME VARCHAR (20), SEX CHAR(1) CHECK (Gender IN ('M' , 'F')), DOB DATE, CLINIC_MEDICAL_SUMMARY VARCHAR (300), MAJOR_ILLNESS VARCHAR (100), CHRONIC_DISEASE VARCHAR (100), PRIMARY KEY (PATIENT_ID));	CREATE TABLE **sdps_clinic.LABTEST(** LABTEST_ID VARCHAR (6), LABTEST_TYPE VARCHAR (60), LABTEST_NAME VARCHAR (60), LABTEST_DATE DATE, PATIENT_ID VARCHAR (6), PRIMARY KEY (LABTEST_ID), FOREIGN KEY (PATIENT_ID) REFERENCES PATIENT (PATIENT_ID)); CHARSET=utf8;

Figure 1. Data definition statements of relational tables used in data sharing.

Figure 2. Creating medical summaries of a particular patient: Translation of databases and manipulation of ontological individuals.

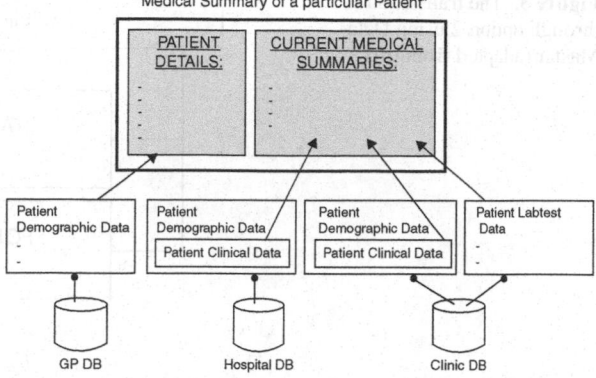

use patient X's demographic data from his/her own database *GP_DB*. We have to take into account that patient clinical and demographic data (a) may be stored under different table/attribute names in these databases, and (b) do not necessarily belong to different data structures within one schema.

For example the sdps_hospital.PATIENT table stores both: *Patient Demographic* and *Clinical Data*, where the latter is stored within the HOSPITAL_MEDICAL_ SUMMARY, MAJOR_ILLNESS, CHRONIC_DISEASE attributes. These three attributes constitutes the *Patient Clinical Data* "block" within the hospital environment as in Figure 2. The same applies to the clinic environment: CLINIC_MEDI-CAL_SUMMARY, MAJOR_ILLNESS, CHRONIC_DISEASE from sdps_clinic. PATIENT is within the *Patient Clinical Data* "block" of the clinic environment. Almost all attributes from sdps_clinic.LABTEST are needed for the *Patient Labtest Data* "block" as in Figure 2.

It is obvious that there are overlapping information/data across these three databases, which may contain a variety of semantic conflicts between them. The issue of resolving them is outside the scope of this paper and we refer readers to our publication which deal separately with semantic conflicts [1 and 2]. However, obtaining MEDICAL SUMMARIES for patient X means that we have to retrieve semantically related data, regardless of their format and location. Retrieving semantically related data may involve resolving semantic conflicts in the first place (if you have any), but may also mean that we want to display overlapping information across the databases (*GP_DB, Hospital_DB* and *Clinic_DB*), because they may give a better picture of the MEDICAL SUMMARIES for patient X.

3   The Translation of Databases

The translation of databases *GP_DB, Hospital_DB* and *Clinic_DB* into ontologies, as specified in 1 from the Introduction, is done through the Protégé 3.4 ontological editing toolkit environment [3]. Protégé 3.4 is an open source tool that works with OWL Full, OWL Lite and OWL DL [4], and provides a plug-in for i) the automatic translation of database schema to ontology and ii) the automatic connection to the

Figure 3. The translation through option 2 using Data-Master (adapted from [6]).

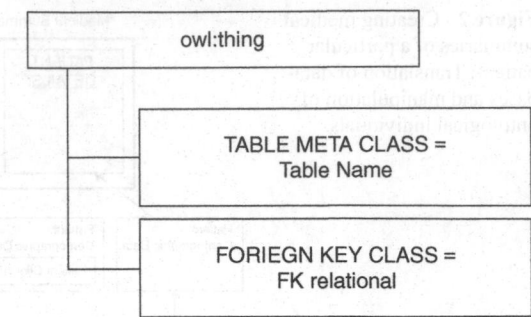

Pellet reasoning engine [5]. The plug-in used for the automatic translation of databases to ontology is DataMaster [6], which allows the importation of database data structures, including their database content into Protégé via a JDBC connector.

The importation is specified according to three choices of translation options:

- option 1: schema structure ontology for Protégé Frames,
- option 2: schema structure ontology for database tables as OWL classes and
- option 3: schema structure ontology for database tables as OWL instances using a predefined Relational.OWL ontology.

A detailed explanation regarding the differences between theses three translation options are in [7]. For the purpose of this paper we briefly outline the differences between options 2 and 3, assuming the content of databases is also being imported via the JDBC connector. Figure 3 illustrates the translation through option 2. Each table within a database schema is translated into an ontological class (TABLE META CLASS) and foreign key relationship between tables within a database schema is translated into the FOREIGN KEY class. Each row of a particular table is translated into an ontological individual. Each ontological individual is connected to a number of datatype properties within an ontology. Each datatype property represents the column (attribute) names in a table. The domain of a datatype property is defined according to which table each column name belongs to. The range of the datatype property is defined according to the data values stored in a row of that table.

Therefore, the sdps_gp.PATIENT table had column names (attributes): PA-TIENT_ID, FIRST_NAME, LAST_NAME and ADDRESS and data values in these columns (for these attributes) are '0002', 'NEMANJA', 'FLEE', and '167 BOULEVARD, LONDON', option 2 would translate the sdps_gp.PATIENT table into the db1:patient ontological class and the row of the table is translated into the db1:patient_Instance_1 ontological individual. Column names (attributes) are translated into datatype properties: db1:patient.PATIENT_ID, db1:patient.FIRST_NAME, db1:patient.LAST_NAME and db1:patient.ADDRESS. The domain for datatype properties is set as follows:

- the domain for the datatype property db1:patient.PATIENT_ID is set as the db1:patient ontological class and the range is set as the data value '0002';

Figure 4. The translation through option 3 using Data-Master (adapted from [6]).

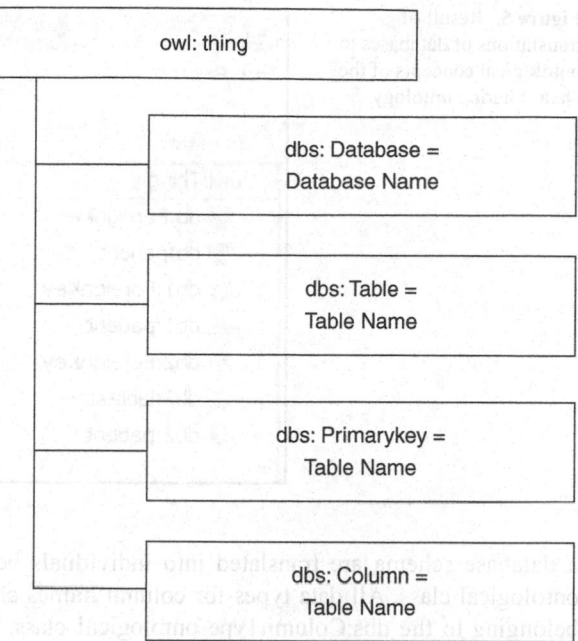

- the domain for the datatype property db1:patient.FIRST_NAME is set as the db1:patient ontological class and the range is set as the attribute 'NEMANJA';
- the domain for the datatype property db1:patient.LAST_NAME is set as the db1:patient ontological class and the range is set as the attribute 'FLEE';
- the domain for the datatype property db1:patient.ADDRESS is set as the db1:patient ontological class and the range is set as the attribute '167 BOULE-VARD, LONDON'.

Option 2 also defines ranges per their corresponding data types and per each column name (attribute): i.e. the data type 'integer' for data value '0002' is translated as the ontological range value of integer for datatype property db1:patient.PATIENT_ID and the data type 'varchar' for data value 'NEMANJA', 'FLEE' and '167 BOULE-VARD, LONDON' are translated as the ontological range type of "string literal" for datatype properties db1:patient.FIRST_NAME, db1:patient.LAST_NAME and db1:patient.ADDRESS respectively.

Figure 4 illustrates translation through option 3. It is clear from the diagram that option 3 carries forward more semantics from the database schema than from database content. Option 3 pre-defines a skeleton of ontological classes in which we translate the semantics from the database schema. For example, the database schema name is translated into an ontological individual belonging to the dbs:Database ontological class. All table names within a database schema are translated into ontological individuals belonging to the dbs:Table ontological class. All primary keys of tables within a database schema are translated into ontological individuals belonging to the dbs:PrimaryKey ontological class. All column names (attributes) within

Figure 5. Result of translations of databases to ontological concepts of the Data_Sharing ontology.

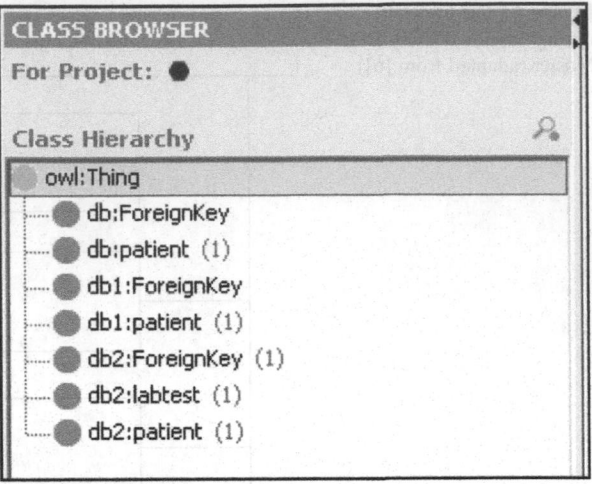

a database schema are translated into individuals belonging to the dbs:Column ontological class. All data types for column names are translated into individuals belonging to the dbs:ColumnType ontological class. There are no provisions for importing the database content.

In our scenario, we are required to bring forward the content of relational tables in terms of displaying MEDICAL SUMMARIES, which consist of data stored across databases *GP_DB*, *Hospital_DB* and *Clinic_DB*. Therefore option 3, which do *not* take into account the content of a database during translation, is automatically eliminated as a solution for us. On top of that, we have found and reported errors in the logical consistency of the ontology produced through translation in option 3. Therefore we advocate option 2 because the translation will then retain the most of the semantics from the databases.

Figure 5 shows a single ontology named Data_Sharing that is used to accommodate the translation of databases *GP_DB*, *Hospital_DB* and *Clinic_DB*. Seven ontological classes are shown in Figure 5 (where "db" denotes *GP_DB*, "db1" denotes *Hospital_DB* and "db2" denotes *Clinic_DB*):

- db:patient and dbForiegnKey ontological classes are translated from the sdps_gp.PATIENT table in the *GP_DB*;
- db1:patient and db1:ForiegnKey ontological classes are translated from sdps_hospital.PATIENT table in the *Hospital_DB*, and
- db2:patient, db2:labtest and db2ForiegnKey ontological classes are translated from sdps_clinic.PATIENT and sdps_clinic.LABTEST tables in the *Clinic_DB*.

Figure 6 also shows a partial view of datatype properties in the Data_Sharing ontology, generated through the translation option 2. The total number of datatype properties in the Data_Sharing ontology corresponds directly to the number of column names (attributes) within the tables in the databases *GP_DB*, *Hospital_DB* and *Clinic_DB*, thus in our example scenario we gain a total of thirty-one datatype properties.

Figure 6. Examples of datatype properties within the Data_Sharing ontology.

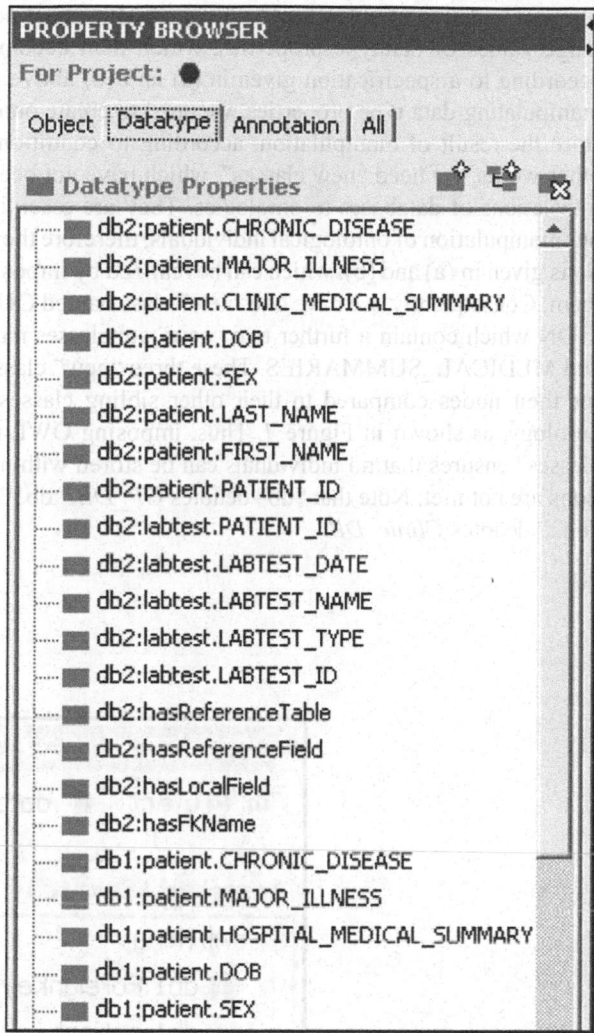

4   Manipulation of Datatype Properties

We manipulate data type properties in order to move individuals around the DATA_ SHARING ontology and achieve data sharing. Manipulation in the context of "creating MEDICAL SUMMARIES for patient X across databases *GP_DB*, *Hospital_ DB* and *Clinic_DB*" from the scenario means that we have to manipulate data within the data "blocks" in Figure 2. The manipulation is as follows:

(a) *Patient Clinical Data* "block" is created from the *Hospital_DB* data, and *Clinic_DB* data, and

(b) *Labtest Data* "block" is created from the *Clinic_DB* data.

The manipulation of ontological individuals shows our power and freedom to move range values of datatype properties, which then become ontological individuals, according to a specification given in (a) and (b) above. However, before we start manipulating data type properties, we need to create ontological classes which can store the result of manipulation, according to conditions given in (a) and (b). In other words, we need "new classes", which have not been created as a result of the translations of databases to ontologies. They are essential for storing the result of our manipulation of ontological individuals, therefore they should conform to conditions given in (a) and (b), which can be realized by imposing OWL restrictions upon them. Consequently, we have one "new" class named OUTPUT_OF_MANIPULA-TION which contain a further two "new" subclasses named PATIENT_DETAILS and MEDICAL_SUMMARIES. These three "new" classes have different symbols for their nodes compared to their other sibling classes within the Data_Sharing ontology, as shown in Figure 7. Thus, imposing OWL restrictions on these "new classes" ensures that no individuals can be stored within them, if the OWL restrictions are not met. Note that "db" denotes *GP_DB*, "db1" denotes *Hospital_DB* and "db2" denotes *Clinic_DB*.

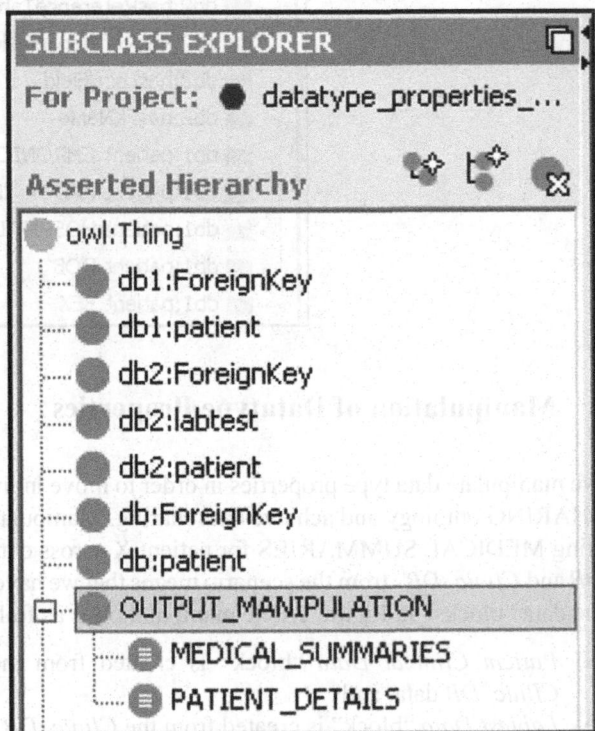

Figure 7. Defining "new" ontological classes for storing results of manipulating ontological individuals in the Data_Sharing ontology.

Figure 8. OWL restriction upon the 'has_patient_details' object property.

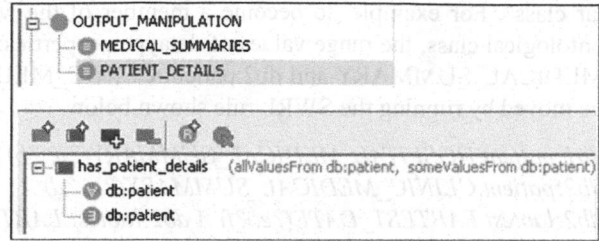

4.1   Owl Restrictions

Conditions, which are listed in (a) and (b) above should be described through object properties, upon which OWL restrictions are set. Object properties, in general, link ontological individuals from one class to another. In our scenario, we restrict the number of ontological individuals belonging to the PATIENT_DETAILS ontological class, by creating object property: has_patient_details as shown in Figure 8. The domain for the object property: has_patient_details is set to the PATIENT_DETAILS ontological class. The range for the object property: has_patient_details is set to the names of ontological classes in which ontological individuals can be connected, i.e. the PATIENT_DETAILS connects ontological individuals from the db:patient ontological class, through the object property: has_patient_details.

We use two OWL restrictions: the *existential restriction* (∃) is used to describe that the PATIENT_DETAILS ontological class has some ontological individuals from the db:patient class. In order to deal with open world reasoning [8] the *universal restriction* (∀) is used to describe that the PATIENT_DETAILS ontological class has only ontological individuals from the db:patient ontological class. Both restrictions are made necessary and sufficient conditions to imply the 'concreteness' of the PATIENT_DETAILS class in the Data_Sharing ontology. The 'concreteness' means that classes in the Data_Sharing ontology will not exist if the necessary and sufficient conditions have NOT been satisfied. Therefore, the PATENT_DETAILS ontological class is defined as: **PATIENT_DETAILS ontological class will store some ontological individuals from the db1:patient ontological class and only from the db1:patient ontological class.** The same applies for the MEDICAL_SUMMARIES class. It will have OWL restrictions which will guarantee that ontological individuals will be drawn from db2:patient and db2:labtest ontological classes.

5   SWRL Rules and the Results of Manipulation of Ontological Indviduals

We run SWRL rules in order to perform the exact manipulation of datatype properties across the Data_Sharing ontology. Range values (i.e. ontological individuals) of datatype properties are used in SWRL rules to become "a member of a particular class". For example, to become a member of the MEDICAL_SUMMARIES ontological class, the range values of dataype properties: db1:patient.HOSPITAL_ MEDICAL_SUMMARY and db2:patient.CLINIC_MEDICAL_SUMMARY must be moved by running the SWRL rule shown below.

db1:patient.HOSPITAL_MEDICAL_SUMMARY(?a, ?b) ∧
db2:patient.CLINIC_MEDICAL_SUMMARY(?c, ?d) ∧
db2:labtest.LABTEST_DATE(?e, ?f) ∧ db2:labtest.LABTEST_ID(?g, ?h) ∧
db2:labtest.LABTEST_NAME(?i, ?j) ∧ db2:labtest.LABTEST_TYPE(?k, ?l)
→ MEDICAL_SUMMARIES(?b) ∧ MEDICAL_SUMMARIES(?d) ∧
MEDICAL_SUMMARIES(?f) ∧ MEDICAL_SUMMARIES(?h) ∧
MEDICAL_SUMMARIES(?j) ∧ MEDICAL_SUMMARIES(?l)

By running the SWRL rule above, we perform manipulation. In general SWRL rules like the one above dictate which ontological indviduals we would like to into which "new" ontological classes. In order to run SWRL rules, we must connect to the JESS engine [9], which performs (i) the conversion of SWRL rule to JESS rules, (ii) the running of Jess rules against the Jess Engine and (iii) the movement of range value (i.e. ontological indiviauls) of datatype properties into "new" onto- logical class, as implied in the SWRL rule. The Pellet reasoning engine is used to check the consistency of the classification made by JESS engine. Therefore, the SWRL rule above consequently moves the following ontological individuals into the MEDICAL_SUMMARIES class:

db1:patient.HOSPITAL_MEDICAL_SUMMARY
db2:patient.CLINIC_MEDICAL_SUMMARY
db2:labtest.LABTEST_DATE
db2:labtest.LABTEST_ID
db2:labtest.LABTEST_NAME
db2:labtest.LABTEST_TYPE

The MEDICAL_SUMMARIES class now contains ontological individuals from databases *Hospital_DB* and *Clinic_DB*.

6 Conclusion

The outcome of this paper may be used in a variety of situations. We can use our process for manipulating ontological individuals as a template for developing a full scale mechanism of data sharing across a more complex environment than the one described in this paper. We can also use it as an addendum to any other ontological solution which resolves semantic conflicts across heterogeneous data sources, such as [1, 2, 10 and 11].

Finally, the steps related to the manipulation of ontological concepts described in this paper, are useful as a "tutorial" for any ontological engineering class, because there are no available sources which can clearly give examples on how to deploy the power of ontological manipulation for achieving a variety of goals.

Our future works includes a full scale implementation of an application which sits on the top of a set of ontologies and delivers data sharing in more complex environments. We need to address the name spaces attached to the ontological individuals generated from data range values through performing SWRL rules manipulation. We should still try to find a way of entailing the translation from database to ontologies through option 3 to retain as much semantics as possible in our ontological models.

References

1. P. Kataria, and R. Juric, "Sharing e-Health Information through Ontological Layering" Proceedings of the 43rd Annual Hawaii International Conference on System Sciences (HICSS 43), 2010.
2. S. Ganguly, P. Kataria, R. Juric, A. Ertas, and M. M. Tanik, "Sharing Information and Data across Heterogeneous e-Health Systems", Tele-Medicine and e-Health Journal, Vol. 15, Issue 5, pp.454–464, 2009.
3. H. Knublauch, R. W. Fergerson, N. Noy, and A. Musen, "The Protégé OWL Plugin: An open Development Environment for Semantic Web Applications", Stanford Medical Informatics, Stanford School of Medicine, available at http://protege.stanford.edu/plugins/owl/publications/ISWC2004-protege-owl.pdf, 2004.
4. P. Hitzler, M. Krotzsch, B. Parsia, P. F. Patel-Schneider, and S. Rudolph, "OWL 2 Web Ontology Language Primer", W3C Working Draft, available at http://www.w3.org/TR/2009/WD-owl2-primer-20090421/ 2009.
5. E. Sirin, B. Parsia, B.C. Grau, A. Kalyanpur, and Y. Katz, "Pellet: A Practical OWL-DL reasoned", Journal of Web Semantics, Vol. 5, Issue 2, pp. 51–53 2004.
6. C. Nyulas, M. O'Connor and S. Tu, "Datamaster – a Plug-in for Importing Schemas and Data from Relational Databases into Protégé", available at http://protege.stanford.edu/conference/20&07/presentations/10.01_Nyulas.pdf, 2007.
7. R. Shojanoori, P. Nemitsas, A. Krishnamurth, "Generation of Ontologies from Relational Databases: An Experience", Proceedings of the 11th International Conference on Integrated Design and Process Technology (IDPT), 2008.
8. A. Rector, N. Drummond, M. Horridge, J. Rogers, H. Knublauch, R. Stevens, H. Wang and C. Wroe, C., "OWL Pizza: Pactical Expeience of Teaching OWL-DL: Common Errors & Common Patterns", Department of Computer Science, University of Manchester, available at http://www.co-ode.org/resources/papers/ekaw2004.pdf, 2004
9. Protégé, "SWRLJessTab", available at http://protege.cim3.net/cgi-bin/wiki.pl?SWRLJessTab, 2009.
10. P. Kataria, R. Juric, and K. Madani, "Go-CID: Generic Ontology for Context Aware, Interoperable and Data Sharing Applications", Proceedings of the 11th International Conference on Software Engineering and Applications (IASTED), 2007.
11. P. Kataria, R. Juric, S. Paurobally, and K. Madani, "Implementation of Ontology for Intelligent Hospital Wards", Proceedings of the 41st Hawaii International Conference on System Science, (HICSS 41), 2008.

Chapter 9
SHARING CANCER-RELATED GENES RESEARCH RESOURCES ON PEER-TO-PEER NETWORK

Chih-Chung Wu, Rong-Ming Chen, and Jeffrey J.P. Tsai

1 Introduction

It has been many years for the computer technology aids in life science research. The bioinformatics researchers spread worldwide and have accumulated volume of research resources. How to integrate their resources and share the resources effectively becomes an important issue. The Internet is the best choice for providing bastion infrastructure to build the shared platform. Network enabled applications such as Client/Server architecture, distributed network system architecture also help people to share their resources. But there is no killer application fit for every bioinformatics researchers to use, because of the diversity of life science. As the result, people still build their application using different programming language on different operating system. It is therefore difficult to integrate these resources. Web Service and XML-based document can resolve this problem. But it is not good enough. How to properly authorize accessing of these services and easily discover these web services become another issue.

To share resources between users more effectively we need better technology. There are some convenient peer-to-peer applications such as ICQ, MSN Messenger, BitTorrent, Emule, which help people to meet and share information with each other. It would be a good idea for bioinformatics researchers to share their resources on peer-to-peer network. In this chapter we proposed system architecture for bioinformatics to develop application based on the free, ready to use JXTA peer-to-peer (P2P) technology. Using this architecture people can easily add the peer-to-peer feature into their application. This feature will let people find or share services more easily.

R.-M. Chen (✉)
Tainan, Taiwan
e-mail: chen.rickyjoseph@gmail.com

S. Suh et al. (eds.), *Biomedical Engineering*,
DOI 10.1007/978-1-4614-0116-2_9, © Springer Science+Business Media, LLC 2011

Figure 1. Relationship
between promoter, TF, and
TFBS. (Redrawn from
"Bioinformatics blog": http://
bioinformaticsblog.blogspot.
com/)

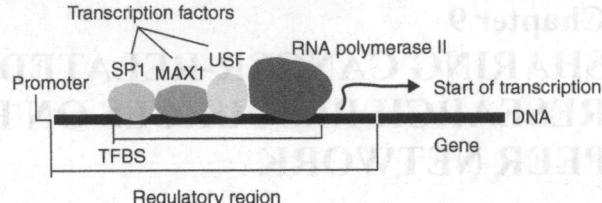

2 Methods

2.1 Recognizing Cancer-related Genes based on Transcription Factor Binding Sites

The purpose of Transcription Factors (TFs) is to regulate the express of other genes. If mutation occurs on promoter region it's possible to cause gene express abnormal. Figure 1 show several transcription factors (SP1, MAX1, USF and basal factors) that are necessary for transcription of some genes [1]. Transcription Factor Binding Sites (TFBS), and Promoter. This method proposed a new approach to predict possible cancer-related genes based on TFBS. By analyzing the occurrence frequencies of these TFBS to investigate the relations of TFBS and possible cancer-related genes, it shows that the TFBS-based approach for predicting possible cancer-related genes is a reliable method to recognize possible cancer-related genes. Table 1 shows the ratio that two TFs exist on the same binding site in known selected cancer-related gene [2].

Table 1. Two TFs that exist in the same binding region and the ratio (%).

TF	AP2alpha	AML1a	C/EBPalpha	C/EBPbeta	MZF-1	CREB	Oct-1	SRY	GATA-1
MZF-1	67	19	38	18					
GATA-1	53	11	37	16	43				
C/EBPalpha	46	21							
Oct-1	43		31	13	33				
SRY	40		27	14	31		30		
YY1	33		26	11	24		20	19	
USF	25		15						20
C/EBPbeta	22	11	18						
CREB	12								
deltaCREB	12						13		
GR alpha	13								
GR beta	13								
Sp1	14								
AML1a			21			19			
AP-1			11						

We implement this method according to finding if they are two TFs exist on the same binding site of the candidate gene. If such TFs exist, it is possible a cancer-related gene.

2.2 Recognizing Cancer-related Genes based on Single Nucleotide Polymorphisms

A Single Nucleotide Polymorphism (SNP) is a single base substitution of one nucleotide with another in DNA sequence. If transcription factor binding site on promoter has SNP it will influence transcription and cause the gene to change abnormally. Table 2 shows an example of nucleotide change on TFBS. This approach predict possible cancer-related gene by examining whether a candidate gene has SNP on TFBS [3].

We implement this method to determine cancer-related gene according to finding if there is SNP on the TFBS of the candidate gene.

2.3 Reliability Verification of Microarray Gene Expression database

cDNA microarray can be used to analyze the gene expression in different cancer cells. Numerous researches have been studied in this area and produced volume of microarray gene expression data. It is usually set α-value to 0.05 for T-test when analyzing the microarray data. This method proposed that it is not appropriate for setting α-value to 0.05 for any T-test. It is better to choose an adaptable α-value for verifying various microarray data.

The steps for determining an adaptable α-value are summarized as follows.

Step 1: Calculate the ratio of samples for over- and under-expressed:

Let OEL# be the number of samples that are over-expressed, UEL# the number of samples that are under-expressed, define r as the ratio of difference between OEL# and UEL# over OEL#:

$$r = \frac{OEL\# - UEL\#}{OEL\#}$$

Table 2. An example for nucleotide change on TFBS.

promoter SEQ	TF name	TFBS SEQ	SNP Allele frequency
CCACAGGTAGAA	AP2alpha	NCAGNTAG	C/G

Table 3. P(k) value of liver tumor.

k	r	P(k)	f(k)	$\sum_{i=0}^{k} f(i)$
0	$0 \le r \le 0.05$		0	0
1	$0.05 \le r \le 0.1$	0.2309	2	2
2	$0.1 \le r \le 0.15$	0.2376	2	4
3	$0.15 \le r \le 0.2$	0.4687	5	9
4	$0.2 \le r \le 0.25$	0.0861	1	10
5	$0.25 \le r \le 0.3$	0.3238	4	14
6	$0.3 \le r \le 0.35$	0.3372	2	16
7	$0.35 \le r \le 0.4$	0.1543	2	18
8	$0.4 \le r \le 0.45$	0.1533	9	27
9	$0.45 \le r \le 0.5$	0.0006	4	31
10	$0.5 \le r \le 0.55$	0.001	6	37
11	$0.55 \le r \le 0.6$	9.15E-05	2	39
12	$0.6 \le r \le 0.65$	4.57E-05	2	41
13	$0.65 \le r \le 0.7$	5.43E-05	3	44
14	$0.7 \le r \le 0.75$	2.92E-05	6	50
15	$0.75 \le r \le 0.8$	1.67E-10	2	52
16	$0.8 \le r \le 0.85$	1.86E-09	4	56
17	$0.85 \le r \le 0.9$	5.45E-16	2	58
18	$0.9 \le r \le 0.95$	8.40E-08	1	59
19	$0.95 \le r \le 1.$		0	59
20	$r = 1$		0	59

Step 2: Create P(k) mapping table as shown in Table 3.

Let T denote the total number of genes which have the value of r lies between 0 and 1, P(k) denote the maximum p-value of genes with r in the interval [0.05k, 0.05(k + 1)] where k = 0, 1, 2, ... 19, f(k) denote the number of genes in that interval, and define α_R as follows.

$$\alpha_k = \left\{ P(k) \middle| k = \min_{n} \left\{ n \middle| \sum_{i=0}^{n} f(i) \ge T/2 \right\} \right\}$$

When we want to determine whether a given gene is a cancer-related gene, we first compute its r value. Next, get the P(R) value from the P(R) mapping table. If the p-value of the given gene is less than α_R, It is possible a cancer-related gene [4].

We implement this method by verifying the given microarray experimental data to find if a candidate gene is a possible cancer-related gene.

2.4   JXTA

JXTA technology is a set of open protocols that enable any connected device on the network, ranging from cell phones and wireless PDAs to PCs and servers, to communicate and collaborate in a P2P manner [5].

The JXTA software architecture is divided into three layers, as shown in Figure 2. The JXTA Core provides the minimal and essential functions for peer-to-peer application. The JXTA Service Layer includes useful but no necessary network services for peer-to-peer application such as searching, indexing, directory, authentication, and PKI (Public Key Infrastructure) services. The Application Layer includes implementation of integrated application, such as our application.

Following are some of the important components we should know [6].

1. Peer, any networked device that implements one or more of the JXTA protocols. They are four kinds of peers, minimal edge peer, full-featured edge peer, rendezvous peer and relay peer. Edge peer can send and receive messages. Rendezvous peer is an infrastructure peer, it aids other peers with message propagation, discovers, route and maintains a topology map of other infrastructure peers. Relay peer is also an infrastructure peer, it aids peers behind firewall or NAT to communication.
2. Peer group, a logical grouping of peers. Peers can join public group without authentication or join private group with authentication and authorization. Peer can join multiple groups.
3. Modules, an abstract presentation of a piece of code made available by a peer or peer group for use by another peer or peer groups.
4. Message, are basic unit of data exchange between peers. The message can be either XML document format or binary data format.
5. Pipe, communication channels created between peers.
6. Advertisements, XML-based messages used to publish information between peers.

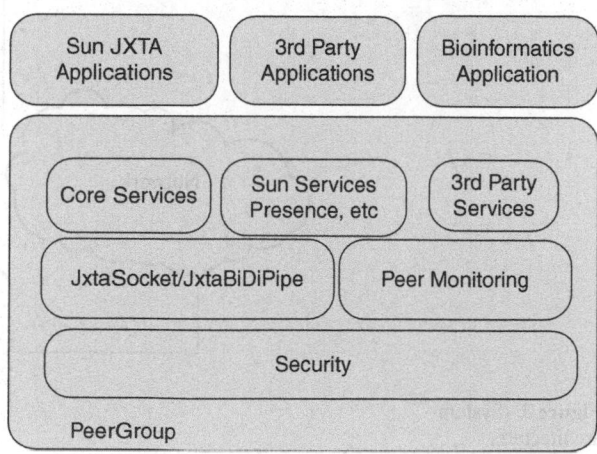

Figure 2. JXTA software architecture.

7. Services, predefined functionality that can be utilized by peers.
8. Security, JXTA provides confidentiality, authentication, authorization, integrity, refutability mechanism for us to develop secure application.
9. ID, is a unique string used for the identification for peers, peer groups, pipes, modules

JXTA is free, mature, open standard software architecture and is independent of programming language, system platform, networking platform. It is appropriate for us to develop the peer-to-peer application.

2.5 System Architecture

The system architecture, as shown in Figure 3, has three major components, namely, web service creation, service advertisement, service consuming. We accomplish our application by doing the following processes in each component:

1. Web service creation: Implement each of the three recognizing cancer-related gene methods as web service and provide web methods for access.
2. Service advertisement: We created and advertise JXTA peer group named 'Bioinformatics', join this group, create and advertise modules, then wait for message. When server peer got request from client peer, it invoke related web service

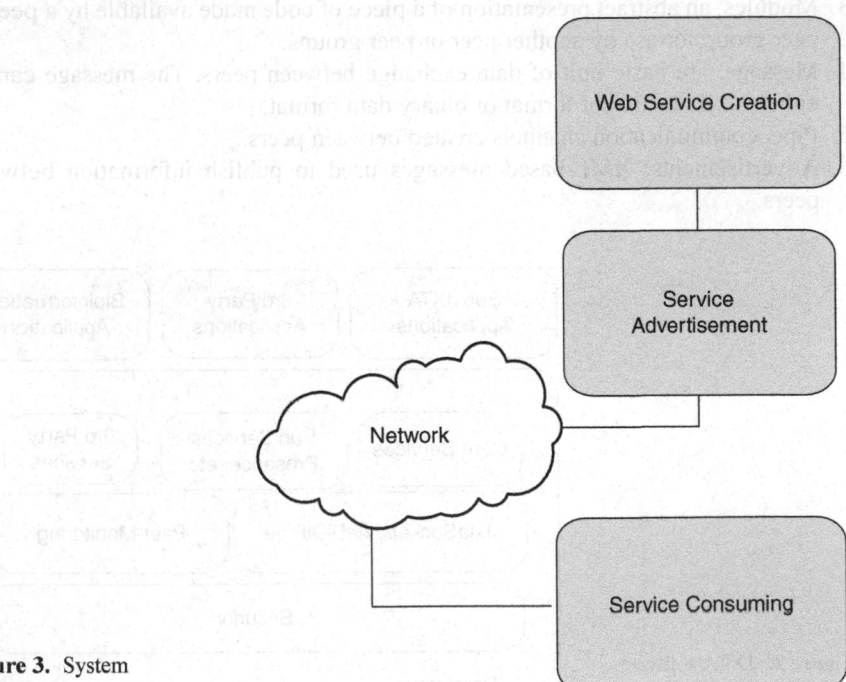

Figure 3. System architecture.

which the client request, return result to the client. Each peer should advertise its basic information and services that want to be published.

3. Service consuming: Create a graphics user interface for user to discover services, input gene symbol or gene name, select and invoke the services, get the results.

3 Results

We implement all the three components of the system architecture and deploy it to different environments to validate the application. Following are four scenarios:

1. Peers in the same local area network: Peers can direct communicate to each other by broadcast, multicast or unicast packets.
2. Peers between subnets on local area: Routers between subnet must enable multicast packet to pass through. Otherwise, we must implement at least one rendezvous peers for communication.
3. Peers cross the Internet: We must implement at least one rendezvous peers on the Internet for communication.
4. Peers behind the firewall or NAT network: We must implement relay peers for communication as shown in Figure 4.

Although we implement bioinformatics methods as web services, it is not necessary to implement the Universal Description, Discovery, and Integration (UDDI) service for discover web service in the system architecture. JXTA peers in the same peer group can find service by the Shared Resource Distributed Index (SRDI).

In our system architecture peer can create web service and ask for authentication, uses JXTA to advertise which service they want to public. When another peer discovers and asks for the service, service provider invokes their service locally or remotely with their credential. Also, we can create private peer group for friend to

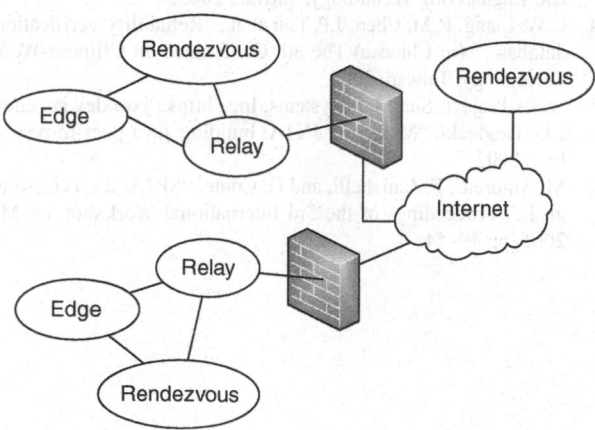

Figure 4. Peers behind the firewall or NAT device.

share private resources. Service provider can decide what to publish, when to publish, and who can access these service.

4 Conclusions

A public JXTA peer group for sharing bioinformatics research resources is introduced in this chapter. Researchers are welcome to discover and consume service in the group or provide services to this group. Using Peer-to-peer network, researchers with the same study area can share resources with each other over the Internet. In future work we will improve the application focus on scalability and reliability. The JXTA technology provides infrastructure for us to build the application easily. More than this, it also provides the possibility for us to build bioinformatics application based on the Service Oriented Architecture and it is also possible to build a grid computing environment based on the JXTA technology [7].

Acknowledgments This work was supported in part from the National Science Council, Taiwan (Project No. 95-2745-E-468-001-URD, Project No. 95-2745-E-468-003-URD, and Project No. NSC 99-2221-E-024-010). We also thank the website for providing the graph of relationship between promoter, TF, and TFBS (http://bioinformatics-blog.blogspot.com).

References

1. M. Tompa et al., "Assessing computational tools for the discovery of transcription factor binding sites," Nature Biotechnology, Vol. 23, No. 1, 2005, pp. 137–144.
2. M.Y. Lee, R.M. Chen, J.P. Tsai et al., "Recognizing cancer-related genes based on transcription factor binding sites," (in Chinese) The 8th Conference on Chinese-Western Medicine and Engineering Technology, Taiwan, 2005.
3. I.C. Wang, R.M. Chen, J.P. Tsai et al., "Recognizing cancer-related genes based on single nucleotide polymorphisms," (in Chinese) The 8th Conference on Chinese-Western Medicine and Engineering Technology, Taiwan, 2005.
4. C.W. Liang, R.M. Chen, J.P. Tsai et al., "Reliability verification of microarray gene expression database," (in Chinese) The 8th Conference on Chinese-Western Medicine and Engineering Technology, Taiwan, 2005.
5. JXTA Project, Sun Microsystems, Inc., https://jxta.dev.java.net/, 2008.
6. J.D. Gradecki, "Mastering JXTA: building Java peer-to-peer applications," Wiley publishing Inc., 2002.
7. M. Amoretti, F. Zanichelli, and G. Conte, "SP2A: a service-oriented framework for P2P-based grids," Proceedings of the 3rd International Workshop on Middleware for Grid computing, 2005, pp.49–54.

Chapter 10
IHCREAD: AN AUTOMATIC IMMUNOHISTOCHEMISTRY IMAGE ANALYSIS TOOL

Chao-Yen Hsu, Rouh-Mei Hu, Rong-Ming Chen, Jong-Waye Ou, and Jeffrey J.P. Tsai

1 Introduction

Cancer is defined as uncontrolled or abnormal cell division. Untreated cancer is generally lethal. Liver cancer is the fifth most common cancer in the world [1] and is much more prevalent in the developing countries, especially Southeast Asia. Chronic infection with hepatitis B virus or hepatitis C virus is a very important liver cancer risk factor in this area. The majority of primary liver cancers arises from liver cells and is called hepatocellular carcinoma (HCC). Treatment of liver cancer patients was limited because the traditional diagnosis is not sensitive enough to detect the early symptoms [2]. Some novel technologies were used to increase the sensitivity and accuracy of cancer detection, typing and treatment.

Microarray is a powerful tool and is generally applied in molecular marker screening and cancer typing. However, it can only give preliminary gene expression information. Quantitative PCR is used to verify the differential gene expressions and immunohistochemistry (IHC) can help us to understand the distribution and localization of biomarker proteins in different parts of a tissue. However, IHC application is limited by its high labor intensity and poorly defined objectives. In this chapter, we introduce an automatic IHC image analysis system to identify the differential expression of marker proteins between cancer and normal cells. IHC images of patients with HCC were analyzed.

2 Overview of the System

A workflow of IHCread is presented in Figure 1. An original image was transformed from the RGB color model to YIQ color model. The I-layer image was selected and transformed into a binary image where the white region (Image In) represented the

R.-M. Chen (✉)
Tainan, Taiwan
e-mail: chen.rickyjoseph@gmail.com

S. Suh et al. (eds.), *Biomedical Engineering*,
DOI 10.1007/978-1-4614-0116-2_10, © Springer Science+Business Media, LLC 2011

Figure 1. Workflow of the automatic IHC image analysis system IHCread.

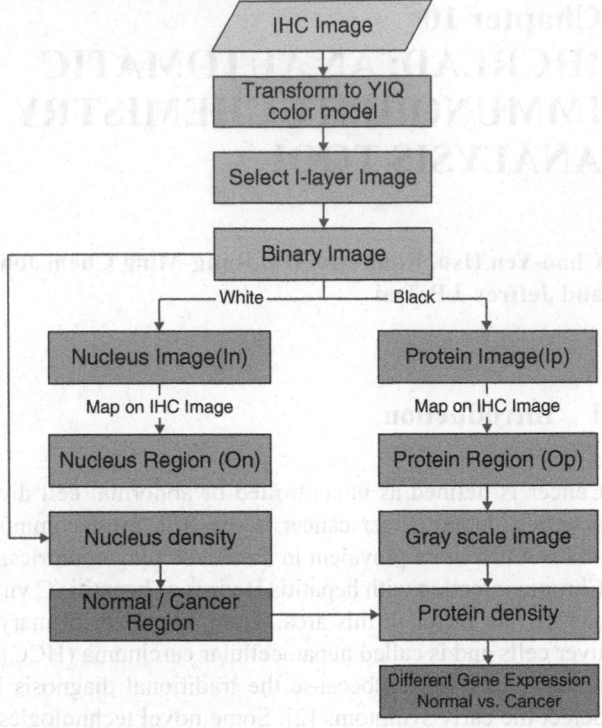

nuclear region which was stained in dark blue on IHC (Image On). The black region (Image Ip) which represented the region hybridized with antibodies was mapped onto the original IHC image to get Image Op. The Image Op was transformed to a Gray scale image. The gray level represents the quantity of the marker protein. The IHC image was segmented into cancerous and normal region according to the density of nucleus and the density of marker protein in each region was calculated to determinate if the marker protein is over or under-expressed in cancer tissue.

3 Methods

3.1 *RGB Color Model*

RGB model is an additive color model in which red, green, and blue light are added together in various ways to reproduce a broad array of colors [3]. In an IHC image, the hybridized proteins are stained in brown color and the nuclei are stained in dark blue color (Figure 2). The aims of our system are: 1) to distinguish the normal and the cancerous region on an IHC image; 2) to quantify the density of marker protein in normal and cancer area, and 3) to determine if the marker protein is over, under or normally expressed in a specific biopsy (patient). To gain our goal, the RGB model is transformed to the YIQ model.

Figure 2. Four IHC image samples, a, b, c, and d were analyzed in this work. In the images a and c, the marker protein was over-expressed in cancerous tissue, and in the image b, the marker protein was under-expressed in cancerous tissue.

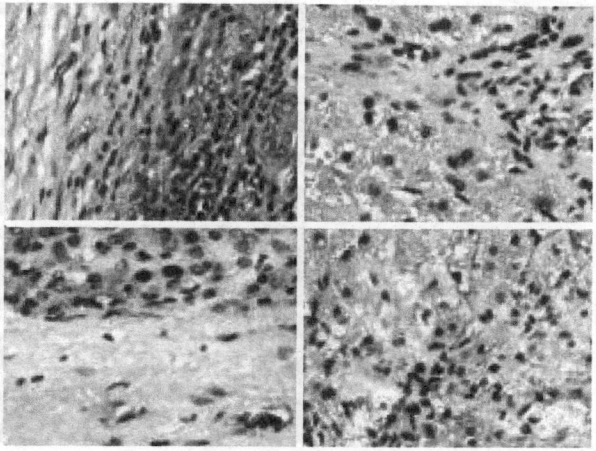

3.2   YIQ Color Model

YIQ is the color space used by the NTSC color TV system. The Y component represents the luma information. I (orange-blue) and Q (purple-green) represent the chrominance information [4]. The conversion from RGB to YIQ is as follows [5]:

$$\begin{bmatrix} Y \\ I \\ Q \end{bmatrix} = \begin{bmatrix} 0.299 & 0.587 & 0.114 \\ 0.596 & -0.274 & -0.322 \\ 0.211 & -0.523 & 0.312 \end{bmatrix} \begin{bmatrix} R \\ G \\ B \end{bmatrix} \tag{1}$$

To get the nuclear area in an IHC image, The I layer of YIQ image were selected which represent the change in the orange-blue range (Figure 3).

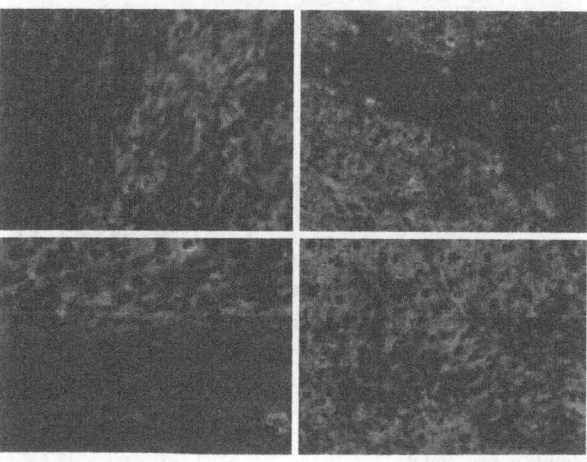

Figure 3. I-layer image of sample a, b, c and d.

Figure 4. Binarized I-layer image of sample a, b, c, and d. The white region (In) corresponds to the nuclei.

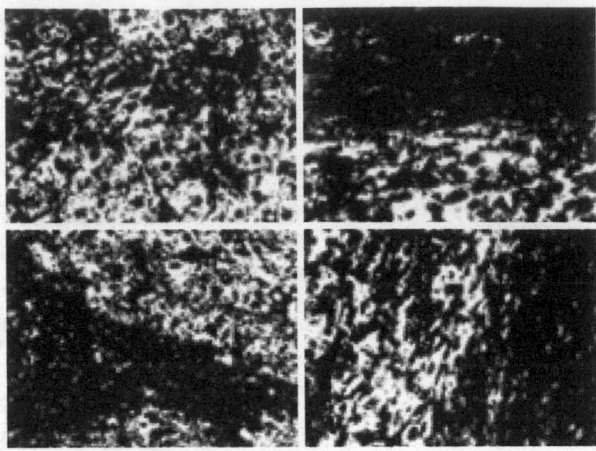

3.3   Binary Image

The Otsu's method [6], which chooses the threshold to minimize the intraclass variance of the black and white pixels, was applied on I-layer image to transform the I-layer image to a binary image (Figure 4).

The white (In) and black (Ip) areas were selected and mapped back to the original RGB image to get the nuclear (On) (Figure 5) and protein region (Op) (Figure 6), respectively.

3.4   Gray Conversion

The Op images were transformed to gray-scale image by the transformation: $0.2989R + 0.5870G + 0.1140B$ [7]. The gray value represents the quantity of immuno-stained marker protein.

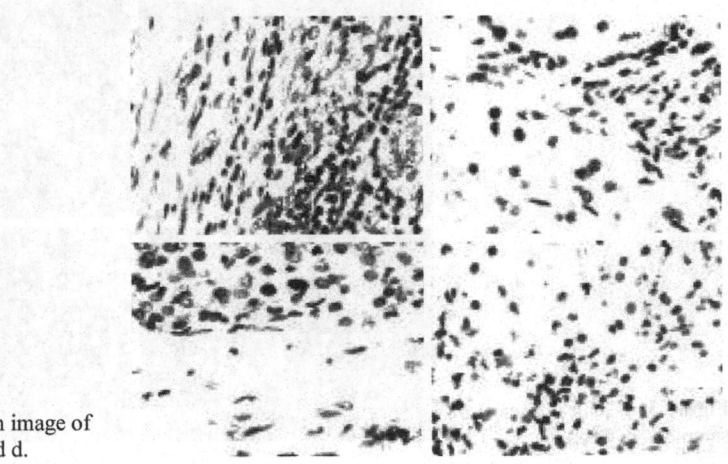

Figure 5. The On image of sample a, b, c, and d.

Figure 6. The Op images of sample a, b, c, and d.

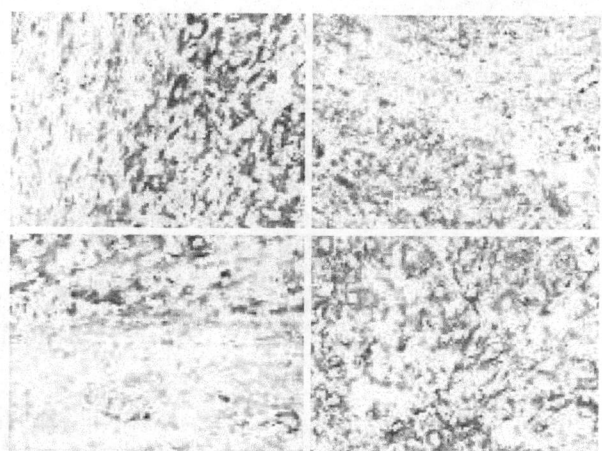

3.5   Cancerous/Normal Region Determination

The binary images were segmented into blocks size of 130x174 pixels. The nuclear density of each block was calculated with the following function:

$$DN_i = \frac{Ci}{A} \; i = 1, \; 2, \; 3, \; 4.....n, \tag{2}$$

where DN_i denotes the nuclear density of each block, C_i denotes the pixel number of the nuclear region (In), n denotes the block number, and A denotes the total pixel number of the blocks.

The cancerous and normal regions were segmented by a threshold ($DisThr$) determined from the average value of nuclear density of total blocks:

$$DisThr = \overline{DNi} = \frac{1}{n} \sum_{i=1}^{n} DNi \tag{3}$$

The results were shown in Figure 7; the cancerous regions were colored in red.

3.6   Protein Density

The density of marker protein in cancerous and normal regions was calculated separately by Equation (4).

$$DP = \sum_{j=1}^{m} \frac{\sum\limits_{(x,y)\in OPj} f(x,y)}{A - C_j}, \tag{4}$$

where DP is the density of marker protein in cancerous region or normal region, OP_j is the block j in image OP and $f(x, y)$ is the gray level at coordinates (x, y).

Figure 7. Results of segmentation of sample a, b, c and d. The cancerous regions were colored in red.

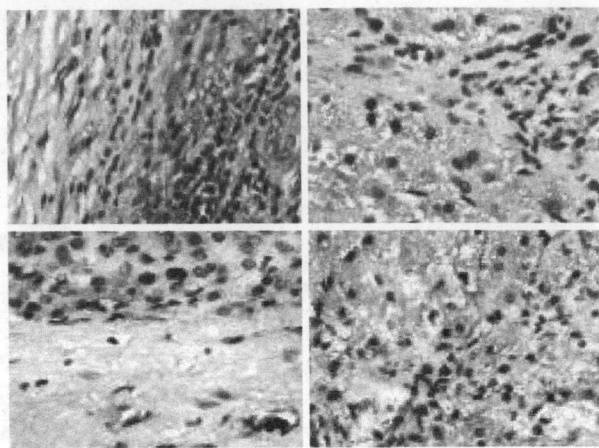

Table 1. Analysis results of IHCread of images a, b, c, and d. The ratio represents the difference of marker protein density between cancerous and normal tissues. A conclusion was made according to the expression level change of marker protein in the cancerous tissue.

Image	Ratio (%)	Conclusion
a	180.8	Over-expression
b	57.7	Under-expression
c	178.1	Over-expression
d	104	No significant differential expression

4   Results

The final results were shown in Table 1. As one can judge by naked eye, the marker protein was over-expressed in cancerous tissue of the images a and c; and under-expressed in the image b, but it's difficult to judge in image d because of the mixed color. The differential gene expression ratio calculated by IHCread showed a similar result (Table 1). This suggests that IHCread is a useful tool for qualitative and quantitative IHC image analysis.

Acknowledgments This work was supported in part from the National Science Council, Taiwan (Project No. 95-2745-E-468-001-URD, Project No. 95-2745-E-468-003-URD, and Project No. NSC 99-2221-E-024-010].

References

1. J. Bruix, L. Boix, M. Sala and J. Llovet, "Focus on hepatocellular carcinoma," Cancer Cell, Vol. 5, No. 3, 2004, pp. 215–219.
2. M.T. Lotze, J.C. Flickinger and B.I. Carr, "Hepatobiliary neoplasm," In: VT DeVita, S Hellman, SA Rosenberg, eds. Cancer: Principles & Practice of Oncology,. 4th ed., Philadelphia, PA: Lippincott Williams & Wilkins", 1993, pp. 883–887.

3. M.F. Cowlishaw, "Fundamental requirements for picture presentation," Proc. Society for Information Display, Vol. 26, No. 2, 1985, pp. 101–107.
4. W.H. Buchsbaum, Color TV Servicing, 3rd edition, Englewood Cliffs, NJ: Prentice Hall. 1975.
5. T. Tanaka, T. Joke, and T. Oka, "Cell nucleus segmentation of skin tumor using image processing," IEEE/EMBS 23rd Annual Conference, Oct. 25–28, 2001.
6. N. Otsu, "A Threshold Selection Method from Gray-Level Histograms," IEEE Transactions on Systems, Man, and Cybernetics, Vol. SMC-9, No.1, 1979, pp. 62–66.
7. J. D. Foley, A. van Dam, S. K. Feiner, and J. F. Hughes, Computer Graphics: Principles and Practice. Reading, MA: Addison-Wesley, 1990.

Chapter 11
A SYSTEMATIC GENE EXPRESSION EXPLORER TOOL FOR MULTIPLE AND PAIRED CHIPS

Jong-Waye Ou, Rong-Ming Chen, Rouh-Mei Hu, Chao-Yen Hsu
and Jeffrey J.P. Tsai

1 Introduction

High throughput microarray technologies were developed nowadays. By using these new technologies, tremendous amount of gene expression data were accumulated in a short time. These data can help us to analyze the complex relationship of gene interactions. DNA microarray is an orderly arrangement of many oligonucleotides or identified sequenced genes printed on an impermeable solid support, usually glass, silicon chips or high-polymer material [1]. DNA microarray is also called DNA chip, gene chip, or biochip. The biochip is an integration of several technologies, including automated DNA sequencing, DNA amplification by PCR, oligonucleotide synthesis, nucleic acid labeling chemistries and bioinformatics. Microarrays have become standard tools for gene expression measurement in biology and biomedicine [2, 3, 4]. The application ranges from gene expression changes of the cell cycle over the classification of disease to drug design.

In this chapter we propose a systematic gene expression explorer (GEE) tool to analyze multiple and paired chips. The GEE tool can help the biologists to analyze expression patterns of individual genes and gene clusters based on microarray data. Typically this involves using various databases, data retrieval, statistical and direct manipulation user interface techniques to identify similar expression patterns. Here all of the data being analyzed is copied to a user's computer and computationally expensive analyses are done there. This feature allows more effective relational database with many alternate views and avoiding excessive system delays. In this tool direct manipulation on a variety of graphic and tabular displays makes it easier for the users to discover interesting gene expressions. They can understand what the abstract graphics data represent in terms of the particular cluster genes expression. The expression profile of a gene is the plot of its normalized intensity as a function of sample patient. Scrollbar lists of gene expression plots may be generated allow-

R.-M. Chen (✉)
Tainan, Taiwan
e-mail: chen.rickyjoseph@gmail.com

S. Suh et al. (eds.), *Biomedical Engineering*,
DOI 10.1007/978-1-4614-0116-2_11, © Springer Science+Business Media, LLC 2011

ing multiple gene symbols to be viewed simultaneously. Scatter plotting, histogram, clustering methods and clustering visualization tools are available for presentation. The users can adjust various parameters such as threshold and ratio of scattering to observe the changes of histogram and scatter plot.

2   Methods

2.1   Multiple Chip Analysis

The flowchart of multiple chip analysis of GEE tool is shown in Figure 1. This tool has three major parts: input block, process block, and output block. The input block deals with the processing of gene expression data. It allows the usage of SMD

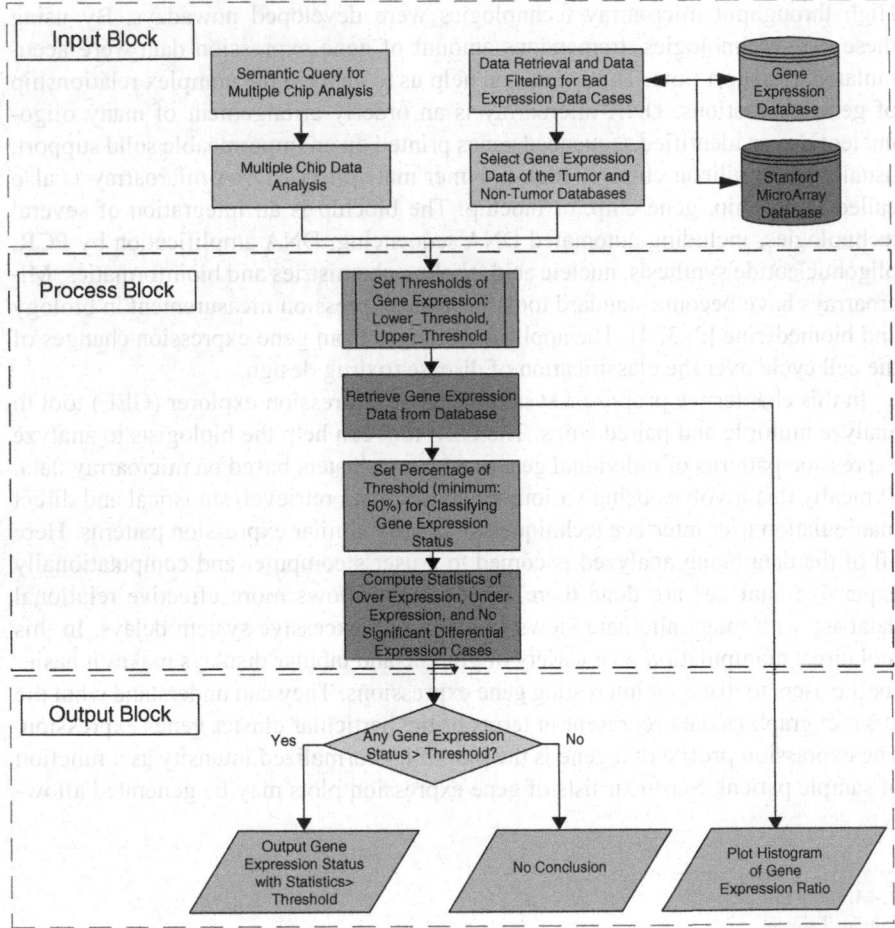

Figure 1. Flowchart of multiple chip analysis of GEE tool.

database or user-defined gene expression data. Input block provides scripts for the creation of the database tables and the conversion of already existing databases into these tables. The process block computes statistics according to user-defined threshold. The output block presents the conclusion of over-expression, under-expression or no significant differential expression, and histogram of expression ratio. Data normalization is an important step in gene expression data analysis. We normalize the gene expression data using one housekeeping gene.

2.2 Paired Chip Analysis

The flowchart of paired chip analysis of GEE tool is shown in Figure 2. This tool presents the compared gene expression of paired chip by scatter plot and frequency histogram of gene expression values. An illustration of scatter plot for the comparison of normalized gene expression ratio between tumor and non-tumor tissues is shown in Figure 3. Scatter plots are useful for obtaining a better understanding of the distribution of genes on different expressed ratio. The dots on the scatter plot represent those genes that are over-expressed, under-expressed, or no significant differential expression based on a user-defined threshold.

2.3 Pathway Map Analysis

The KEGG pathway database represents networks of molecular interactions and reactions in the cell in a graphical manner [5]. The available pathways provide key information of the functional and metabolic systems within a living cell. We use this database and color differential gene expression on the pathway shown on the current web, allowing the exploration of functional relationships between genes. The human gene activity, described by numbers in database, is used for connecting KEGG maps to the probes on the chip. As a gene activity can be catalyzed by more than one gene the pathway view shows different expression values for each different gene. For genes that exist in the microarray database, all KEGG maps associate with this gene symbol. On the other hand, all genes given on a specific map can be used as a gene subset in analysis. All genes on a KEGG map can be selected by clicking mouse on the gene names to get the detailed information about the corresponding genes on the web through the Internet.

3 Results

We have developed the GEE Tool to facilitate the visualized analysis for gene expression of cDNA microarrays. With this program it is possible to analyze the expression of individual genes, to plot the histogram of gene expression for multiple

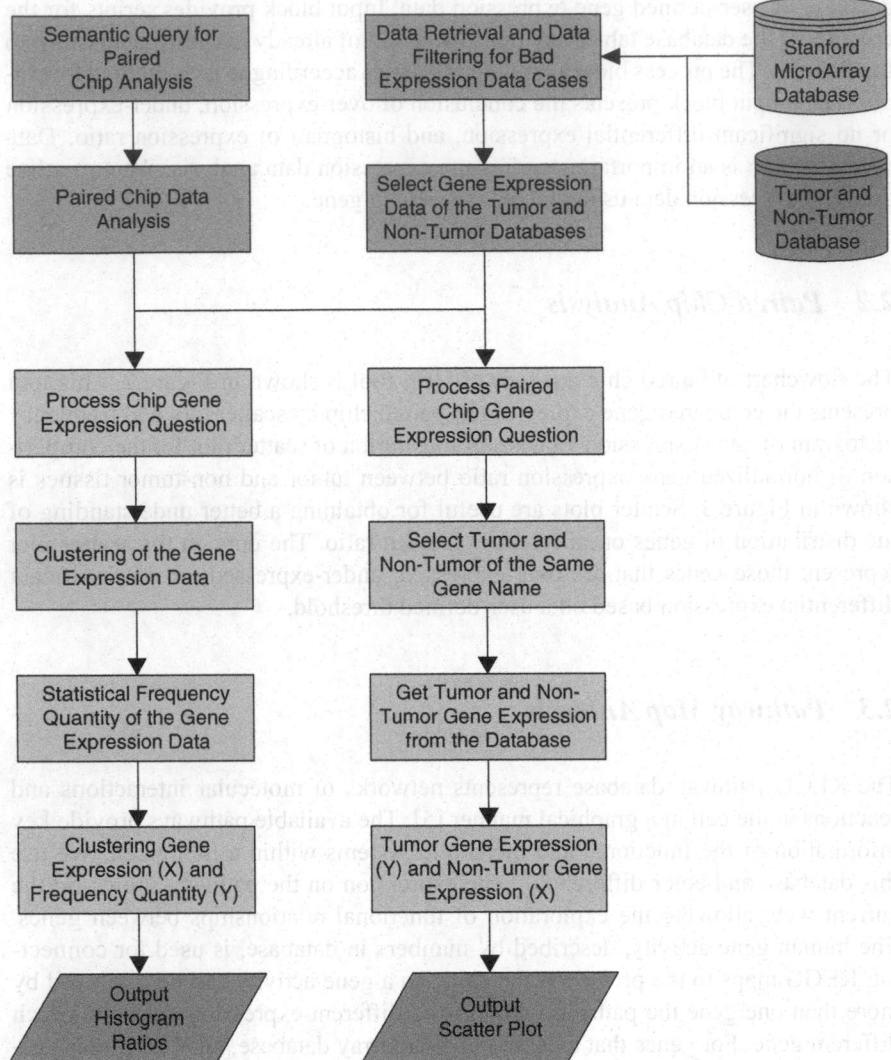

Figure 2. Flowchart of paired chip analysis of GEE tool.

chips. The graphical user interface of multiple chip analysis is shown in Figure 4. The user interface provides a set of pull-down menus which invoke various analysis operations of gene expression. These menus are fully documented in the system reference manual.

With this program it is also possible to analyze the frequency histogram for gene expression of paired chip and to analyze the scatter plot of paired gene expression ratios. Different user-defined threshold will result in different gene ex-

Figure 3. Paired chip analysis results.

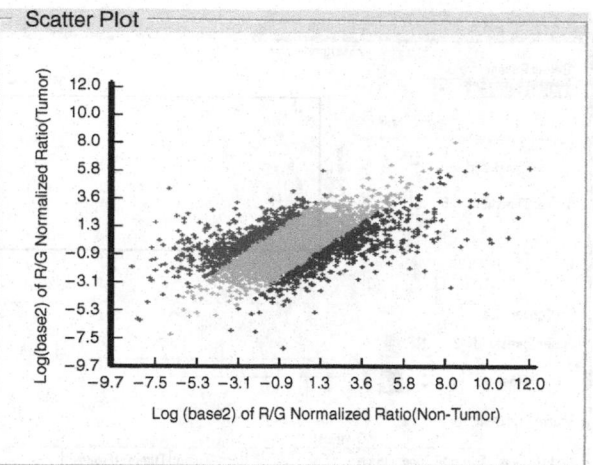

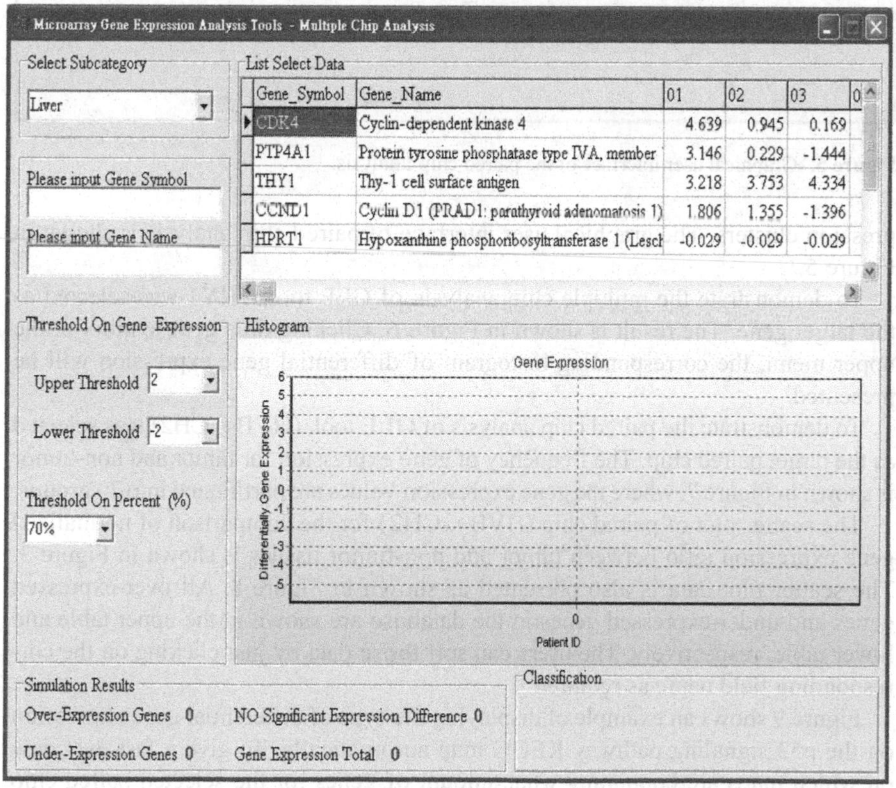

Figure 4. Graphical user interface of the multiple chip analysis.

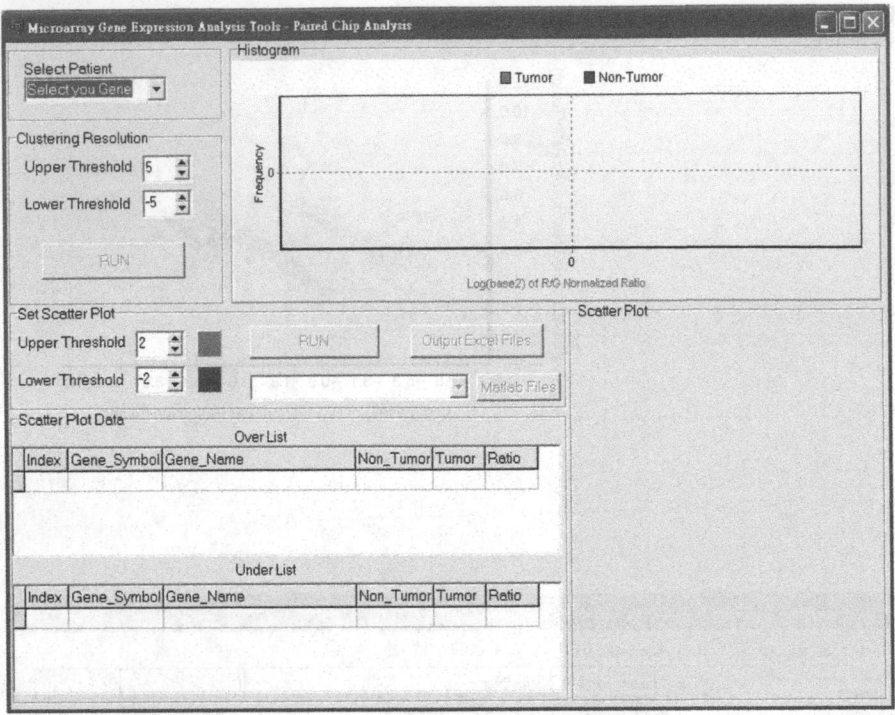

Figure 5. Graphical user interface of the paired chip analysis.

pression clusters. The graphical user interface of paired chip analysis is shown in Figure 5.

To demonstrate the multiple chip analysis of GEE tool, THY1 was selected as the target gene. The result is shown in Figure 6. Clicking on a gene symbol in the upper menu, the corresponding histogram of differential gene expression will be presented.

To demonstrate the paired chip analysis of GEE tool, (HVB+, CH2) was selected as the target paired chip. The frequency of gene expression for tumor and non-tumor is shown in Figure 7, where the gene expression values are partitioned into 20 groups.

The scatter plot of paired chip (HVB+, CH2) for the comparison of normalized gene expression ratio between tumor and non-tumor tissues is shown in Figure 3. The scatter plot data is also presented as shown in Figure 8. All over-expressed genes and under-expressed genes in the database are shown in the upper table and lower table, respectively. The users can sort those data by just clicking on the corresponding field name as required.

Figure 9 shows an example of displaying the type of differential gene expression on the p53 signaling pathway KEGG map automatically. To give a fast overview on which maps are containing what amount of genes for the selected paired chip set, an overview bitmap are displayed. Different colored gene symbols shown in the pathway denote different gene expression types. A gene symbol with pink color

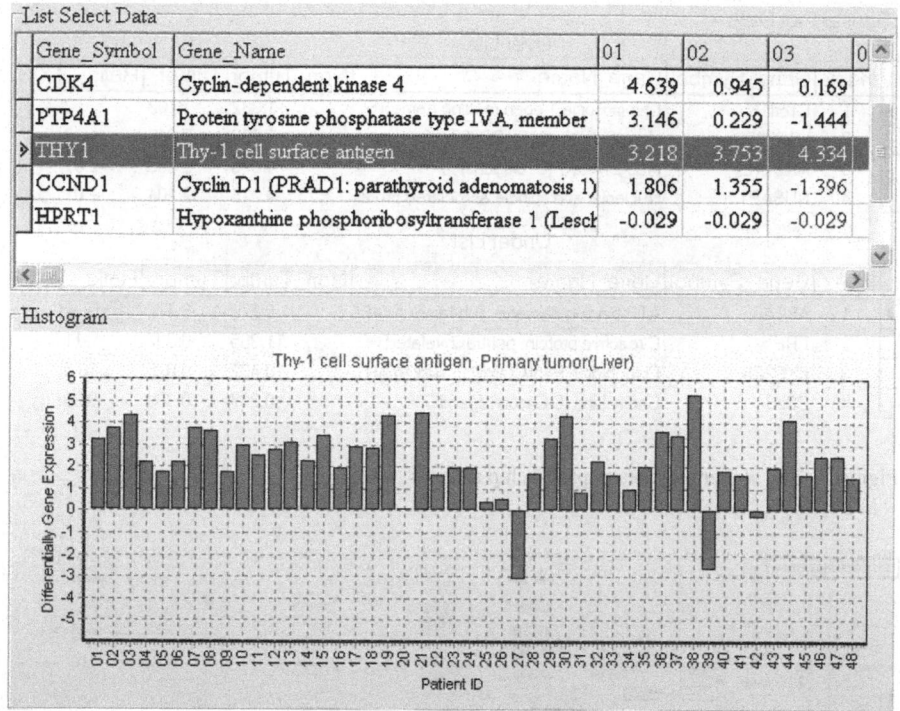

Figure 6. Demonstration of multiple chip analysis using THY1.

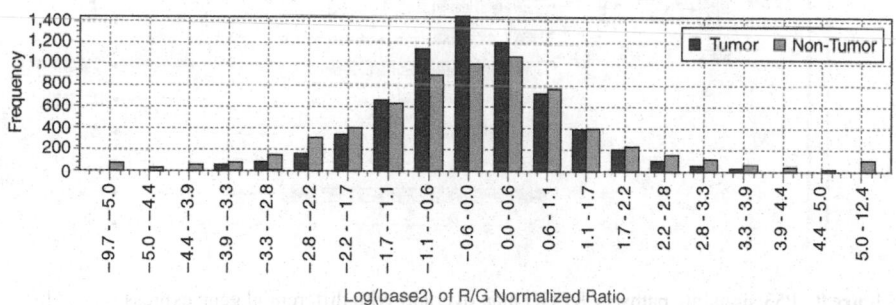

Figure 7. Demonstration of paired chip analysis using (HVB+, CH2).

in the pathway is used to denote that the corresponding gene is over-expressed. A gene symbol with blue color is used to denote that the corresponding gene is under-expressed and the one with yellow color is used to denote that the corresponding gene has no significant differential expression. However, if there are genes on the pathway map that can not be found in the microarray data, the corresponding gene symbol will be marked as white color. Finally, the users can retrieve more information on the genes by clicking mouse on the gene name.

Scatter Plot Data

Over List

Index	Gene_Symbol	Gene_Name	Non_Tumor	Tumor	Ratio
1	C1orf16	Chromosome 1 open reading frame 16	-1.382	0.62	2
2	LOC90268	Hypothetical protein BC007706	-1.356	0.646	2
3	MAB21L1	Mab-21-like 1 (C. elegans)	-4.375	-2.368	2.01
4	NUSAP1	Nucleolar and spindle associated protein	-4.214	-2.204	2.01

Under List

Index	Gene_Symbol	Gene_Name	Non_Tumor	Tumor	Ratio
1	ABCA6	ATP-binding cassette, sub-family A (ABC	8.278	-3.472	-11.8
2	CRP	C-reactive protein, pentraxin-related	11.365	0.397	-11
3	CYP2C8	Cytochrome P450, family 2, subfamily C,	10.462	-0.025	-10.5
4	CCL3	Chemokine (C-C motif) ligand 3	10.297	0.309	-9.99

Figure 8. The scatter plot data of paired chip analysis.

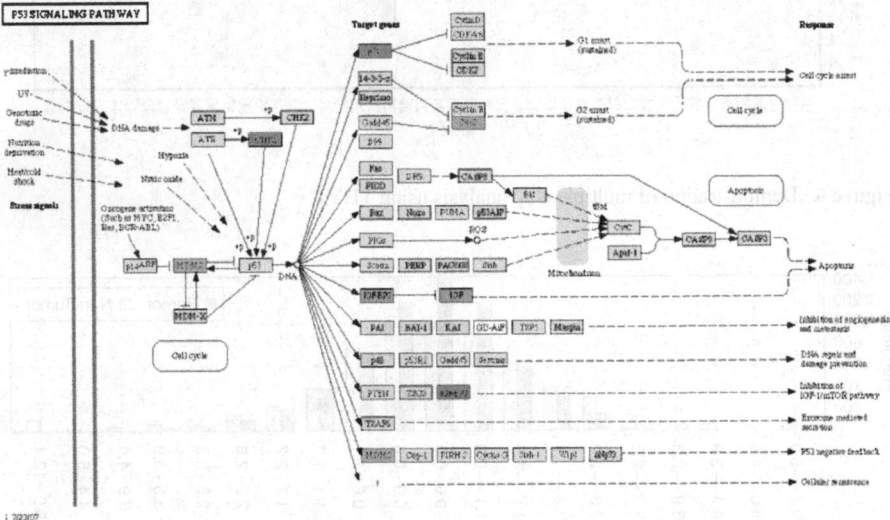

Figure 9. P53 signaling pathway KEGG map overlaid with differential gene expression results.

4   Conclusions

The GEE Tool is developed as a microarray data analysis application with emphasis
on direct graphical and tabular manipulation of the data. It helps the users to organize and analyze microarray gene expression data for multiple and paired chips. We
can use the GEE tool to analyze gene expression patterns among different biochips.
We can identify cluster of genes with similar expression patterns. In addition, we

are able to analyze cDNA arrays from different sources. The GEE is implemented with C++ language and hence can be executed on different computational platforms. The GEE tools can be run as stand-alone software on the user's computer system where the primary microarray data resides.

In summary, the GEE tool is an easy to use application for analyzing microarray gene expression data using techniques of direct manipulation, data filtering, statistical tests, and clustering. The analysis results are presented in various graphic and table formats. Tables may be exported to text files or be used to access databases. It is also made freely available to the microarray research community for use with their own arrays as well as for use with data from commercial arrays. Future enhancements will include new statistical and direct-manipulation techniques, data caching, additional support for microarray databases and simplified configuration for the other arrays.

Acknowledgments This work was supported in part from the National Science Council, Taiwan (Project No. 95-2745-E-468-001-URD, Project No. 95-2745-E-468-003-URD, and Project No. NSC 99-2221-E-024-010]. We also thank SMD (Stanford MicroArray Database) and KEGG (Kyoto Encyclopedia of Genes and Genomes) for their valuable databases.

References

1. P. Chavan, K. Joshi, and B. Patwardhan, "DNA microarrays in herbal drug research," Evidence-based Complementary and Alternative Medicine, Vol. 3, No. 4, 2006, pp. 447–457.
2. P.F. Lemkin, G.C. Thornwall, K.D. Walton et al., "The microarray explorer tool for data mining of cDNA microarrays: application for the mammary gland," Nucleic Acids Research, Vol. 28, No. 22, 2000, pp. 1452–1459.
3. J. K. Choi, U. Yu, O. J. Yoo, and S. Kim, "Differential coexpression analysis using microarray data and its application to human cancer," Bioinformatics, vol. 21, 2005, pp. 4348–4355.
4. J. A. Warrington, R. Todd, and D. Wong, Microarrays and Cancer Research. Eaton Publishing Company, 2002.
5. Kyoto Encyclopedia of Genes and Genomes (KEGG), 2010, Available from: <http://www.genome.ad.jp/kegg>.

Part II
Technology and Techniques

Chapter 12
ON SOLVING SOME HETEROGENEOUS PROBLEMS OF HEALTHCARE INFORMATION SHARING AND INTEROPERABILITY USING ONTOLOGY COMPUTING

Ching-Song Don Wei, Jiann-Gwo Doong, and Peter A. Ng

1 Introduction

A Clinical Information System (CIS) (or interchangeably, called Healthcare Information System (HIS)) is usually implemented using a Windows-based client-and-server architecture. The development of this system relies heavily on the use of a medical database as a backend and a graphic user interface (GUI) for data input and output as a frontend. The data in the frontend consists of multiple GUI's fields of various forms, which compose structured data and free-text medical reports on patients. These free text medical reports are mostly made of the complicated part of unstructured text data. Usually, the free-text medical reports include reports on patient intake, examination, and discharge, which could be an input from a keyboard, a handwriting device, a voice microphone, transcription service or others. All the data, including structured and unstructured data, are stored in the backend database. Generally a relational database (RDB) is adopted and its schema was well designed by applying normalization rules so that a flat data view in GUI is implemented with multiple tables and constraints to maintain the data integrity and consistency.

1.1 Dealing with Clinical Free Text

Much research is done in developing methods that can automatically map a clinical free text to concepts of standardized medical knowledge source such as UMLS Metathesaurus. Most of these systems employ various methods based on string matching, linguistic processing and utilization of part of speech tagging followed

P. A. Ng (✉)
Fort Wayne, IN, USA
e-mail: npg@ipfw.edu

S. Suh et al. (eds.), *Biomedical Engineering*,
DOI 10.1007/978-1-4614-0116-2_12, © Springer Science+Business Media, LLC 2011

by identification of noun phrases. The Metathesaurus also provides a method which can be used to index text or to map terms from one vocabulary to another. It then discovers a set of UMLS concepts in a body of the index text and provides the ranking on these concepts in the set.

Some of medical language extraction and encoding systems, such as MedLEE [1], can be used to translate the medical text, written in a free-text and unstructured format, into a text of a standard and structured format. Once the unstructured data is transformed into a standard one, together with structured medical data, a clinical data repository is formed by applying the concept extraction, information categorization and information retrieval for integrating these data.

1.2   Inability of Relational Database in Medical Applications

Most of the CIS's provide users with versatile functionalities such as searching, creation, updating, and deletion of data records of a medical database, which is normally a RDB satisfying normalization forms, to support the data manipulation. A RDB requires data to be specified in multiple tabular forms and conformed to a predefined schema and constraints. This data requirement promotes data integrity but discourages any changes on irregular data or data that evolves rapidly. In addition, the normalized RDB schema does not really reflect data format in the GUI view of the information. For example, the patient information may be stored in three tables such as a patient table, a doctor table, and an insurance table. In this patient table, there are foreign keys referring to the doctor table and insurance table which enforce the referential integrity. However, the Relational Database Management System (RDBMS) will continue to be a dominant information management system, which manages all the critical enterprise data, although it is difficult to support data sharing and data exchange in large-scale databases.

Furthermore, since each CIS is usually designed for a specific purpose and there is no unified term and structure among different CIS databases, it is very challenging to enable a system to interact with other systems of the same sort (such as, for the same purpose, area of application, etc.). It is a much more difficult task to provide interactions among different CIS's, even within the same healthcare organization. Even though many of the CIS's share similar GUI forms, but each of them may use a different database system and/or schema that form a heterogeneous environment within a healthcare organization.

1.3   Data Heterogeneity Problem

For the structured and unstructured data to be available and interoperable among healthcare organizations, the well-known problem of data heterogeneity with the distributed database systems has to be solved. This problem can be classified as:

system heterogeneity, syntactic heterogeneity, structural heterogeneity and semantic heterogeneity. The system heterogeneity considers applications and data that
may reside in different hardware and operating systems. The syntactic heterogeneity addresses data source which may use different representations and encodings
[2, 3]. For achieving syntactic interoperability, compatible forms of encoding and
access protocols could be used to allow information systems to communicate each
other [4]. The structural heterogeneity states that different information systems may
use different document layouts and formats, data structures, data models and schemas to store data. Finally, the semantic heterogeneity considers the intended semantic meaning associated with the content of an information item.

1.4   Sharing and Interoperability among Medical Database Systems

The ability of data sharing and interoperability between database systems is a critical
problem in large healthcare organizations. For managing distributed, heterogeneous
and autonomous databases and providing the data interoperability, the federated
database system (FDBS) [5, 6] and mediation [7] are the two dominant approaches
in the 90s. In supporting interoperability of autonomous hospital databases, a FDBS
[8] uses an unbalanced hierarchical structure to locate data in any federation within
the system. But, this requires human assistance to define a hierarchy of terms, based
on the user interests and applications, and to create metadata about a database (schemas and instances) to be shared with other federations within the system.

An architecture [9] is proposed for integrating heterogeneous data source applying an eXtensible Markup Language (XML) global schema [10]. Then, XQuery is
used to specify data retrieval and translated to SQL for each local data resource [11].
The use of a standard-compliant XML for storing the structured and unstructured
data on a GUI form provides a way for data sharing and data integration [12] and
therefore makes the data interoperability possible [13]. The GUI form is similar
to document data. In order to maintain and obtain semantic interoperability, each
value control must associate with its context information [14]. The context information is a metadata specifying the meanings, properties (such as quality, precision
and source) and organization of its associated value of a value control. The use of
XML for specifying the mapping description eliminates the need of a specialized
parser and a specialized description language [15]. In [16], the extended semantic
value model, each data value contained in a value control is associated with a set
of attributes and a context. The semantic association trees, which are of predefined
semantic tree structures, describe the framework of property and semantic relations.
Instead of storing them into a database, the contents of GUI forms and their associated controls, called data source, are exported as a file of a semi-structured XML
format. The exported XML file contains only de-normalized data and is ready and
easier to be shared with other systems. Even though XML is a standard-compliant
format and widely accepted by healthcare applications, its capability of expres-

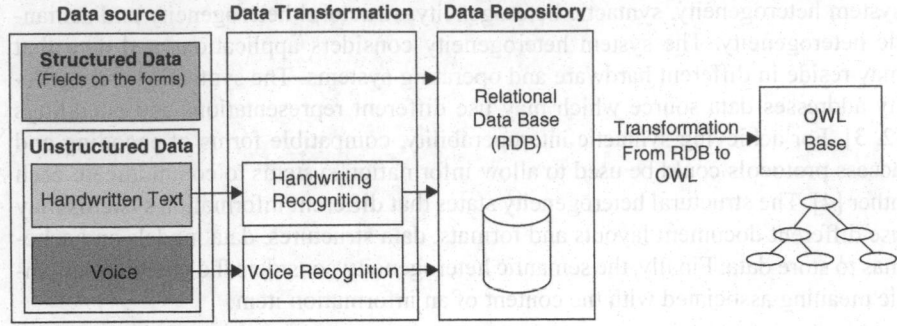

Figure 1.1 An Overall Data Transformation and Process.

siveness for describing complete semantic structure is still not enough to make it machine readable.

The Semantic Web initiative [17] has provided a new approach for data interoperability. The data only can be shared and interchangeable if the data value and its semantic are well described. With Ontology Web Language (OWL) [18], the syntactic and semantic information of the structured and unstructured data can be well described by converting the data stored in the underlying RDB to a flat OWL file which is similar to the ideal data source of XML semi-structured data and forms. OWL provides a promising way for data sharing and interoperability if one can devise a component to mediate over the syntactic and semantic heterogeneities and then map the data from source end to the target end. The data process is depicted as in Figure 1.1.

1.5 Semantic Web Service and Healthcare Data Interoperability

In the healthcare domain, data sharing and interoperability issues exist not only within a hospital but also across enterprises and community boundaries. The first scenario describes a hospital that contains disparate applications and RDBs. Since the databases are within an organization, a shared ontology can be generated by ontology merging processes to resolve the data interoperability. On the other hand, the Semantic Web initiatives can target the solutions for providing the data sharing and interoperability among hospitals, laboratories and governments [19]. Most of the HIS's serve only one special department for a specific purpose, even within a healthcare organization. A patient's information could be spread out over a number of health information systems in different departments which are interoperable. The design described in the section 1.4 is for an ideal platform to resolve the data sharing and interoperability among heterogeneous systems within an organization such as a completely integrated HIS systems or a platform for data mapping between two applications with heterogeneous database systems.

Figure 1.2 Web Service Process for Healthcare Information Systems.

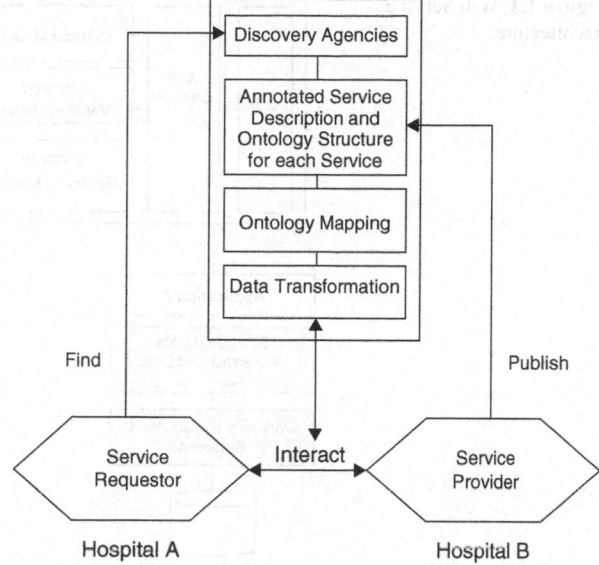

However, many complicated scenarios may occur, when information sharing and interoperability are necessarily applicable among different organizations, such as clinic offices, hospitals, and laboratories. For example, a patient may be transferred from one hospital to another hospital for better treatment. It would be an ideal situation if the latter hospital is able to acquire the treatment history, laboratory results, and others from the former hospital via the internet. For this case, the Web Service model provides the healthcare providers with an ideal platform which supports the data sharing and exchanging. For a data requestor to use services remotely supplied by different providers on the internet, Web Service requires neither a specified programming language nor a dedicated supporting platform [20]. All the services are designed to wrap and expose existing services and resources for promoting interoperability among diverse systems. In addition, Semantic Web Service uses ontology to annotate the description of a Web Service to be machine readable and understandable. Web Service also resolves the nature of dynamic changes in data and structures of a database system. In addition to the annotation language for Web Service Description, the techniques of the ontology mapping, merge, and the data transformation are also required. Figure 1.2 depicts the Web Service processes between two hospitals. Figure 1.3 shows Web Service architecture for a proposed system.

In the remaining of this paper, we will consider two cases: data sharing and interoperability within a hospital, and then across enterprises and communities. Both use different architectures. The former case uses a simple architecture with ontology merge. The latter requires a unified ontology or ontology mapping. Section 2 gives an overview of related works and describes a scenario of a hospital computing environment. Section 3 focuses on the conversion between database and ontology. Finally the experiment and conclusion are summarized in Section 4 and 5.

Figure 1.3 Web Service
Architecture.

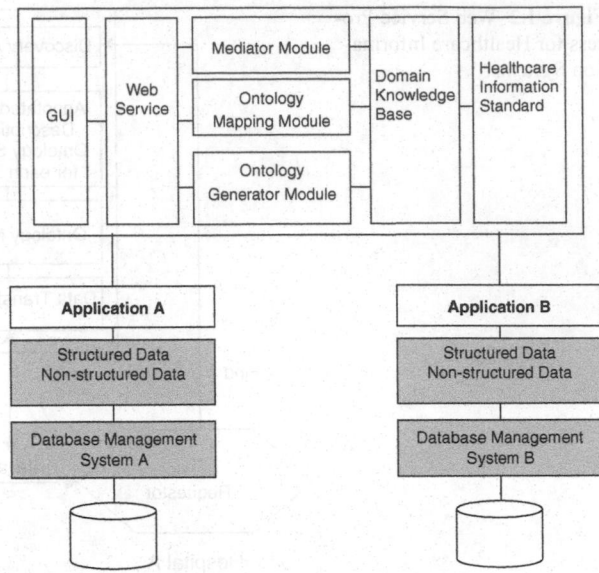

2 Related Work

In this section, several related works on ontologies for dealing with semantic interoperability in healthcare systems or in general applications are reviewed. [21] Proposed a hybrid approach to supporting data sharing and semantic interoperability using ontologies and a central access to objects of database federations. Each federation has an object catalog which serves as a mediator to the federated databases as the underlying data repositories. The object catalog contains an ontology for each data source and provides service to access and search for federated data and metadata. It also stores instances of reference and a reference ontology that represents the global conceptual schema of a federation. At both the structural and the semantic levels, the ontology of each database schema describes mappings from the schema to the elements of reference ontology. In this approach, the use of ontology to serve as a mediator is a new and interesting method. However, as described in Section 1, federated database systems have not shown a promising result in solving data heterogeneity problem since 90s. The maintenance of a federated database system to cope with its local schema evolution is a challenging problem.

In [22], the Semantic Web Service approach is proposed to overcome the difficult problem of interoperability in healthcare industry. The health information standards, such as HL7 [23], are used as domain knowledge and expressed in an OWL format to facilitate the ontology mapping. There are two important ontologies: the Functionality Ontology and Service Message Ontology. The Functionality Ontology classifies coarse-grained Web Services in healthcare domain and the Service Message Ontology annotates finer granularity services to retrieve meaningful health data components. Both the Functionality Ontology structure and Message Ontology structure can be built based on the health standards. Two different clinic systems which use different health standards can share the information through the ontology mapping and reasoning. Using Semantic Web Service is an ideal platform for achieving the interoperability and data sharing for the clinic information systems. But there is an assumption in this approach: the clinic information systems have to use one of the healthcare information standards. However, in reality, many of the clinic information systems do not use any of the existing information standards and therefore the existing mapping rules may not be applicable. Furthermore, many of the local clinic information systems do not need to use such a huge standard that has all kinds of concept structure.

An ontology-based OntoGrate system [24] employs machine learning and data mining techniques to provide users with meaningful mapping rules, and capability for interaction with domain experts to justify the rules and promote its effectiveness. It consists of an ontological language, Web-PDDL, which incorporates database schemas for defining both structure and semantics, and an inference engine – OntoEngine. OntoEngine checks the consistency and redundancy of mapping rules and conducts optimized query answering and data translation as well. Its goal is to validate semantic structure created by OntoGrate using Web-PDDL. The system provides interactive tools for the domain experts to perform final validation process. A drawback is that it is difficult to generate ontological structure manually from the database schemas of a large database system. A reliable automatic tool for converting database to ontology is required.

An ontology-based knowledge sharing system, called OntoShare [25] uses the Resource Description Framework (RDF) to specify the ontology and form a Web-based virtual community. The shared concepts described in ontology are constructed upon the RDF-annotated information resource for providing the information to their users. OntoShare supports, to a certain degree, ontology evolution based on the usage information shared among the users and the assigned concepts for this information.

The system proposed in [26] uses Semantic Web tools and reasoning mechanisms to achieve data sharing and interoperability in a heterogeneous hospital computing environment. Initially, the system constructs a source ontological model via its instances, classes, attributes, relationships and events. A source ontology, which is described using the D2RQ [27] language, represents the semantics of source data. Then, it generates a target ontology by merging the source ontologies that present the common knowledge in the domain, without taking into consideration the health information standards such as HL7. The product of TopBraid Composer is used for creating the source ontology and target ontology [28]. However, users may not have any control over the process of the automatic creation of source ontology. If, in the process of creating a source ontology, it could not be validated by users, the later merging process for source ontologies may not be accurate. The task of merging source ontologies includes the activities for determining the domain and scope of each ontology, defining the ontology class and its scope, and defining the class properties and internal taxonomy structure. Additional reasoning rules are added to the target ontology for handling the interoperability by resolving heterogeneous data structure, and the data sharing by resolving the data linking issues. The reasoning rules are described by Semantic Web Rule Language (SWRL) [29]. Except the creation of source ontology, this framework demonstrates a thin structure and promising design.

An enhanced framework is under investigation by taking the existing works into consideration. This framework includes (1) an automatic approach to converting a database schema into a source ontology that can maintain most of the semantics of source data, (2) an ontology merging process which merges source ontologies to form a target ontology that represents a global-shared concept structure, and (3) a semantic Web Service that provides a Web-accessible interface for accessing the shared data. In the remaining paper, we shall discuss the database conversion component that takes both the database schemas and data into consideration for securing their semantics during the creation of source ontology.

3   Mapping between Relational Database and Ontology

Ontologies are formal and explicit specifications of a shared *conceptualization* which consists of objects, concepts, and other entities in some areas of interest and their relationships among them [30]. It provides an approach to describing a shared and reusable knowledge about a specific domain. In this paper, the ontology is used to describe a formal description of concepts in a given domain of disclosure.

3.1   Sharing and Interoperability among Medical Database Systems

The concept of interest is described by classes, properties of each concept, and restrictions on properties. Classes are the main theme of most ontologies which describes concepts in the domain. Normally, classes are interpreted as sets that contain individuals. Sometimes classes are also called concepts. Concepts can be organized to form a superclass-class (or class-subclass) hierarchy, which is also known as taxonomy. Individuals represent objects (also known as instances) in a given domain. The concept's individuals are equivalent to the class instances in object-oriented terms. Properties describe relationships between two individuals. Relationships characterize binary relations on individuals which link two individuals together. It means that the relationships link the domain's individuals to the range's individuals. Relationships can have inverses. For example, the inverse of hasSibling is isSiblingOf. There are three types of relationships: object relationships, data-type relationships, and annotation relationships. An object relationship links an individual to another individual. A data-type relationship links an individual to a XML schema data-type value or a Resource Description Framework (RDF) literal [31]. Annotation relationships are used to add information to classes of concepts. For example,

Figure 3.1 A Patient, John Smith with Various Relationships in an Ontology.

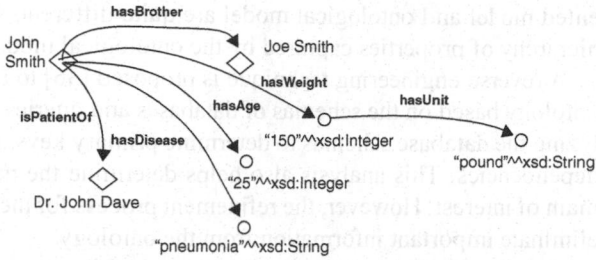

the concept Patient contains all the individuals which are patients in a given clinic domain. Figure 3.1 describes a patient, John Smith, and his properties: is 25 years old, weighs 150 pounds, and has a brother whose name is Joe Smith, has disease pneumonia, and is a patient of Dr. John Dave. The properties of hasBrother John and isPatientOf are called object property. The properties of hasWeight, hasAge, and hasDisease are called data-type property. As shown in this figure, the relationships can be limited to having only single value but they can be either transitive or symmetric.

Many research works have been published in the area of generating ontology from database schema. A detailed ontology development guide in [32] describes a process for creating an ontology for a specific concept. In this paper, for the application of data sharing and exchanging among clinical databases, both source and target databases will be converted into ontology for further processing.

3.2 Transforming a Relational Database to an Ontology

While ontology aims to express the global semantics of the real world explicitly, a relational database system, which provides only a general approach to storing and accessing data, primarily concentrates on expressing the local semantics of the real world implicitly and its schemas are not generally reusable by other database systems. Some of the semantics of their data have to be defined in GUI forms. However, it is possible to discover and capture the semantic information about data (i.e., the meanings of data) from the business logic rules (located at the business logic tier) and GUI forms of the database application in addition to database schemas. Then ontology could be used to represent the semantics for a database including schemas and business logic rules. Having this idea in mind, we will discuss the mapping from a database schema to an ontology in the following section. This process only extracts the data schema from the database. The rest of the meanings of the data should be added manually from the business logic rules and application program.

Both the methods of extracting semantics from database schemas [33] and of transforming a relational data model to an object-oriented model [34] were addressed. But these two methods did not have sufficient information for constructing ontologies from databases. Even though they are quite alike, both the object-oriented model and ontological model are quite different, such as the constraints and hierarchy of properties captured by the ontological model [35].

A reverse engineering technique is proposed [36] to construct a domain specific ontology based on the schemas of databases and queries. The process involves analyzing the database schemas to determine primary keys, foreign keys, and inclusion dependencies. This analysis also helps determine the main entities within the domain of interest. However, the refinement process for the created ontology may also eliminate important information from the ontology.

Based on the degree of semantic recovery, this reverse engineering technique can be classified into three types:

- Direct transformation: A direct generation process is proposed that transforms each table to a class and each attribute to a property [37, 38]. If a table has foreign keys which reference to other tables, then for each foreign key a new property is added to the class corresponding to the reference table whose value is an instance of the class representing the referenced table. In a relational OWL project, based on OWL, a presentation format for data and schema information is proposed [39]. The representation for schemas includes tables, columns, data types, primary keys, foreign keys, and the relations among each other. Since this direct transformation does not fully recover the entity of a database schema and its relationship to other entities, the transformed ontology does not have sufficient semantic information to support semantic interoperability application.

- Semantic recovery: According to [16], the GUI forms of a CIS system contain the full semantic information that can be used for interoperability. The direct transformation transforms only the database schemas to classes in OWL without recovering the entity and relationship information, i.e. the semantic information. In order to recover the semantics, the reverse engineering of normalization process should be applied. A tool called DB2OWL [40] could be used to recognize the concepts (classes) from tables, the object properties from integrity constraints and data type properties from non-key columns. The algorithm only uses a key-based approach to generate the ontology components and the mapping documents. However, this approach does not identify the hierarchy of classes and the cardinality information.

- Complete semantic recovery: According to [35], the learning process generates, as its result, complete semantic information including classes, properties and property characteristics, class hierachy, cardinality, and instances. However, it does not consider many-to-many relationship and does not describe the mapping description document. In this paper, we will examine and adopt this method with an extension of a mapping method.

3.2.1 Definition of Relational Database

Throughout the remaining Section 3, let R, R_1, R_2, ..., R_i and R_j denote the relations in a given database. Let $A = \{A_1, A_2, ..., A_n\}$ denote a set of attributes with their corresponding domain $D = \{D_1, D_2, ..., D_n\}$. Then, let $Si = \{A_1: D_1, A_2: D_2, ..., A_n: D_n\}$ be a relational schema. In the relational model [41], each relation R_i, defined by a relational schema Si, is a set of mapping from the attribute names to their corresponding domains. This means, a relation $R_i(A_1, A_2, ..., A_n) = \{<a_1, a_2, ..., a_n> |$ for each $1 \le k \le n$, a_k is an element of $A_k: D_k\}$. In [35], a series of functions is defined as follows. The function $attr(R_i)$ returns the attributes for a relation R_i. The function $dom(A_i)$ returns the allowable values of attribute A_i. For the key functions, the function $pkey(R_i)$ returns the primary key of the relation R_i and the function $fkey(R_i)$ returns the foreign keys of the relation R_i. The inclusion dependence and equivalence are two important relational features in learning the ontology from a given relational database when considering the values (tuples) in relations. We shall give the definitions of inclusion dependence and equivalence.

Given relations R_1 and R_2 in a database, let $A_1 = \{A_1, A_2, ..., A_m\}$, $A_1 \subseteq attr(R_1)$ and $A_2 = \{B_1, B_2, ..., B_m\}$, $A_2 \subseteq attr(R_2)$. We give the following definitions:

Definition 1 – Inclusion dependence. Let $t(A_1)$ and $t(A_2)$ represent the tuple values of A_1 and A_2 respectively. If, for each $t(A_1)$ in R_1, there exists $t(A_2)$ in R_2 such that $t(A_1) = t(A_2)$, then A_1 and A_2 are of inclusion dependency, denoted as $R_1(A_1) \subseteq R_2(A_2)$. The inclusion dependency is a generalization of referential integrity.

Definition 2 – Equivalence. If there exists $R_1(A_1) \subseteq R_2(A_2)$ and $R_2(A_2) \subseteq R_1(A_1)$, then A_1 and A_2 are of equivalence, denoted as $R_1(A_1) = R_2(A_2)$.

In addition to the inclusion dependence and equivalence, the relational model also has integrity constraints such as entity constraint and referential integrity.

3.2.2   Learning Rules

In this paper, we consider a relation containing either data or no data. The proposed method [35] considers the values in relations without taking many-to-many relationship into consideration. The DB2OWL method [40] does not consider the values in relations. For generalizing these methods, we propose the following learning rules for creating the classes (concepts), properties, hierarchy, cardinality and instances from a relational database.

3.2.2.1   Rules for Creating Classes

In the process of database design, a relation (also called table) may split into multiple relations simply for the sake of different categories of data. During normalization processes, the original relations may be decomposed into sub-relations iteratively. Therefore a class may be derived from information of a relation or information spreading across multiple relations in a database.

1) *Process I. Integrate information scattered in several relations into one class which describes a concept or an entity in the real world.*

a. *Rule 1. Check the table data for table integration.*

if $A_1 = pkey(R_1)$, $A_2 = pkey(R_2)$, ..., $A_i = pkey(R_i)$, where $dom(A_1) = dom(A_2)$ $= ... = dom(A_i)$, and $R_1(A_1) = R_2(A_2) = ... = R_i(A_i)$, then the information spread across $R_1, R_2, ..., R_i$ should be integrated into an ontological class C_i.

When a primary key of a relation has more than one attribute, let $pkey(R_i) = A_i$ which contains a list of $k > 1$ attributes from $attr(R_i)$. $A_i \subseteq attr(R_i)$. $dom(A_i) = dom(A_{i1}) \times dom(A_{i2}) \times ... \times dom(A_{ik})$, for $A_i = \{A_{i1}, A_{i2}, ..., A_{ik}\}$. For this case, $R_i(A_i) = R_i(A_{i1}, A_{i2}, ..., A_{ik})$ is a relation of k columns of primary key values. The above-mentioned Rule 1 can be extended into multiple columns.

When multiple relations are used to describe one same entity, the information should be integrated into one ontological class.

Example 1: As in a sample of database schemas described in Table 1, the relation Patient and the relation Patient_Detail should be integrated into one ontological class Patient because Patient(PatientID) = Patient_Detail(PatientID). Therefore, the class Patient is created via OWL ontology, and the class Patient is as follows.

<owl:Class rdf:ID="Patient"/>.

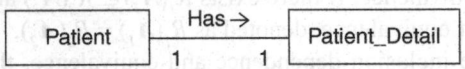

Table 1. A Sample of Relational Database Schemas.

Relation	Primary Key	Foreign Key
Patient (*PatientID* string, FirstName, string, LastName string, Address string, Gender char, DOB datetime)	PatientID	NA
Patient_Detail (*PatientID* string, InsuranceCompany string, InsurancePolicy string, emailAddress string)	PatientID	PatientID *references to* PatientID *in Patient relation*
Physician (*PhysicianID* string, FristName string, LastName string, Address string, Gender char, DepartmentID string)	PhysicianID	DepartmentID
Staff (*StaffID* string, FristName, string LastName string, Address string, Gender char, JobCode string,DepartmentID string)	StaffID	JobCode, DepartmentID
IsPatientOF(*PatientID* string, StaffId string)	PatientID, StaffID	PatientID, StaffID
Department (*DepartmentID* string, DepartmentNAme string)	DepartmentID	NA
Job (*JobCode* string, JobTitle string)	JobCode	NA

Assume that PhysicianID is a subset of StaffID. Since Physican(PhysicianID) is a subset of Staff(StaffID), then the class Staff, which is a superclass of a subclass Physician, is created via OWL ontology. This will be addressed in the later section. As well, we will also address the relationship between classes, such as IsPatientOf later.

b. *Rule 2. Check the table structure for table integration.*
   Assume that $attr(R) = pkey(R) = fkey(R) = \{A_i, A_j\}$ where $A_i \cap A_j = \varnothing$, and $A_i \cup A_j = attr(R)$. If A_i refers to R_i via $pkey(R_i)$ and A_j refers to R_j via $pkey(R_j)$, then the many-to-many relationship should be expressed by object properties. The creating process will be described in Rule 6 of *Process III.*
   All attributes of R, $attr(R)$, can be divided into two disjoint subsets of A_i and A_j, each of them participating in a referential constraint with R_i and R_j, repectively. When a relation R is used only to relate two other relations R_i and R_j in a many-to-may relationship, then the relation R does not need to map to a class.

Example 2: In the sample of relational database schemas described in Table 1, the relation IsPatientOf consists of two attributes, PatientID and PhysicianID. These two attributes are primary keys and also are foreign keys of the relation IsPatientOf. The attribute PatientID is referring to the relation Patient. The attribute PhysicianID, a subset of the attribute StaffID, is referring to the relation Physician. The purpose of the relation IsPatientOf is solely to create the many-to-many relationship between the two relations, Patient and Physician. That means, a patient may have more than one physician and a physician may have more than one patient.

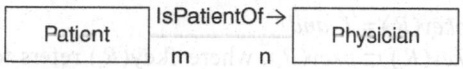

2) *Process II. Create ontological classes for the relations, otherwise.*

   a. *Rule 3. Create an ontological class C_i for the relation R_i.*
      An ontological class C_i is created for the relation R_i, if R_i is not used to relate one relation to the other relation as in Rule 1 or if there does not exist a relation that can be integrated with R_i as in Rule 2. But it meets one of the following conditions:

      Condition 1. $|pkey(R_i)| = 1$;
      Condition 2. $|pkey(R_i)| > 1$, and there exists an A_i, where $A_i \in pkey(R_i)$ and $A_i \notin fkey(R_i)$.

      In other words, if a relation R_i is used to describe an entity instead of the many-to-many relationship between relations or if it is used to describe partial information of an entity and should be integrated with the other relation, then R_i can be mapped into one ontological class.

Example 3: Consider the sample of relational database schemas given in Table 1. Using OWL ontology, create four ontological classes, Physician, Staff, Department

and Job for the relations Physician, Staff, Department and Job, respectively, since
these relations fail to meet Rule 1 and 2 and each of them has only one primary key.
The four corresponding classes are as follows.

```
<owl:Class rdf:ID=" Physician"/>.
<owl:Class rdf:ID=" Staff"/>.
<owl:Class rdf:ID=" Department"/>
<owl:Class rdf:ID=" Job"/>
```

3.2.2.2   Rules for Creating Properties and Property Characteristics

3) *Process III. Identify the relationships between relations and create proper object
   properties accordingly.*

   a. *Rule 4. Check tables that implement the foreign key relationship.*
      Let $Ai \subset attr(R_i)$ *and* $Aj \subset attr(R_j)$. *Let* $Ai = fkey(R_i)$, $Aj = pkey(R_j)$, *and* $Ai \not\subset$
      $pkey(R_i)$. *Let* $R_i(Ai) \subseteq R_j(Aj)$. Then, create an object property OP based on Ai
      : Let the ontological classes C_i and C_j be created for the relations R_i and R_j,
      respectively. An OP is created for C_j, where the domain and range of OP are
      C_i and C_j.
   b. *Rule 5. Check a table that implements the foreign key relationship with partial
      primary key.*
      Let C_i and C_j be the classes corresponding to R_i and R_j, respectively. For the
      relations R_i and R_j, which meet the following two conditions:

      Condition 1. $|pkey(R_i)| = 1$, *and*
      Condition 2. $fkey(R_i) \subset pkey(R_i)$, where $fkey(R_i)$ refers to R_j and $fkey(R_i) \subset$
      $attr(R_i)$.

      Then, create two ontological object properties, called the "has-part" and "is-
      part-of" object property. C_i and C_j are the domain and range of "is-part-of"
      object property, respectively. C_j and C_i are the domain and range of "has-
      part" object property, respectively. These two properties are of inverse
      properties.

Example 4: In the sample of relational database schemas described in Table 1,
consider the relationships "work_for" and "has_Employee" as the "is-part-of" and
has-part object properties, respectively. Create both the is-part-of object property
"work_for" and the has-part object property "has_Employee". The relations Staff
and Department are the domain and the range of the is-part-of object property
"work_for", respectively. The relations Department and Staff are the domain and
the range of the has-part object property "has_Employee". The corresponding OWL
description is as follows:

```
<owl:ObjectProperty rdr:ID="work_for">
        <rdfs:domain rdf:resource="# Staff"/>
        <rdfs:range rdf:resource="# Department"/>
```

```
</owl:ObjectProperty >
<owl:ObjectProperty rdr:ID="has_Employee">
        <rdfs:domain rdf:resource="# Department"/>
        <rdfs:range rdf:resource="# Staff"/>
</owl:ObjectProperty >
```

c. *Rule 6. Check tables that only implement many-to-many relationship.*
Let C_i and C_j be the ontological classes for the relations R_i and R_j, respectively. Assume that $attr(R)$ = pkey(R) = $fkey$(R) = $\{A_i, A_j\}$ where $A_i \cap A_j = \emptyset$, and $A_i \cup A_j = attr(R)$. If A_i refers to R_i through $pkey(R_i)$ and A_j refers to R_j through $pkey(R_j)$ as described in *Rule 2*. Then, based on (A_i, A_j) create an object property OP_i for C_i, where C_i and C_j are its domain and range, respectively. Likewise, based on (A_j, A_i), create another object property OP_j for C_j, where C_j and C_i are its domain and range, respectively.

d. *Rule 7. Check tables that indicate the relationship between two relations.*
Let C_i and C_j be the ontological classes for the relations R_i and R_j, respectively. Assume that A_i = pkey(R_i), A_j = pkey(R_j), where $A_i \cap A_j = \emptyset$; and $A_i \cup A_j = fkey(R)$, where $fkey(R) \subset attr(R)$. If A_i refers to R_i through $pkey(R_i)$ and A_j refers to R_j through $pkey(R_j)$. Then, based on the semantic of R, create two object properties OP and OP', where C_i and C_j, respectively, are its domain and range of OP, and C_j and C_i, respectively, are its domain and range OP'.

Example 5: In the sample of relational database schemas given in Table 1, the relation Staff identifies the relationship between the two relations Department and Job. According to the semantic of the Staff table, an object property"has_job" is created and is given as follows. Its domain and range are Department and Job, respectively.

```
<owl:ObjectProperty rdf:ID="has_job">
        <rdfs:domain rdf:resource="# Department">
        <rdfs:range rdf:resource="# Job"/>
<owl:ObjectProperty>
```

4) *Process IV. Identify the data type properties from the attributes that are not converted in Process III.*

a. *Rule 8. Create data type property for each attribute in relation which cannot be converted into object type property.*
Let C_1, C_2, ..., C_i be the corresponding ontological classes for the relations R_1, R_2, ..., R_i of a given database. If any attribute A of a relation R_k, $1 \leq k \leq i$, cannot be used to create object property as described in Process III, then it will be used to create a data type property P for the corresponding ontological class, C_k of the relation R_k. The domain and range of the data property P are C_k and dom(A), respectively, where $P \in DP(C_k)$, the set of data type properties of C_k, and $A \in$ attr(R_k).

3.2.2.3   Rules for Creating Class/Property Hierarchy

5) *Process V. Identify the class hierarchy in ontological structure.*

    a. *Rule 9. Check the inheritance relationship of two relations to determine the hierarchy relationship of their corresponding classes.*
       If $A_i = pkey(R_i)$, $A_j = pkey(R_j)$, and $R_i(A_i) \subset R_j(A_j)$, then the class/property of the R_i is a subclass/subproperty of the class/property of R_j.
       Since the classes and properties can be organized in a hierarchy, this rule is to determine the inheritance relationship between classes or properties. If two relations in a given database have their inheritance relationship, then their two corresponding ontological classes or properties can be organized in a hierarchy.

Example 7: Given the sample of relational database schemas as shown in Table 1, the ontological class Physician is a subclass of the class Staff according to Rule 9. The following ontology description describes this hierarchical relationship.

```
<owl:Class rdf:ID="Physician">
        <rdfs:subClassOf rdf:resource="# Staff"/>
</owl:Class>
```

3.2.2.4   Rules for Learning Cardinality

6) *Process VI. Identify the cardinality constraints.*
    The cardinality specifies the constraint among classes and is one of the OWL properties. This property can be learned from the constraint of attributes in relations [35]. Let R_i be a relation and $A \in attr(R_i)$.

    a. *Rule 10. Create minCardinality and maxCardinality of a property P.*
       If $A = pkey(R_i)$ or $A = fkey(R_i)$, then the minCardinality and maxCardinality of the property P corresponding to A is 1.
    b. *Rule 11. Create minCardinality of a property P when A is NOT NULL.*
       If A is set as NOT NULL, then the minCardinality of the property P corresponding to A is 1.
    c. *Rule 12. Create maxCardinality of a property P when A is UNIQUE.*
       If A is set as UNIQUE, then the maxCardinality of the property P corresponding to A is 1.

3.2.2.5   Rules for Learning Instances

7) *Process VII. The instances in ontological class C_i will be created from the tuples of its (i.e., C_i) corresponding relation R_i.*

a. *Rule 13. Create instances.*
For an ontological class C_i, which corresponds to a number of relations, says, $R_1, R_2,..., R_j$, in database, then every tuple $t_i \in R_1 \times R_2 \times... \times R_i$ can be an instance of C_i.

Given a relational database, we could apply Rule 1 to Rule 13 to construct automatically its corresponding ontology by means of the OWL ontology. The conversion between database and ontology can be achieved with the intent that the converted system can support data sharing and interoperability as well as dealing with the semantic heterogeneity problem.

3.2.3   Mapping Document Generation

During the learning process, a DR2Q [27] document is automatically generated to record the mapping relationships between generated ontology components and the original database components. DR2Q mapping language is a declarative mapping language that includes a full description of relational schema and OWL ontologies. In DR2Q document, a construct of dr2q:Database defines a JDBC or ODBC connection to a local relational database and specifies the type of the database columns used by D2RQ. A construct of dr2q:ClassMap represents a class or a group of similar classes of an OWL ontology. Property Bridges relate database table columns to Resource Description Framework (RDF) properties [25, 31, 42].

4   Experiments

The sample of database schema and data as described in Table 1 is used to create an ontology using various algorithms and tools. Our evaluation focuses on the proper creation of classes, object properties for foreign keys, data type properties for attributes, hierarchy of classes, and instances in the ontology. The tools or algorithms include Datamaster [37], [35], [39], Topbraid [28], and the method proposed in this paper. The more information is captured by the ontology, the more semantics of information is preserved during the conversion. For evaluating these algorithms, all rules are used for creating the ontology manually. For evaluating these tools, the sample database is inputted into the tools for generating the ontology. The proposed method is found to restore more semantics of data and schema. Currently the proposed method has been implemented in Visual Studio 2008 with Oracle server.

5   Conclusion

The Healthcare Information Exchange and Interoperability (HIEI) between providers (hospitals and medical group practices), independent laboratories, pharmacies, payers, and public health departments has been assessed and recognized as a highly promising benefit [43]. Interoperability is a basic requirement for ensuring that the widespread Electronic Medical Record (EMR) adoption could yield the social and economic benefits for us. Without interoperability, the EMR adoption will strengthen the information vaults that exist in today's paper-based medical files, resulting in even greater proprietary control over health information and, with it, control over patients themselves [44].

In this paper, we proposed using of ontology computing to resolve the issues regarding healthcare information sharing and interoperability. Our discussion starts with the conversion of database to ontology. A set of learning rules is presented for constructing automatically an OWL ontology from both database schemas and data. According to our experiment and analysis, the proposed method can capture the semantics of database schemas and data information more efficiently. This method is integrated into a healthcare information system that promotes information interoperability and data exchange using ontology. Since the construction of an ontology from a database plays an important role for information interoperability, there is a need to improve the existing methods. Our future research includes improvement of learning rules based upon the feedback from the use of healthcare information system and the evaluation of the representation language between database and OWL ontology.

Currently, we are also investigating the generation of a shared global ontology within a healthcare organization and the application of Web Service in reconciling heterogeneous attributes and their semantics. The representation and mapping of semantics will be elaborated to develop a standard framework for the environment of healthcare. Since the healthcare organizations are not expected to use standard format, a mechanism is devised for organizations to describe their formats and for the mediator to generate semantic mapping and merging. We are implementing a prototype for healthcare organizations, including database conversion, generation of shared ontology, and Web Services to study the efficiency and effectiveness of the interoperability of the systems within the organization or across organizations and to investigate other issues such as security, privacy, network paradigm, and knowledge base integration.

Reference

1. C. Friedman, L. Shagina, Y. Lussier, and G. Hripcsak,"Automated Encoding of Clinical Documents based on Natural Language Processing," Journal of the American Medical Informatics Association, Vol. 11, No. 5, 2004, pp. 392–402.
2. W. Kim and J. Seo,"Classifying Schematic and Data Heterogeneity in Multidatabase Systems," IEEE Computer, Vol. 24, No. 12, 1991, pp. 12–18.
3. V. Kashyap and A. Sheth, "Semantic Heterogeneity in Global Information Systems: The Role of Metadata, Context and Ontologies," in Cooperative Information Systems, M. Papzoglou and G. Schageter (eds.), Academic Press, San Diego, 1996, pp.139–178.
4. J. Cardoso and A. P. Sheth,"The Semantic Web and Its Application," in Semantic Web Services, Processes and Applications, Jorge Cardoso and Amit P. Sheth (eds.), Springer, 2006, pp. 5–36.

5.   A. P. Sheth and J. A. Larson,"Federated Database Systems for Managing Distributed, Hetero-geneous, and Autonomous Databases," ACM Computing Surveys, Vol. 22, No. 3, September 1990, pp.183–236.
6.   R. M. Colomb,"Impact of Semantic Heterogeneity on Federating Databases," The Computer Journal, Vol. 40, 1997, pp. 235–244.
7.   G. Wiederhold,"Mediation to Deal with Heterogeneous Data Sources," Proceedings of the Interoperating Geographic Information Systems Conference, Zurich, Switzerland, March 10–12, 1999. LNCS 1580, Springer Verlag.
8.   A. Zisman and J. Kramer, "Supporting Interoperability of Autonomous Hospital Database: a Case Study," Proceedings of the 1st East-European Symposium on Advances in Databases and Information Systems, St. Peterburg, 1997, pp. 285–294.
9.   I. Manolescu, D. Florescu and D. Kossmann,"Answering XML Queries over Heterogeneous Data Source," Proceedings of the 27th International Conference on VLDBs, Roma, Italy, 2001, pp. 241–250.
10.   T. Bray, J. Paoli, C. M. Sperberg-McQueen, and E. Maler (eds.), Extensible Markup Language (XML) 1.0 (Second Edition). W3C Recommendation, October 2000. Latest version is available at http://www.w3.org/TR/REC-xml/
11.   XQuery,2007. http://www.w3.org/TR/xquery/.
12.   J. Ostell,"Databases of Discovery", ACM QUEUE, Vol. 3, No. 3, 2005, pp. 40–48.
13.   L. Seligman and A. Rosenthal,"XML's Impact on Databases and Data Sharing," Computer, Vol. 34, No. 6, 2001, pp. 59– 67.
14.   E. Sciore, M. Siegel and A. Rosenthal,"Using Semantic Values to Facilitate Interoperability among Heterogeneous Information Systems," ACM TODS, Vol. 19, No. 2, 1994, pp. 254–290.
15.   H. Karadimas, F. Hemery, P. Roland and E. Lepage,"DbMap: Improving Database Interoperability Issues in Medical Software using a Simple, Java-XML based Solution," Proceedings of AMIA Symposium, Vol. 7, 2000, pp. 408–412.
16.   C-S. D. Wei, S. Y. Sung, S. J. Doong and P.A. Ng, "Integration of Structured and Unstructured Text Data in a Clinical Information System," Transactions of the SDPS: Journal of Integrated Design & Process Science, Vol. 10, No. 3, September 2006, pp.61–77.
17.   W3C,"Semantic Web Activity," available at http://www.w3.org/2001/sw/.
18.   M.K. Smith, C. Welty and D.L. McGuinness,"OWL Web Ontology Language Guide. W3C Recommendation," February 2004. Latest version is available at http://www.w3.org/TR/owl-guide/.
19.   L. Russel,"New Directions in Semantic Interoperability," The Semantic Interoperability Community of Practice (SICoP) Spec. Conference 2, Building Knowledge-bases for Cross-Domain Semantic Interoperability, April 25, 2007.
20.   L. Baresi, E. Di Nitto and C. Ghezzi,"Toward Open-World Software: Issues and Challenges," IEEE Computer, Vol. 39, No. 10, October 2006, pp. 36–43.
21.   D. F. Brauner, M. A. Casanova, and C. J. P de Lucena,"Using Ontologies as Artifacts to Enable Databases Interoperability," University of Hamburg: Workshop on Ontologies as Software Engineering Artifacts, Hamburg, Germany, July 6, 2004.
22.   A. Dogac, G. Laleci, S. Kirbas, S. Sinir, A. Yildiz and Y. Gurcan,"Artemis: Deploying Semantically Enriched Web Services in the Healthcare Domain," Information Systems, Vol. 31, No. 4, 2006, pp. 321–339.
23.   Health Level 7, Health Level 7 (HL7), 2006, http://www.hl7.org.
24.   D. Dou, P. LePendu, S. Kim and P. Qi,"Integrating Databases into the Semantic Web through an Ontology-based Framework," Proceedings of 3rd International workshop on Semantic Web and Databases, 2006. pp. 54.
25.   J. Davies, A. Duke and Y. Sure,"OntoShare: Using Ontologies for Knowledge Sharing," Proceedings of International Workshop on Semantic Web, at 11th International WWW conference, Honolulu, Hawaii, USA, May 7–11, 2002.
26.   P. Kataria, N. Koay, R. Juric, K. Madani and I. Tesanovic,"Ontology for Interoperability and Data Sharing in Healthcare," Proceedings of the 4th IASTED International Conference on

Advances in Computer Science and Technology, Langkawi, Malaysia, April 2–4, 2008, pp. 323–328.

27. DR2Q,2009. http://www4.wiwiss.fu-berlin.de/bizer/d2rq/spec/index.htm.

28. TopBraid,2010, http://www.topquadrant.com/products/TB_Composer.html.

29. SWRL,2004, http://www.w3.org/Submission/SWRL/.

30. T.R. Gruber,"A Translation Approach to Portable Ontology Specifications," Knowledge Acquisition, Vol. 5, No. 2, 1993, pp. 199–220.

31. G. Klyne, J.J. Carroll, and B. McBride (eds.),"Resource Description Framework (RDF) Concepts and Abstract Syntax," W3C Recommendation (February 2004). Latest version is available at http://www.w3.org/TR/rdf-concepts/.

32. N. F. Noy and D. L. McGuinness,"Stanford Knowledge Systems Laboratory Technical Report KSL-01-05" and"Stanford Medical Informatics Technical Report SMI-2001-0880," March 2001.

33. H.L.R. Chiang, T.M. Barron and V.C. Storey,"Reverse Engineering of Relational Databases: Extraction of an EER Model from a Relational Database," Data & Knowledge Engineering, Vol. 12, No. 2, 1994, pp. 107–142.

34. M. Vermeer and P. Apers,"Object-Oriented Views of Relational Databases Incorporating Behaviour Database," Proceedings of the 4th International Conference on Database Systems for Advanced Applications (DASFAA), Singapore, April 11–13, 1995.

35. M. Li, X.Y. Du and S. Wang,"Learning Ontology from Relational Database, " Proceedings of the 4th International Conference on Machine Learning and Cybernetics, Guangzhou, China, August 18–21, 2005, pp. 3410–3415.

36. V. Kashyap,"Design and Creation of Ontologies for Environmental Information Retrieval," Proceedings of the 12th Workshop on Knowledge Acquisition, Modeling, and Management, Alberta, Canada, 1999.

37. C. Nyulas, M. O'Connor and S. Tu,"DataMaster-a Plug-in for Importing Schemas and Data from Relational databases into Protégé," Proceedings of the 10th International Protégé Conference, Budapest, Hungary, July 15–18, 2007.

38. DataGenie,2007. http://protege.cim3.net/cgi-bin/wiki.pl?DataGenie.

39. C.P. de Laborda and S. Conrad,"Relational.OWL - A Data and Schema Representation Format Based on OWL," Proceedings of 2nd Asia-Pacific Conference on Conceptual Modelling, Newcastle, Australia. CRPIT, 43. Hartmann, S. and Stumptner, M., Eds. ACS, 2005, pp. 89–96.

40. R. Ghawi and N. Cullot,"Database-to-Ontology Mapping Generation for semantic Interoperability," The 3rd International Workshop on Database Interoperability (InterDB 2007), held in conjunction with VLDB 2007, Vienna, Austria.

41. E.F. Codd,"A Relational Model of Data for Large Shared Data Banks," Communications of ACM, Vol. 13, No. 6, June 1970, pp. 377–387.

42. Maedche, A., Motik, B., Silva, N. and Volz, R. (2002). MAFRA - A MApping FRAmework for Distributed Ontologies, Proceedings of the 13th European Conference on Knowledge Engineering and Knowledge Management (EKAW), Siqüenza, Spain, pp. 235–250.

43. J. Walker, E. Pan, D. Johnston, J. Adler-Milstein, D.W. Bates and B. Middleton, "The Value of Health Care Information Exchange and Interoperability," Health Affairs (The Policy J. of the Health Sphere), Vol. 10 January, 2005, pp. 10–18.

44. D. Brailer,"Interoperability: The Key to the Future Health Care System," Health Affairs (The Policy J. of the Health Sphere), Vol. 10 (January), 2005, pp. 19–21.

Chapter 13
CLINICAL APPLICATIONS AND DATA MINING

David E. Robbins and Marco Chiesa

1 Introduction

As the quantity of medical patient data available to the number of physicians increases, Electronic Healthcare Record (EHR) systems become a necessity for providing more reliable and better quality healthcare [1]. The benefit of using EHR's is dramatic enough that several nations, including the United States, have enacted legislation to provide strong incentives encouraging the use of Electronic Healthcare Records (EHR), as well as penalties for failing to use them [2]. These factors combine to make the adoption of EHR over paper-based systems inevitable. As the use of EHR increases, biomedical practitioners should be expected to make the best possible use of the wealth of computable patient data EHR systems will contain.

Simultaneously, an increasing awareness of the need to integrate the research and practice aspects of biomedicine has driven the development of two integration paradigms: research-based practice (RBP) and practice-based research (PBR) [3]. In RBP, emphasis is placed on ensuring biomedical practitioners can incorporate up-to-date medical knowledge in their treatment of patients. Contrastingly, PBR emphasizes the need for practice to inform and drive research; without the practice of medicine, biomedical research as a field would be much smaller. RBP and PBR share a common need, however: generating usable knowledge for one application from data originally generated or collected for a different application.

Data mining provides the solution. Applying data mining tools to EHR systems can allow medical practitioners deeper insight into the patients they serve, while simultaneously providing biomedical researchers with the knowledge required to best serving the practice of medicine. Biomedical researchers use data mining tools to enhance their studies of molecular biology by sifting through large amounts of genomic or proteomic data. Through data mining applications in clinical decision support and literature mining, up-to-date medical research can be applied to medical practices efficiently. Finally, as education and research move into the combinatorics

D. E. Robbins (✉)
Birmingham, AL, USA
e-mail: robbinsd@uab.edu

S. Suh et al. (eds.), *Biomedical Engineering*,
DOI 10.1007/978-1-4614-0116-2_13, © Springer Science+Business Media, LLC 2011

and integration era, where knowledge is generated through automated, systematic, meta-fusion processes [4], applications of clinical data mining become increasingly important.

This chapter provides an introduction to the clinical applications of data mining. We begin with an overview of the data mining field: overall goals, terminology, techniques, and limitations. Next, we briefly examine current applications of data mining to the biomedical field as a whole. An in-depth review of clinical data mining to decision support, group identification, and event detection follows. After a discussion of the current challenges in applying data mining tools to clinical data, we provide concluding remarks.

2   Data Mining Overview

Modern enterprises generate large amounts of electronic information during the course of daily processes. This information contains a wealth of hidden knowledge buried within it. The use of data mining techniques can reveal previously unrealized patterns, trends, and associations controlling the enterprise's operational environment. Regardless of the type of environment (business, healthcare, academic, etc.), the general goal is the same. It is to discover and leverage the value contained in a data store through the application of data mining techniques.

Knowledge Discovery can be defined as the non-trivial extraction of implicit, previously unknown and potentially useful information from data [5]. The use of data mining methods makes up only one stage in the knowledge discovery process. Also, a single application of a data mining technique is not always sufficient. It is common for the data mining stage of the knowledge discovery process to be iterative. Each use of a different data mining method may reveal something new.

2.1   The Knowledge Discovery Process

A general representation of the knowledge discovery process appears in Figure 1. Typically, multiple data sources are integrated into a unified data model. This model is recorded in a data store. Each information source may have separate prerequisite data extraction and transformation steps. Next, the information will normally

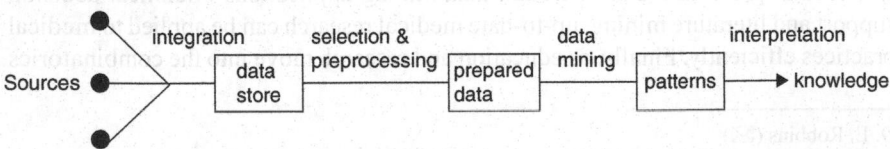

Figure 1. The knowledge discovery process, incorporating data mining.

undergo some preprocessing to clean up or normalize the data values. Any data records with missing parameter values could also be handled at this time. Further, it is common to reduce the number of parameters of interest for each data record through feature selection. After these preprocessing activities, the prepared data is ready to be used as input to the data mining stage. One or more data mining techniques would then be applied, hopefully revealing patterns, trends, or connections in the data. The system operator or a domain expert would then interpret the results from this process. The intent of this procedure is to gain new knowledge about the entities under observation.

2.2 Data values

Since quality data is required for data mining, it makes sense to consider what type of data values are commonly dealt with. Each data record will have a number of parameters. The value for each parameter will represent a feature or characteristic of the entity being considered. For example, if we have a data store containing patient data records, then each patient record might have parameters for the patient's date of birth, weight, admitting diagnosis, allergies, etc.

A record's parameter values will commonly be one of two types. Categorical parameters can only take on one of several discrete values (Boolean parameters are a special case of this). The unit nurse assigned to a patient would be an example of this. Continuous parameters on the other hand can take on any value (possibly within some valid range) along the dimension of interest. The patient's weight is an example of this type of parameter.

2.3 Data types

Regardless of the type of data records and parameters being investigated, the data set can be said to fall into one of two categories. With labeled data sets, each data record contains a specially designated parameter. The intention when using labeled data sets is to take the rules and patterns, which were discovered through the use of data mining and use them to predict the value of the 'special' parameter for data records you haven't seen yet. Data mining performed on labeled data sets where the specially designated parameter is categorical is referred to as classification. It is referred to as regression or numerical prediction if the designated parameter is continuous or more specifically numerical. Overall the knowledge discovery process is referred to as supervised learning when using labeled data sets because each data sample used during training was labeled with the expected output value.

If the data set does not have a specially designated parameter, then it is an unlabeled data set. The goal when using unlabeled data sets is simply to discover whatever useful relationships you can from the data. Common procedures when using unlabeled data sets include looking for relationships that exist between a record's parameter values. It may be possible to predict the likelihood that a parameter will take a particular value based on the value of other parameters in the same data record. These relationships are called association rules. Another common procedure when using unlabeled data sets is to look for clusters in the data records. This involves creating a distance measure to be used to compute 'nearness' between any two data records (points). This is easy to imagine when there are a maximum of three parameters (dimensions) per data record (point), but can also be easily extended to case with a larger number of parameters (dimensions). The knowledge discovery process used with unlabeled data sets is generally called unsupervised learning.

2.4   Data mining techniques

2.4.1   Classification

The data mining technique of classification involves the use of supervised learning when the specially designated attribute is categorical. Classification can be carried out several ways. Some common ones are decision tree induction [6], nearest neighbor [7], naïve Bayes [8], and support vector machines [9]. The intention is to assign new data points to one of several predefined categories or classes. The general approach is to start with a set of training data that has already been labeled with the correct class. Each technique then uses a different procedure to generate a model, which can be used to assign the appropriate label when classifying new data.

2.4.2   Numerical prediction

Supervised learning conducted on continuous attributes is referred to a numerical prediction or regression. Common approaches used are linear regression and multivariate linear regression. The intention is to find the function that models the training set data with the least error. This function can then be used to predict the value of the target attribute for new data [10].

2.4.3   Association Rules

When dealing with data for which there is no attribute that has special importance, one technique for revealing knowledge is to look for association rules. It is usually possible to find relationships between the values of a data set instance's attributes [11]. While the number of possible relationships will be large, most will be of little practical value. A measure of interestingness can be used to identify the subset of more useful relationships. Two common such measures are support and confidence. Support is a measure of the coverage of an association rule over the data set since obviously a rule that only applies to a few cases is of less general use. The value is calculated as a ratio of the number of data instances that are covered by the rule to the total number of data instances. Confidence is a measure of the predictive accuracy of an association rule. This can be calculated as the ratio of the number of instances for which the association rule was correct to the total number of data instances [5].

2.4.4 Clustering

Clustering is an unsupervised learning technique involving the use of a distance measure to find groupings of similarity in the data records. Common algorithms are: k-means, hierarchical, and density [12]. Clustering is also an application of knowledge extraction to a situation where there is no special attribute for the data instances. Here, however, the data instances will be split into groups based on some measure of similarity. Sometimes this is done to gain a greater understanding of the instances of the data set. For instance, to reveal classes which can then be used in classification. Sometimes, however, this is done purely as an intermediate step where the clusters themselves provide some usefulness. For example: data compression [10].

3 Applications of Data Mining to Biomedicine

Clinical decision support represents the original application of data mining techniques to biomedicine. Image and signal analysis also played a key role in introducing biomedical applications of data mining. In recent years, however, these fields have been largely eclipsed by a growing interest in molecular biology, and in particular genomics and proteomics. Epidemiology, biosurveillance, and literature mining have also seen increased interest. Figure 2 presents a concept map depicting the relationships between various biomedical applications of data mining.

3.1 *Molecular Biology*

Molecular biology's applications of data mining center on three data sets: gene sequences, gene expression data, and protein expression data [13]. Although initial work focusing on individual data sets provided excellent results, genomic and proteomic data mining is approaching boundaries as the search spaces involved grow. Literature mining provides a potential method for overcoming these barriers as relationships between genes and proteins are extracted from biomedical publications. Combing this meta-analysis with consensus algorithms and Bayesian priors represents a recent development in the application of data mining to biomedicine.

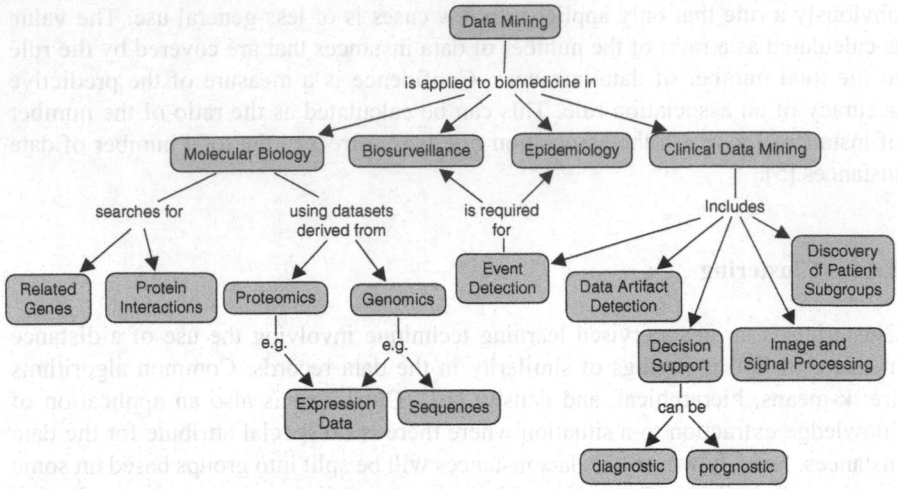

Figure 2. Biomedical Applications of Data Mining.

4   Clinical Data Mining

Clinical data, generally patient data gathered in clinical settings, represents the key source for knowledge discovery in the practice of biomedicine. Data mining tools may be applied to clinical data to provide both predictive and descriptive knowledge about individual patients, specific groups of patients, and of a local, regional, or national population as a whole.

4.1   Clinical Decision Support

For individual patients, clinical data mining is most often used to drive a clinical decision support system. This support may include real-time signal and image analysis. Clinical decision support systems generally make use of data mining tools on their patient models (Figure 3). The measured health parameters of an individual patient, possibly including historical values, can be mined for patterns consistent with either known diseases or probable causes.

The knowledge derived for data mining a patients measured health parameters may be used to provide either a diagnosis or extended into a prognosis. Interest in prognostic (predictive) data mining has increased in recent decades, although diagnostic (descriptive) data mining represented the initial focus of clinical decision support. This emphasis on the predictive elements of clinical decision support has led to the introduction of machine learning [14] and artificial intelligence concepts [15] in support of clinical data mining.

Figure 3. Concept map depicting the patient model of a clinical decision support system.

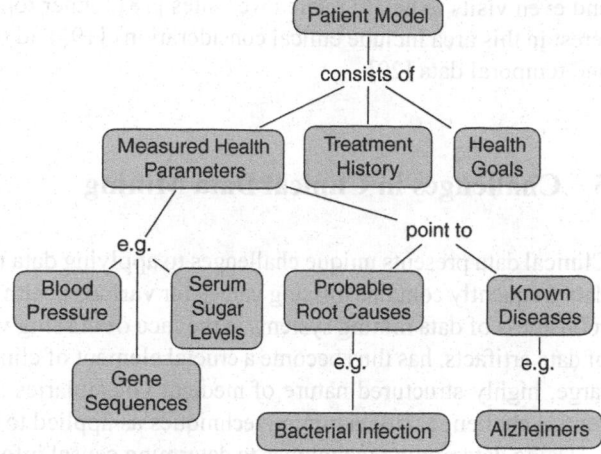

4.2   Subgroup Identification and Classification

Clinical data sets consisting of numerous patients allow data mining tools to discover both homogenous subgroups of patients and events affecting groups of patients. Using large-scale clinical data to derive conclusions regarding the causal effect of therapies represents a seemingly obvious but generally biased and invalid biomedical application of data mining tools [13]. Apparent casual effects in the clinical data may, however, be used to indicate a need for a controlled study of the phenomena discovered.

Identifying homogenous subgroups of patients provides insights into potential risk factors for known diseases or conditions, e.g. infections in children with febrile neutropenia [16]. If all the members of a group share many common traits, subgroup identification may also point to a need for further research into the causal effect of a therapy within a certain patient subgroup (e.g. patients with high blood pressure). Looking for subgroups in patient data represents a combination of both the clustering and classification techniques of data mining.

4.3   Epidemiology and Biosurveillance

Event detection in clinical data sets comprises two primary applications: the tracking of diseases at population levels (epidemiology) and the early detection of biological and chemical attacks (biosurveillance). The knowledge generated by these fields is most applicable to policy makers and government agencies required to respond to outbreaks and potential attacks [17]. Data used for clinical event detection has recently been expanded to include a large variety of sources, including hospital discharge records, prescription and over the counter drug sales, emergency calls, and even visits to health related web sites [18]. Other topics of increased recent interest in this area include ethical considerations [19] and the increased use of spatial and temporal data [20]

5 Challenges in Clinical Data Mining

Clinical data presents unique challenges to applying data mining techniques. Patient data frequently contains missing values for various health parameters. Interest in the robustness of data mining systems in the face of missing values, and in the detection of data artifacts, has thus become a crucial element of clinical data mining. The very large, highly structured nature of medical vocabularies and ontologies also represents a challenge to data mining techniques as applied to patient data.

Using data mining techniques to determine causal information regarding courses of treatments and therapies from clinical data can be misleading. Due to the uncontrolled nature of the patient sample, the knowledge acquired from this form of clinical data mining is neither unbiased nor does it demonstrate patient exchangeability. As a result, biomedical applications of the subfield of literature mining are receiving increased attention, e.g. [21, 22].

Medical data sets tend to come from either research or clinical settings. A data set that has been collected to further medical research efforts will usually be of high quality. There will be few data entry errors or missing values, and a large number of parameter values will be recorded for each record. Unfortunately, this type of data set will probably not contain a large number of records. Data sets that have been generated from normal operations in a clinical setting, on the other hand, will usually contain a very large number of data records. The downside to real-world data collection, however, is that the data will be lower quality (errors, missing values, etc.). This type of data set will often have temporally related sub-records to chronicle a patient's course of treatment throughout their stay and possibly later re-admission. The patient records may contain only simple textual parameters like the results of laboratory tests or they may also contain medical images or electronic health records (EHR) [23].

In dealing with the quality of the data set, handling missing parameter values requires special attention. Missing values in medical data sets may occur as the result of simple data entry mistakes just as with data sets generated from business operations. However, missing data values may also occur for reasons unique to the clinical setting. For instance, if a clinician decides that a certain physical or laboratory exam is not need to diagnose a patient, then the associated parameters would be blank for that patient record. As another example, progress in medical science continually leads to the development of new examinations. As these new procedures enter standard use, parameters to represent their results will appear in newer patient records. Older patient records, however, will not have these parameters [23]. In regards to dealing with missing values, the correct approach will vary. It may be best just to discard the entire record. Another approach is to replace the blank value with the average (for numeric parameters) or most frequent value (for categorical parameters) for that parameter.

A central concern when working with medical data sets that is not usually an issue when dealing with business data is privacy [19]. Since medical records are related to human subjects privacy protection must be taken very seriously. The Health Insurance Portability and Accountability Act (HIPAA) require protecting patient privacy and ensuring the security of medical data [24]. HIPAA violations can result in monetary penalties. An additional difficulty found in medical data sets concerns medical coding. It is common in health care to make use of codes to refer to diagnoses and procedures. Also, there is more than one standard set of codes. These codes will need to be transformed appropriately when extracting data from the source system. In most if not all cases coded values will indicate a categorical parameter.

6   Conclusion

Data mining tools represent a crucial element of an automated knowledge discovery process. Allowing for the discovery of patterns and the prediction of parameter values, these tools give key insights into the abundance of data generated by modern clinical practice. These insights, in turn, drive clinical decision support systems, inform epidemiology and biosurveillance studies, and provide direction to biomedical research as a whole.

As clinical practice moves into the 21st century, the tools provided by clinical data mining will become increasingly important. These tools allow for improved patient care through decision support and subgroup identification. They provide government officials and policymakers with the data needed to protect their populations from outbreaks and attacks. Finally, they provide researchers with the clues needed to ensure the science of medicine best serves the practice of medicine.

References

1. L.C. Burton, G.F. Anderson and I.W. Kues,"Using electronic records to help coordinate care," *The Milbank quarterly*, vol. 82, pp. 457–81, Jan , 2004
2. *Health Information Technology for Economic and Clinical Health (HITECH) Act*, Public Law No. 111-5, Title XIII, 2009, pp. 226–79. Available: http://www.gpo.gov/fdsys/pkg/PLAW-111publ5/pdf/PLAW-111publ5.pdf
3. Epstien, *Clinical Data-Mining: Integrating Practice and Research*, New York: Oxford University Press, 2009.
4. Ertas, T. Maxwell, V.P. Rainey, and M.M. Tanik,"Transformation of higher education: the transdisciplinary approach in engineering," *IEEE Transactions on Education*, vol. 46, May. 2003, pp. 289–295.
5. D. Hand, H. Mannila and P. Smyth, *Principles of Data Mining*, Boston: The MIT Press, 2001.
6. F. Esposito, D. Malerba, G. Semeraro, and J. Kay,"A comparative analysis of methods for pruning decision trees," *IEEE Transactions on Pattern Analysis and Machine Intelligence*, vol. 19, May. 1997, pp. 476–493.

7. J.-G. Wang, P. Neskovic, and L.N. Cooper,"An adaptive nearest neighbor algorithm for clas-sification," *2005 International Conference on Machine Learning and Cybernetics*, IEEE, 2005, pp. 3069–3074.

8. R. Abraham, J.B. Simha, and S.S. Iyengar,"A comparative analysis of discretization methods for Medical Datamining with Naive Bayesian classifier," *9th International Conference on Information Technology (ICIT'06)*, IEEE, 2006, pp. 235–236.

9. P.S. Kostka and E.J. Tkacz,"Feature extraction for improving the support vector machine biomedical data classifier performance," *2008 International Conference on Technology and Applications in Biomedicine*, IEEE, 2008, pp. 362–365.

10. P. Tan, M. Steinback, and V. Kumar, *Introduction to Data Mining*, Addison- Wesley, 2006.

11. S. Brin, R. Motwani, and C. Silverstein,"Beyond market baskets," *Proceedings of the 1997 ACM SIGMOD international conference on Management of data - SIGMOD '97*, New York, New York, USA: ACM Press, 1997, pp. 265–276.

12. A.K. Jain, M.N. Murty, and P.J. Flynn,"Data clustering: a review," *ACM Computing Surveys*, vol. 31, Sep. 1999, pp. 264–323.

13. N. Peek, C. Combi and A. Tucker,"Biomedical data mining," *Methods Inf Med*, vol. 48, no. 3, pp. 225–228, Mar, 2009.

14. L. Nanni, A. Lumini, and C. Manna,"A Data Mining Approach for Predicting the Prgnancy Rate in Human Assisted Reproduction," *Advanced Computational Intelligence Paradigms in Healthcare 5*, S. Brahnam and L. Jain, eds., Berlin / Heidelberg: Springer, 2011, pp. 97–111.

15. L. Nanni, S. Brahnam, A. Lumini, and T. Barrier,"Data Mining Based on Intelligent Systems for Decision Support Systems in Healthcare," *Advanced Computational Intelligence Para-digms in Healthcare 5*, S. Brahnam and L. Jain, eds., Berlin / Heidelberg: Springer, 2011, pp. 45–65.

16. Z. Badiei, M. Khalesi, M.H. Alami, H.R. Kianifar, A. Banihashem, H. Farhangi, and A.R. Razavi,"Risk factors associated with life-threatening infections in children with febrile neu-tropenia: a data mining Approach," *Journal of Pediatric Hematology/Oncology*, vol. 33, Jan. 2011, p. e9-e12.

17. H. Rolka and J. O'Connor,"Real-Time Public Health Biosurveillance," *Infectious Disease In-formatics and Biosurveillance*, R. Sharda, S. Voß, C. Castillo-Chavez, H. Chen, W.B. Lober, M. Thurmond, and D. Zeng, eds., Springer, 2011, pp. 3–22.

18. E. Koski,"Clinical Laboratory Data for Biosurveillance," *Infectious Disease Informatics and Biosurveillance*, C. Castillo-Chavez, H. Chen, W.B. Lober, M. Thurmond, and D. Zeng, eds., Berlin / Heidelberg: Springer, 2011, pp. 67–87.

19. J. Collmann and A. Robinson,"Designing Ethical Practice in Biosurveillance," *Infectious Disease Informatics and Biosurveillance*, R. Sharda, S. Voß, C. Castillo-Chavez, H. Chen, W.B. Lober, M. Thurmond, and D. Zeng, eds., Springer, 2011, pp. 23–44.

20. T. C. Chan and C. C. King,"Surveillance and Epidemiology of Infectious Diseases using Spa-tial and Temporal Clustering Methods," *Infectious Disease Informatics and Biosurveillance*, R. Sharda, S. Voß, C. Castillo-Chavez, H. Chen, W.B. Lober, M. Thurmond, and D. Zeng, eds., Springer US, 2011, pp. 207–234.

21. L.J. Jensen, J. Saric, and P. Bork,"Literature mining for the biologist: from information re-trieval to biological discovery," *Nature reviews. Genetics*, vol. 7, Feb. 2006, pp. 119–29.

22. V.P. Gurupur and M.M. Tanik,"A System for Building Clinical Research Applications using Semantic Web-Based Approach," *Journal of medical systems*, Feb. 2010.

23. S. Tsumoto, "Problems with mining medical data," *Computer Software and Applications Conference, 2000. COMPSAC 2000. The 24th Annual International*, pp.467–468, 2000.

24. E. Poovammal and M. Ponnavaikko, "Task Independent Privacy Preserving Data Mining on Medical Dataset," *Advances in Computing, Control, & Telecommunication Technologies, 2009. ACT '09*, pp.814–818, Dec. 2009.

Chapter 14
IMAGE PROCESSING TECHNIQUES IN BIOMEDICAL ENGINEERING

Reza Hassanpour

1 Introduction

Advances in biomedical imaging technology have brought about the possibility of non-invasive scanning the structures of the internal organs and examining their behavior in healthy and disease states. 2D/3D ultrasound imaging, high resolution multi-slice computed tomography, and endoscopic imaging provide valuable information about the functioning of the human body. These advances have made the medical imaging an essential component in many fields of biomedical research such as generating 3D reconstructions of viruses from micrographs or studying regional metabolic brain activities. Clinical practice also benefits from the data provided by biomedical imaging modalities. Detection and diagnosis of cancer for instance, is carried out by using multi-slice computer tomography or magnetic resonance imaging. These benefits on the other hand, have triggered an explosion in the amount of biomedical images obtained daily from the image acquisition modalities. Automatic processing and interpretation of these images through image processing methods therefore has become unavoidable. The analysis and interpretation of complicated or unexpected results require deep understanding of the underlying theory and methods involved in image processing beside the medical physics. Biomedical image processing is an interdisciplinary field combining biomedical engineering and computer science. The first class of image processing operations for biomedical application are the fundamental techniques intended to improve the accuracy of the information obtained from the imaging modality. These techniques which include adjusting the brightness and contrast of the image, reduce image noise and correcting for imaging artifacts, generally involves only basic arithmetic operations.

The second class of common image processing operations that are often used to reduce or enhance image details are called convolution operations. Image smoothing and sharpening algorithms are the two most common convolution based algorithms.

R. Hassanpour (✉)
Ankara, Turkey
e-mail: reza@cankaya.edu.tr

S. Suh et al. (eds.), *Biomedical Engineering*,
DOI 10.1007/978-1-4614-0116-2_14, © Springer Science+Business Media, LLC 2011

Extracting information in a higher level by segmentation and classification requires emphasizing distinguishing features while suppressing minor differences. These types of operations are always preceded by convolution operations for enhancing discriminating properties such as organ edges, while reducing noise or texture effect. Applicability and usefulness of many of these algorithms are directly related to the characteristics of the image accusation modality. Hence, in the following section, we introduce the main features of biomedical imaging systems to emphasize the differences between biomedical image processing and the image processing in general. We further introduce the imaging processing tools and techniques applied to the biomedical images.

2   Biomedical Imaging

Biomedicalimaging is a discipline that involves the use of technology to take images from the inside of the human body in as non-invasive way as possible. These images can provide an insight into the human body and are used in diagnostics and in routine healthcare. The study of the biomedical images can be used to identify unusual cases inside the body, such as broken bones and tumors. The type of the information being sought determines the required technology and imaging technique. In the following sections the most important techniques are introduced.

2.1   *Projectional Radiography*

In x-ray imaging or radiography, an image as a 2D projection of a 3D body is produced by letting x-ray beams pass through the body. [1, 2]. This mode of imaging is referred to as projection or planar radiography. The x-ray beams are attenuated by a factor depending upon the attenuation factor of the material between the beam source and the sensing surface. Generally the material type changes along the x-ray beam path and hence the attenuation is obtained by the integral of the linear attenuations factors along the x-ray path as shown in equation 1.

$$\int_{path} \mu(x,y)ds = \ln\left(\frac{N_i}{N_o}\right) \tag{1}$$

Where $\mu(x,y)$ is the attenuation factor at coordinate x and y of the cross-sectional plane, N_i is the number of input photons and N_o the corresponding photons exiting the object. The integration is along the x-ray path and the beam is assumed to be mono-energetic. As an example, projectional radiograph of lungs is used for detecting lesions in the lungs, or fracture of the ribs or the spinal column.

2.2   *Magnetic Resonance Imaging (MRI)*

A nucleus with an odd number of protons or an odd number of neutrons has an inherent nuclear spin. These atoms have magnetic moment with a vector of random orientation. In the presence of a magnetic field however, some of these atoms orient along the axis of the magnet and precess about it with a rate proportional to the power of the magnetic field (Figure 1).

Next an RF electromagnetic pulse is applied to perturb the aligned atoms and make their precession axis deviate from the field axis. When the RF perturbation is removed, the perturbed atoms return to their unperturbed states emitting energy in the form of RF signals. An RF detector system is used to detect the RF signals emitted from the perturbed atoms. The RF signal measured outside the body represents the sum of the RF signals emitted by active atoms from a certain part or slice of the body. Two important measured parameters are the time constant of longitudinal relaxation named T_1, and the decay time constant of transverse magnetization or T_2. Longitude relaxation is the relaxation of the component of the magnetization vector along the direction of the external magnetic field. Transverse magnetization applying a 90° RF pulse causes to orient the net magnetization vector in the plane perpendicular to the external magnetic field. MR images are characterized with a lower resolution and quality compared to CT images. They also suffer from artifacts such as non-uniform distribution of gray levels values at the regions close to RF detector. However, since MRI imaging does not expose the patient to ionizing radiation, it does not pose safety hazards and is preferred particularly during the first stage of examinations. MRI is also used in studying the functions of organs (fMRI) by measuring the amount of change in oxygen [3].

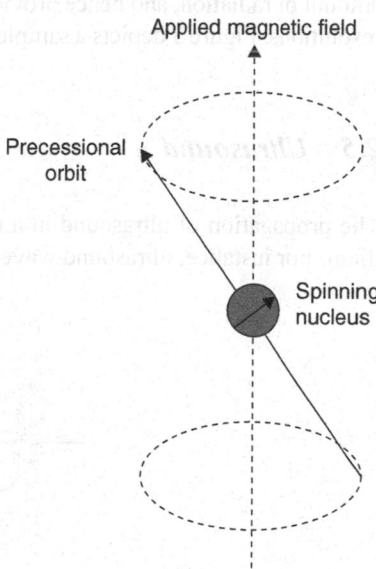

Figure 1. Precession
Vector and Magnetic Field
Directions.

2.3 Nuclear medicine

In nuclear medicine imaging, the gamma ray photons emitted from a radiopharmaceutical are detected and measured by gamma cameras. The radiopharmaceutical is administered into the patient's body either by injection, inhalation, or orally depending on the type of the examination. The radiopharmaceutical is then absorbed by the organ of interest. The localization of radiopharmaceutical in the organ is observed and examined by the practitioners. Positron Emission Tomography (PET) can produce 3-dimensional tomography images using the pairs of gamma rays emitted by the radiopharmaceutical material. Nuclear medicine is used in examining the physiological functions of organs such as lungs, heart, liver, and spleen.

2.4 Computed Tomography (CT)

Computer Tomography (CT) is based on the same principal as projectional radiography. The x-ray beam is detected by a series of surrounding detectors around the patient. The attenuation of each point is computed by solving simultaneously the set of linear equations obtained from each sensor [4]. Inverse Radon transform is used for fast computations. More recent technologies are based on Helix CT scanners where the patient or the beam gun moves along the axis passing through the patient body from head to feet while rotating around the patient's body. In this way the x-ray beam gun travels in a helix shaped trajectory. Figure 2 depicts a simple view of the helical computed tomography scanner.

The main advantage of helical CT technology is in exposing the patient to less amount of radiation, and hence providing the possibility of taking images with higher resolutions. Figure 3 depicts a sample CT image taken from the abdomen of a patient.

2.5 Ultrasound

The propagation of ultrasound in a medium is a function of the density of the medium. For instance, ultrasound wave travels with a speed of 330m/s in air, 1540 m/s

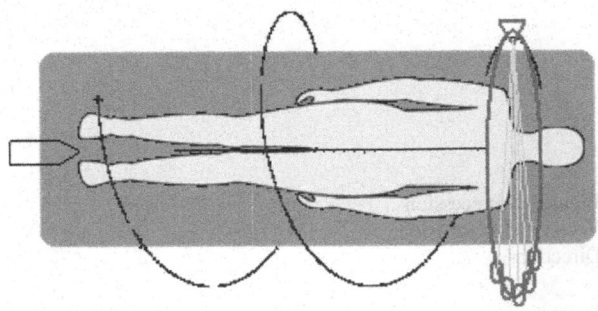

Figure 2. Helical Computed
Tomography Scanner.

Figure 3. Sample Abdominal CT Image. Reprinted with the Courtesy of Hacettepe University Hospital.

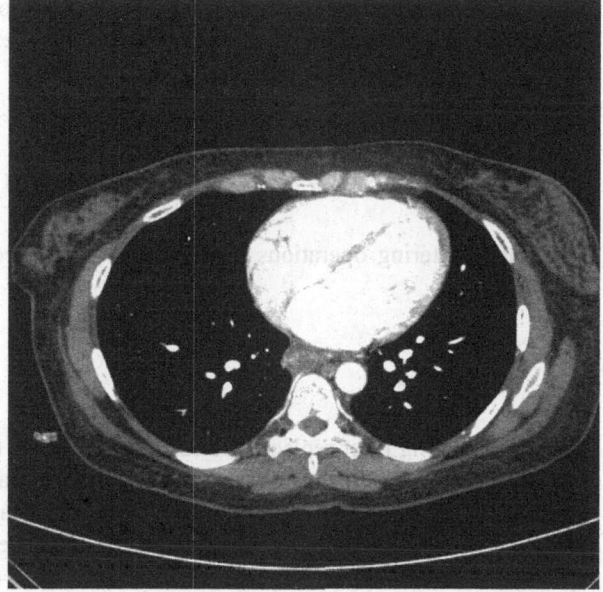

in soft tissue, and 3300 m/s in bone. This property of ultrasound is used for determining the tissue type in human body. Ultrasonography is carried out in different modes. In amplitude mode (A mode) the amount of attenuation in the signal together with the propagation delay time are measured. Propagation delay time and echo depth are measured in M mode, which is used for studying motion as in cardiac valve motion analysis. Ultrasonography images suffer from quality and resolution; however, non-ionizing nature of these images make them widely used biomedical-imaging tool [5].

3   Image Enhancement

Despite the recent improvements in biomedical imaging, the image acquisition process is not completely error free. Low resolution, sampling and quantization errors, physiological artifacts due to patient movements, and artifacts due to the physical properties of the imaging system such as beam hardening are some of the sources for image acquisition impairments. Reducing the effect of these impairments by pre-processing the images can improve the results of the higher level processing stages such as organ segmentation.

3.1   Noise Removal

Noise is the result of a perturbation in the signal. Removing noise is almost impossible however; reducing the effect of noise can be done if a good guess of the noise type is made. Spatial filtering which refers to the operations performed on the pixels of an image in a neighborhood called a window is generally used for noise reduction. Equation 2 shows the result of applying the spatial filter $w(x,y)$ of size a×b to the image $f(x,y)$.

$$g(x, y) = \sum_{s=-a/2}^{a/2} \sum_{t=-b/2}^{b/2} w(s,t) f(x+s, y+t) \qquad (2)$$

Two spatial filtering operations namely mean and median filters are introduced below.

3.1.1 Mean Filter

Mean or averaging filter replaces the current pixel with the average of pixel values in its neighborhood. The mean filter is a low-pass filter and can reduce the effect of noise which is generally of high frequency content. Mean filters blur the edges and corners of the objects which correspond to the high frequency contents. Figure 4 depicts the result of applying mean filter of size 5×5 to a CT image.

3.1.2 Median Filter

Spurious noise known as salt and pepper noise shows itself as randomly occurring black and white pixels. Applying low-pass filters such as averaging filter results in blurring the boundaries. A more suitable filter in this case is the median filter. Median filter replaces a pixel by the middle value of the pixels in its neighborhood, after sorting them numerically [6]. The idea behind the median filter is that a single pixel that is different from its neighbors will fall either to the beginning or the end of the sorted list. On the other hand, a pixel has equal number of darker and lighter neighbors on average. Median filter preserves the edge pixels however, tends to round the sharp corners of objects. Figure 5 depicts a CT image after superimposing salt and pepper noise, and the result of applying median filter.

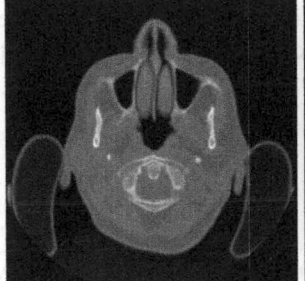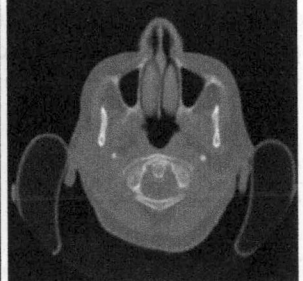

Figure 4. Original CT image
(left), Result of apply-
ing the mean filter (right).
Reprinted with the Courtesy
of Hacettepe University
Hospital.

Figure 5. Noisy CT image (left), Result of applying the median filter (right). Reprinted with the Courtesy of Hacettepe University Hospital.

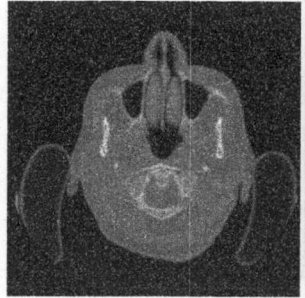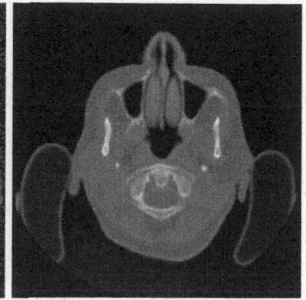

3.2   *Histogram Transformation*

The histogram of an image is a function mapping a gray level to the number of pixels having that gray level value [6]. The number of gray levels in an image is a dependant on the quantization type, and the physical properties of the imaging tools. However, the distribution of the gray levels is an important factor in the quality of the image and can be improved through post-processing. Let us assume $p(i)$ denotes the probability of the occurrence of gray level i in the image. The cumulative distribution of $p(i)$ is given by Equation 3.

$$CDF(i) = \sum_{j=0}^{i} p(j) \tag{3}$$

A histogram transformation which converts the cumulative distribution of the image into a linear distribution is called histogram equalization transform. The transformed gray level values of the pixels are given as shown in Equation 4.

$$y = T(x) = CDF(x) \tag{4}$$

Figure 6 depicts a dark image and its histogram. The equalized histogram and the corresponding image are depicted in Figure 7. As it is clear from the equalized histogram, the exact linearization of the histogram is not possible when the histogram values are discrete. However, a close to linear equalization is always possible. In biomedical applications, especially in normalized images such as normalized CT

Figure 6. A sample CT image and its histogram. Reprinted with the Courtesy of Hacettepe University Hospital.

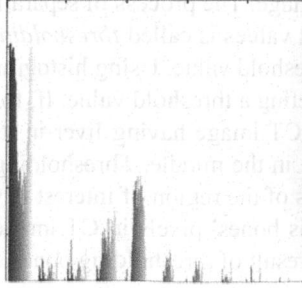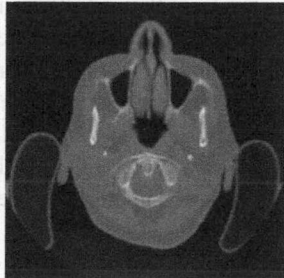

Figure 7. The equalized histogram and the corresponding image.Reprinted with the Courtesy of Hacettepe University Hospital.

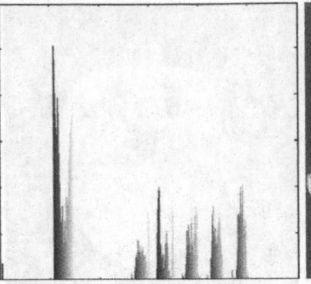 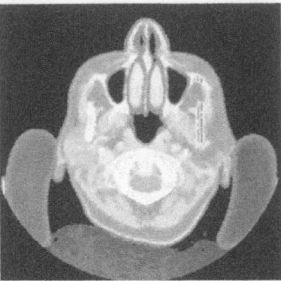

images, applying histogram equalization can result in disturbing the measured values. It should be noted that in normalized CT images, pixel values reflect the material type independently from the imaging modality settings.

4   Detection of Regions of Interest

Enhancing the features that a physician or a radiologist is interested in can greatly facilitate the detection, diagnosis, or interpretation processes. As mentioned in section 2, biomedical imaging modalities are capable of producing images with much higher number of measured levels than the distinguishing capability our visual system. This can lead to cases of unnoticed details in biomedical images. Thorough scanning of the images in different windowing adjustments can be tedious and time-consuming. Enhancing features like corners, edges, and straight lines can be of large importance in the screening of images. These features are also used during the higher level processes such as segmentation, or registration.

4.1   *Thresholding*

The images produced by biomedical imaging modalities reflect the physical properties of tissue or the object. Similar objects, therefore, are expected to be represented using close pixel values [7]. This feature provides a method to separate an object or tissue from the rest of the image. The process of separating the region of interest in an image based on the pixel values is called *thresholding,* with the most important step being selection of a threshold value. Using histogram analysis is the most generally used method for selecting a threshold value. If, for instance, the image being processed is an abdominal CT image having liver in its largest cross-section; the histogram will show a peak in the middle. Thresholding is especially useful when the range of intensity values of the region of interest is very different from the rest of the image. An example is bones' pixels in CT images. Figure 8 depicts an abdominal CT image and the result of thresholding bones.

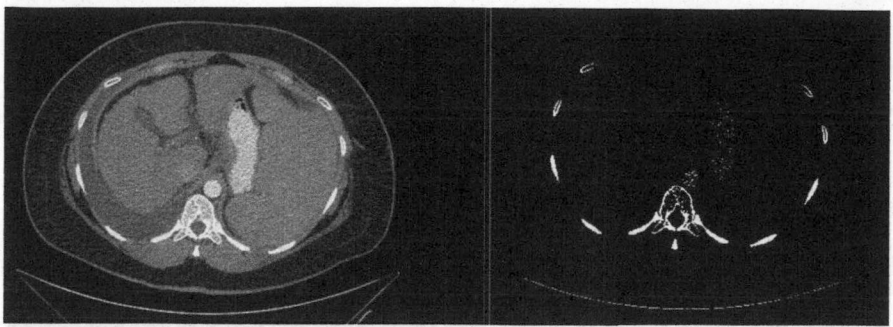

Figure 8. An abdominal CT image (left) and the corresponding thresholded bone areas.Reprinted with the Courtesy of Hacettepe University Hospital.

4.2   Edge Detection

One of the approaches to extract the region of interest is using the boundaries of the objects. Boundaries define the shape of objects and help the viewer to detect and analyze them. Edges correspond to high frequency content of the image so the edge detection approaches are, in fact, high pass filters. First derivative, zero crossing of the second derivative, and gradient operators are the most common approaches used for edge detection. An important problem in high-pass filter operators is their sensitivity to noise. A smoothing operation therefore is combined with edge detection operation to suppress the noise effect. Figure 9 depicts a Sobel filter for edge detection in x and y directions. The size of the window determines the amount of smoothing applied. Laplacian of Gaussian (LoG) is a second derivative edge detection which applies the Gaussian smoothing for noise effect reduction [8]. Specifying the Gaussian function by:

$$g\left(x,y\right) = -e^{\left(-\frac{x^2+y^2}{2\sigma^2}\right)} \tag{5}$$

Then LoG is given by:

$$\nabla^2 g\left(x,y\right) = LoG\left(r\right) = -\frac{r^2 + 2\sigma^2}{\sigma^4} e^{\left(-\frac{r^2}{2\sigma^2}\right)} \tag{6}$$

Figure 9. Horizontal (left) and vertical (right) Sobel operator masks.

1	2	1
0	0	0
−1	−2	−1

1	0	−1
2	0	−2
1	0	−1

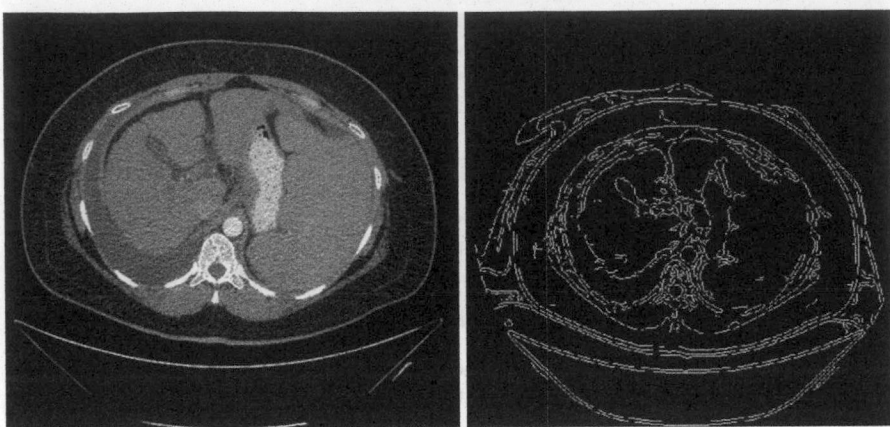

Figure 10. An abdominal CT image (left), the result of Canny edge detection (right). Reprinted with the Courtesy of Hacettepe University Hospital.

Where $r = \sqrt{x^2 + y^2}$. Canny edge detector operator is a first derivative based edge detector which combines edge gradients applied in horizontal, vertical, and diagonal directions. Canny operator combines Gaussian smoothing with gradient operation for edge detection, however it eliminates the blurring effect of Gaussian smoothing by suppressing non-maxima values in gradient results. Figure 10 depicts the result of applying Canny edge detector to an abdominal CT image.

4.3   Straight Line Detection

Simple shapes such as curves or straight lines can be used as features for detecting higher level structures. These shapes are composed of edge points, therefore detecting them is always performed after an edge detection step. Detecting curves or lines may fail if some of the edge pixels are missing or the geometric properties of the shapes are slightly violated. A voting method for counting the number of pixels satisfying the geometric requirements of the line or curve can solve the above mentioned problems. The detection procedure through voting which is referred to as Hough transform, is performed in parameter space. In case of line detection, the slope of the line and its intersection with x axis is used as parameters. The maxima of an accumulator in the parameter space represent the parameters of the detected lines. One problem with using slope and intercept points as parameters is that the vertical lines cannot be represented as their slope tends to infinity. Other sets of parameters such as the distance of the line to the origin (τ), and its angle with the detected line (θ) can be used [6]. The equation of the line becomes as:

$$y = \left(\frac{\cos \theta}{\sin \theta} \right) x + \frac{\tau}{\sin \theta} \tag{7}$$

Figure 11 depicts a line in the parameter space.

Figure 11. The Hough transform parameter space.

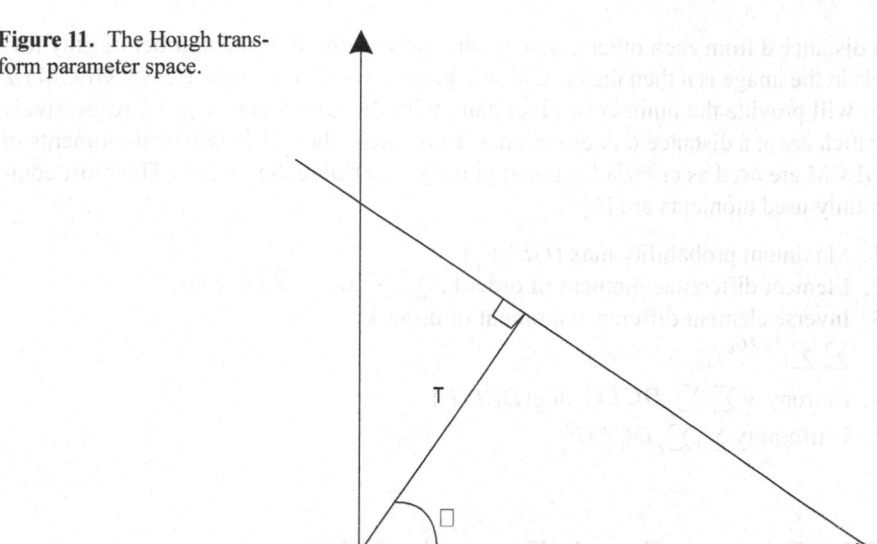

Origin

5   Texture Analysis

Texture has been described by repeating similar structures in a surface which are considered as a descriptive property of that surface. The texture types can be classified by five different properties in the psychology of perception: coarseness, contrast, directionality, line-likeness and roughness [ref]. The regularity of the repetition of the texture structure elements is also a criterion for classifying textures into two broad group of (quasi-)regular and irregular categories. Regular repetition of texture structures can be described using statistical measures such as mean, variance and entropy. These statistical measures can be also analyzed in the frequency domain. Regular textures, however, are more common in biomedical images. Examples are the fibrous structure of muscles, repeated skin patterns, and blood vessels present in medical images. Hence we review the stochastic texture analysis methods in the following sub-sections.

5.1   Statistical Texture Analysis

Texture is described by repeating similar structures, as mentioned earlier. Therefore, the statistical distribution of gray levels in the image space can be used for quantitatively measuring textures. The most commonly used method for statistical analysis of textures is gray level co-occurrence matrix (GLCM) proposed by Haralick et al. [6,9]. GLCM measures the probability of the occurrence of gray level pairs (l_1, l_2) at a distance d from each other given by direction vector $\vec{v}$. If the number of gray levels in the image is n then the GLCM will have a size of n×n and the entry GLCM (a, b) will provide the number of pixel pairs with the gray levels a and b respectively which are at a distance d in direction $\vec{v}$ from each other. The statistical moments of GLCM are used as criteria for quantitatively describing the texture. The most commonly used moments are [6]:

1. Maximum probability max ($GLDC_{ij}$)
2. Element difference moment of order k, $\sum_i \sum_j (i - j)^k DCLG_{ij}$
3. Inverse element difference moment of order k,
   $$\sum_i \sum_j {}^{DCLG_{ij}} /_{(i-j)^k} \quad i \neq j$$
4. Entropy $-\sum_i \sum_j DCLG_{ij} \log(DCLG_{ij})$
5. Uniformity $\sum_i \sum_j DCLG_{ij}^2$

5.2 Frequency Domain Texture Analysis

If a texture is spatially periodic, its power spectrum will show peaks at the corresponding frequencies. These peaks can be used as characteristics of the corresponding texture. Coarse or fine texture types that correspond to low and high frequencies are also detectable by analyzing the power spectrum of the image [9]. Power spectrum of an image is defined as $|F|^2$ where F is the Fourier transform. Frequency features can be analyzed in radial and angular bins. The radial bins consider the coarseness of the texture and measure the energy on a circular strip as shown in Figure 12.

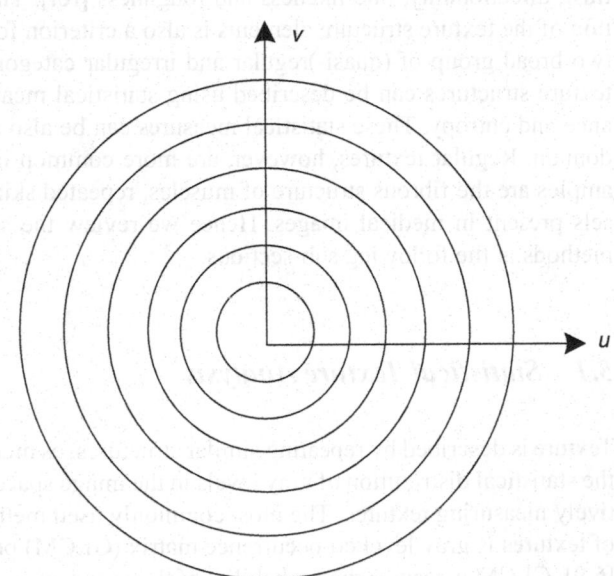

Figure 12. Radial bins of power spectrum.

Figure 13. Angular bins of
power spectrum.

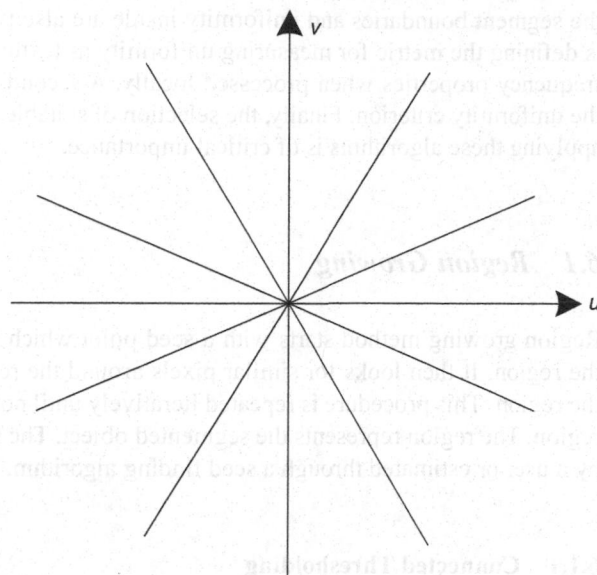

The radial bin features are calculated by:

$$V_{r1,r2} = \iint |F(u,v)|^2 du\, dv \tag{8}$$

Where r_1 and r_2 are the inner and outer radius of the radial bin respectively.

The angular features are found in a similar way by:

$$V_{\theta1,\theta2} = \iint |F(u,v)|^2 du\, dv \tag{9}$$

Where θ_1 and θ_2 are the angles limiting the angular strip. Figure 13 depicts the angular bins in power spectrum.

A combination of both angular and radial partitioning can provide information about the coarseness and the direction of the texture elements.

6   Segmentation

Precise detection of the boundaries of objects in an image makes representation, analyzing, and visualization of the images easier. Segmentation is the process of partitioning an image into disjoint sets of pixels, where each set is the representative of an object and is called a segment. A meaningful segmentation of an image requires consideration of the properties which distinguish objects in the image. These properties can be based on the discontinuities such as edges, or uniformities such as the internal region of an object or organ. A hybrid method requiring discontinuity at the segment boundaries and uniformity inside are also possible. An important task is defining the metric for measuring uniformity as texture elements that show high frequency properties when processed locally. A second factor is the threshold for the uniformity criterion. Finally, the selection of suitable initial or seed points when applying these algorithms is of critical importance.

6.1   Region Growing

Region growing method starts with a seed point which is the initial definition for the region. It then looks for similar pixels around the region and appends them to the region. This procedure is repeated iteratively until no more point is added to the region. The region represents the segmented object. The seed point can be provided by a user or estimated through a seed finding algorithm.

6.1.1   Connected Thresholding

Segmentation by region growing can be greatly simplified if a thresholding step is performed first. As the result of the thresholding is a binary image, the region growing will attract the foreground pixels connected to the seed point. Region growing after thresholding or connected thresholding eliminates the pixels from other objects or organs having the same intensity or texture features but being apart from the object of interest. By marking the foreground pixels, it would be possible to start new region growing processes and detect all connected components. A clue about the approximate location or size of the object of interest will then help to choose the right component.

6.1.2   Splitting and Merging

Finding a seed point in region growing is essential for the success of the segmentation. If the whole image is to be segmented into connected regions with similar properties, split and merge algorithm can be used [9]. Assuming the image is a square matrix of pixels, the splitting and merging of the regions are repeated recursively. A function H(R) determines whether a region R is homogeneous or not and returns true if all neighboring pairs of points in R are such that $f(x) - f(y) < T$ where T is a threshold, and false otherwise. The split and merge algorithm is as follows:

Algorithm: Split and Merge

1. Pick a region R and apply homogeneity property test using H. If H(R) = false then split the region into four sub-regions. If joining any pair of the new sub-regions gives a region with $H(R_i \cup R_j)$ = true then merge them into a single region. Stop when there is no further region to split or merge.
2. If there are any neighboring regions Ri and Rj (perhaps of different sizes) such that $H(R_i \cup R_j)$ = true, merge them.

6.2   *Level Sets*

The 3D object is assumed to be cut by a plane and the boundary is the intersection of this plane and the 3D surface. Any change in the boundary can be described as a movement in the normal direction to the cutting plane [10]. The 3D surface and the cross sectional curve are called the level set function Φ and the zero level set x, respectively. The level set function is then evolved under the control of a differential equation. It is clear from the definition that all points on the zero level set x(t) are obtained when the level set function is zero and hence should satisfy the following equation.

$$\Gamma\left((x),t\right)=\phi\left(x,t\right)=0 \tag{10}$$

Here Γ is the contour extracted by setting the level set function equal to zero. Considering Φ (x, t)=0 and differentiating both sides we get

$$\phi_t + \nabla\phi(x(t),t).x'(t)=0 \tag{11}$$

The term $x'(t).\nabla\phi/|\nabla\phi|$ determines the propagation speed of the contour. Level sets can be used for describing any arbitrarily complex shape and modeling the topological changes such as merging and splitting implicitly. Figure 14 depicts liver area segmented in an abdominal CT examination.

The two main drawbacks of the level set method are selecting a proper function for controlling the evolving speed of the boundary, and finding a proper seed or initial point. The evolving speed should be adjusted so that it has the maximum value of 1 at uniform regions inside the region of interest (anatomical structures for instance) and approaches zero at the boundaries. The initial points to start evolving of the level set towards the boundaries is estimated from the previous steps in multiple slice image sets, and from the a priori information such as approximate size or location of the region of interest at the single image cases.

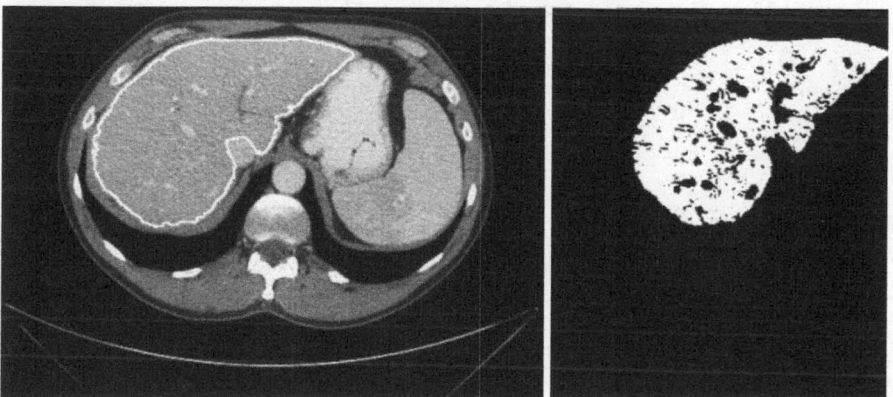

Figure 14. Liver area segmented in an abdominal CT image. Reprinted with the Courtesy of Hacettepe University Hospital.

7   Concluding Remarks

This chapter has focused on the techniques used in the processing of biomedical images. These techniques have been effective in many applications and are commonly used in practice. Typically, the techniques presented in this chapter are the most common methods used in image enhancement and noise removal, feature extraction and low level processing, and segmentation and high level processing. Further processing with more advanced methods and higher complexity may be required for extracting much more detailed information.

References

1. Cho ZH, Jones JP, and Singh M. Foundations of Medical Imaging. Wiley, New York, NY, 1993.
2. Huda W and Slone R. Review of Radiologic Physics. Williams and Wilkins, Baltimore, MD, 1995.
3. Huettel S.A., Song A.W., and McCarthy G., Functional Magnetic Resonance Imaging, Sinauer Associates, 2004.
4. Herman, G. T., Fundamentals of computerized tomography: Image reconstruction from projection, 2nd edition, Springer, 2009.
5. Novelline, Robert, Squire's Fundamentals of Radiology, 5th edition, Harvard University Press, 1997.
6. Gonzalez R.C. and Woods R.E. Digital Image Processing. Prentice Hall, 3rd edition, Upper Saddle River, NJ, 2007.
7. Sezgin M. and Sankur B., Survey over image thresholding techniques and quantitative performance evaluation, Journal of Electronic Imaging 13(1), 146–165, 2004.
8. Haralick R. and Shapiro L. Computer and Robot Vision, Vol. 1, Addison-Wesley Publishing Company, 1992.
9. Ballard D. H., Brown C. M., Computer Vision, Prentice Hall; first edition,1982.
10. Sethian J., Level Set Methods and Fast Marching Methods: Evolving Interfaces in Computational Geometry, Fluid Mechanics, Computer Vision, and Materials Science, Cambridge University Press, Cambridge, UK, 2003.

Chapter 15
PROBABILISTIC TECHNIQUES IN BIOENGINEERING

S. Ekwaro-Osire and T. H. Jang

1 Introduction

Research in biomechanics endeavors to develop scientific knowledge on injury in order to prevent accidental injury and minimizes the damage caused by the accidents. Accidental injuries can happen to anyone with devastating effects on the individual's quality of life. Injury biomechanics is one of the most important fields for understanding accidental injury of the human body. Evaluation of the mechanism of injury is vital for a complete understanding and treatment of injuries. Knowledge of the injury mechanism can also be critical to choosing the treatment and management of injuries.

Biomechanical research has been conducted using various methods, of which the computational modeling method is one of the most widely. Deterministic analysis generally has been conducted in various bioengineering fields to understand human body mechanism and to verify the related injuries. In this type of analysis, deterministic values were assigned to define the mechanical properties of biological structures. In assigning material properties, the best numerical fit of a material property set was selected after comparing response with experimental results.

In computational analysis, however, one of the most important points is that the human body is a complex biological structure, and there are inevitably some elements of uncertainty in the basic design parameters, such as material properties, injury tolerances, and loadings. Uncertain parameters may come from measurement error, environment effects, and so on [1]. Furthermore, these uncertainties can produce significant unexpected influences in the analysis of biological structures and it is important to evaluate the influence of such uncertainties in analyzing bioengineering structures [2].

The probabilistic structural analysis is the way to overcome the difficulties associated with uncertainties in bioengineering structures. For the realistic and accurate analysis, probabilistic analysis method should be employed to consider uncer-

S. Ekwaro-Osire (✉)
Lubbock, TX, USA
e-mail: stephen.ekwaro-osire@ttu.edu

S. Suh et al. (eds.), *Biomedical Engineering,*
DOI 10.1007/978-1-4614-0116-2_15, © Springer Science+Business Media, LLC 2011

tainties (or variability) contained in biological structures. The probabilistic injury analysis is also vital to reducing the likelihood of occupant injury by identifying and understanding the primary injury mechanisms and the important factors leading to injuries. This analysis method can be applied in broad purpose of applications that minimize the probability of injury. The objective of this chapter is to provide an overview of the probabilistic techniques in bioengineering.

2   Probabilistic Analysis

2.1   Uncertainty

The high performance computers combined with probabilistic theories make it possible to find numerical solutions of analyses of complex biological systems involving uncertainty (or randomness). This approach has been introduced and broadly employed in science and various engineering fields. Uncertainty may be due to errors in physical observation, statistical approximations, and definition of modeling parameters [3]. An accurate representation of uncertainties for a given system is critical because the uncertainty may yield different interpretations of the response of given systems [4]. In probabilistic analysis, reliable data is required to make accurate estimation of the uncertainties.

2.2   Random Variable

Any possible outcome of a random experiment can be identified numerically using random variables [5]. Random variable is the real numerical valued function defined over a sample space and the value of a random variable represents a distinct event. As shown in

Figure 1, random variable is the function, $X(s)$ that transforms the random phenomenon over the sample space, S, into the real number, x. A random variable may be discrete if it can assume a finite number of possible value or continuous if it can assume any value in some interval or intervals of real number. Random variables are defined by a mean, standard deviation and probabilistic density function (PDF).

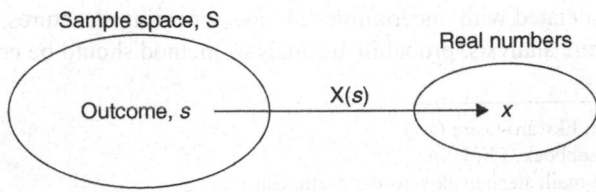

Figure 1. Depiction of a random variable.

2.3   Probability of Injury

An injury function, Z(X), can be defined for a biological structure as

$$Z(X) = T - R(X) \qquad (1)$$

where T is an *ad hoc* injury tolerance, $R(X)$ is the structure response due to various loads, and X is the vector of random variables. It is difficult to define an injury tolerance because of uncertainty in biological structure and the errors involved in extracting the material mechanical factors [6]. Often an injury tolerance value is selected based on the prevailing PDF in a given research. Using the injury function, the limit state function, $g(X)$, for the injury can be defined as

$$g(X) = Z(X) - z_0. \qquad (2)$$

The value z_0 is a particular value of $Z(X)$. The function $g(X)$ delineates the boundary between the safe and unsafe states. The limit state function also referred to as the performance function; define the boundary between the safe and failure regions in the design parameter space. This function plays a central role in the development of reliability analysis methods. An explicit form of limit state function can be extracted for simple problems. In the real world, however, it is impossible to extract the explicit form of this function. Therefore, the function is often expressed in an implicit form and numerical or approximation methods are used for the analysis.

Figure 2 depicts the concept of limit state function. If the limit state value is less than zero, the cervical spine is considered injured. Therefore, the probability of injury, p_f, can be formulated as [3]

$$p_f = P[g(X) \le 0]$$
$$= \int \cdots \int_{\Omega} f(X)dX \qquad (3)$$

where $f(X)$ is the joint probability density function and Ω is the injury region. The probability of injury is the probability of $g(X)$ less than zero.

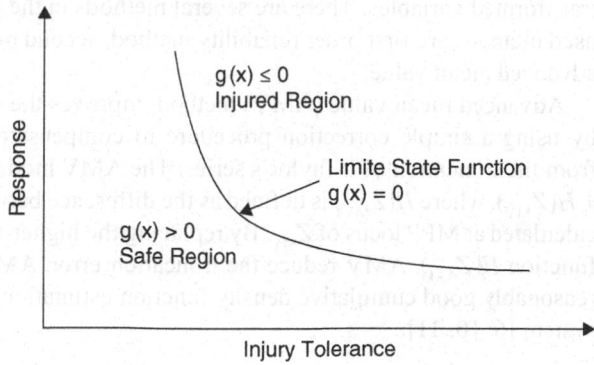

Figure 2. Concept of limit
state function.

2.4   Probabilistic Analysis Method

Direct numerical integration may be used if the performance function (limit state function) is given in simple closed form. In reality, however, it is impossible to obtain an exact deterministic equation of performance function. Even if the closed-form exists, the multiple integral is extremely complicated and it is sometimes impossible in joint PDF. In this case, the probability of failure needs to be estimated by analytical approximation [7].

Monte Carlo Simulation is a widely used method in probabilistic analysis. The general concept of the method is to solve mathematical problems by the simulation of random variables. This method uses randomly generated samples of the input variables according to their PDF for each simulation, and estimates response statistics and reliability after numerous repetitions of the deterministic analysis. Finally, the probability of failure is estimated as $P_f = N_f / N$, where N_f is the number of failed simulation and N is the total number of simulation. Accuracy of the method depends on the number of samples [3].

Response Surface Method (RSM) constructs a polynomial closed-form approximation for limit state function through a few deterministic analysis and regression analysis of these results. Goal of the RSM is to find a predictive equation relating a response to a number of input variables [8]. Once the closed-form equation is obtained, the equation is used to investigate the response instead of repeating the deterministic analysis. The closed-form equation can be used in numerical integration, Monte Carlo, and limit state approximation to determine the probability of failure.

Limit State Approximation can be divided into the mean value (MV) method and the most probable point (MPP) method. It is known that the MV is relatively inaccurate for high reliability and for highly nonlinear performance function [8]. The MPP method is used as a means of reducing the number of g-function evaluations from that of Monte Carlo [9]. In MPP method, the design variable distributions are transformed into standard normal distribution and then the most probable point is identified. The MPP is the closest point to the origin in the transformed space. Finally a polynomial approximation to the limit state function is developed around the MPP and probability of failure is computed using the newly defined g-function and the transformed variables. There are several methods in the group, but the most widely used methods are first order reliability method, second order reliability method, and advanced mean value.

Advanced mean value (AMV) method improves the mean values (MV) method by using a simple correction procedure to compensate for the errors introduced from the truncation of a Taylor's series. The AMV model is defined as $Z_{AMV} = Z_{MV} + H(Z_{MV})$, where $H(Z_{MV})$ is defined as the difference between the vales of Z_{MV} and Z calculated at MPP locus of Z_{MV}. By replacing the higher-term of MV by a simplified function $H(Z_{AM})$, AMV reduce the truncation error. AMV solution has provided a reasonably good cumulative density function estimation of engineering application system [6, 10, 11].

2.5 Sensitivity analysis

The sensitivity analysis can be considered from two points of view. The first is the change in the probability relative to the change in the distribution parameters, such as mean and standard deviation [12]. The two sensitivity coefficients are

$$s_\mu = \frac{\partial p_f}{\partial \mu_i}(\sigma_i)$$

$$s_\sigma = \frac{\partial p_f}{\partial \sigma_i}(\sigma_i)$$ (4)

which is probabilistic sensitivity with respected to mean, μ_i and standard deviations, σ_i, respectively.

The second is determination of the relative importance of the random variables. It is important to know which material properties have great influence on the probabilistic response.

The probabilistic sensitivity is

$$\alpha_i = \frac{\partial \beta}{\partial u_i}$$ (5)

which measures the change on the safety index, β, with respect to the standard normal variables u_i. The sensitivity factors satisfy the following rule $\alpha_1^2 + \alpha_2^2 + \cdots + \alpha_n^2 = 1$ which implies the each α_i^2 is a measure of the contribution to the probability and higher α indicate the higher contribution.

Sensitivity analysis is performed to determine influence of a particular random variable on the system response. It is important to know which problem parameters are the most important and the degree to which they control the design. Based on the sensitivity analysis, it is possible to ignore the number of random variables, which do not have a significant effect on reliability; thus researchers can save a great amount of computational effort [13, 14].

2.6 Probabilistic analysis techniques in bioengineering

Probabilistic analysis techniques have been broadly used in various bioengineering research fields by integrating with theories and computational tools. Various analysis techniques are applied to account for uncertainty (variability) contained in geometry, material properties, and loadings [15]. These techniques have been applied in kinematics, joint mechanics, musculoskeletal modeling and simulations, and so on. These methods have been also employed to prosthetic design, structural reliability, novel clinical technique, and prediction of injury.

Dopico-González et al. [16] studied an uncemented hip replacement by integrating the Monte Carlo simulation with a realistic finite element mode, and they analyzed the percentage of bone volume that exceeded specified strain limits and the maximum nodal micromotion. Probabilistic finite element analysis was conducted for an uncemented total hip replacement considering variability in bone-implant version angle [17]. Based on the research, they proposed a potential for application in matching patient to an implant.

Probabilistic techniques have also been applied in the area of total knee replacement. Pala et al. [18] contributed to the problem of polyethylene wear in total knee replacement. The Monte Carlo and AMV probabilistic methods were used the probabilistic component of the analysis. Their development has potential for clinical applications involving implant positioning and addressing problems in knee arthroplasty patients.

South west research institute (SwRI) developed probabilistic analysis software called NESSUS. They conducted stochastic crashworthiness FE analysis by combing NESSUS with commercial software LS-DYNA and Hybrid III dummy [19]. Based on the sensitivity analysis, they carried out re-design analysis and improve the reliability of vehicle crashworthiness. The probability of injury to the neck of naval aviators due to high acceleration maneuvers is also investigated by SwRI. Neck injury of female and male aviators was investigated during the ejection using SwRI-developed advanced probabilistic analysis techniques [20]. This method was employed to verify how the biological variability affects predictions of system performance and functionality.

Pérez et al. [21] used a probabilistic technique based on a probabilistic cumulative damage approach and on stochastic finite elements by combining the B-model based on the Markov chain with finite element model of the proximal part of the femur. They considered three random variables (muscle and joint contact forces at the hip, cement damage and fatigue properties of the cement) to calculate the probability of failure of a cemented hip implant. Jang et al. [22] used random field analysis in analyzing the cervical spine. Based on the sensitivity analysis, the most sensitive parameters (disc annulus and nucleus pulpous) were specifically considered as random fields. Then, each random field parameters were discretized into a set of random variables. In the random field analysis, in stead of single random variable, the sets of random variables were defined for the sensitive parameters for increasing the analysis confidence [23].

In addition, probabilistic techniques are used in design and development of novel prosthesis. Rohlmann et al. [24] conducted probabilistic finite element for developing an artificial disc on lumbar spine and considered uncertainties of the input parameters implant position, implant ball radius, presence of scar tissue, and gap size in the facet joints. Failure production of graft for aortic aneurysm and mineral-collagen composite bone are also the application of the probabilistic techniques in bioengineering engineering [25, 26].

3 Conclusion

The objective of this chapter is to provide an overview of the probabilistic techniques in bioengineering. The basic concept and theory behind the majority of these techniques is presented. In order to account for uncertainty (variability) contained in biological system, probabilistic analysis methods are being increasingly applied in bioengineering and these techniques are the way to overcome inaccuracy caused form the uncertainty factors. These techniques have been applied in kinematics, musculoskeletal modeling and simulation, and so on. These methods could have contribution in implant device design, cure and treatment techniques, structural reliability, and prediction and prevention of injury.

References

1. C.-K. Choi and H.-C. Noh, "Weighted Integral SFEM Including Higher Order Terms," Journal of Engineering Mechanics, Vol. 126, No. 8, 2000, pp. 859–866.
2. R. Ishida, "Stochastic Finite Element Analysis of Beam with Statistical Uncertainties," AIAA Journal, Vol. 39, No. 11, 2001, pp. 2192–2197.
3. A. Haldar and S. Mahadevan, "Probability, Reliability, and Statistical Methods in Engineering Design," John Wiley, 2000.
4. S.-K. Choi, R. Grandhi, and R. A. Canfield, "Reliability-Based Structural Design," Springer Verlag, 2007.
5. A. H.-S. Ang and W. H. Tang, "Probability Concepts in Engineering," 2nd ed, Wiley, 2006.
6. B. H. Thacker, D. P. Nicolella, S. Kumaresan, N. Yoganandan, and F. A. Pintar, "Probabilistic Injury Analysis of the Human Cervical Spine," BED Bioengineering Conference ASME, Vol. 50, 2001, pp. 879–880.
7. E. Vanmarcke, "Random Field," MIT, 1983.
8. M. W. Long and J. D. Narciso, "Probabilistic Design Methodology for Composite Aircraft Structure," FAA, Springfield Virginia DOT/FAA/AR-99/2, 1999.
9. D. P. Nicolella, B. H. Thacker, G. M. Jones, R. C. Schenck, M. Simonich, and C. M. Agrawal, "Probabilistic Design of Orthopaedic Implants," Advance in Bioengineering Winter Annual Meeting, New Orleans, Louisiana, Vol. 26, 1993, pp. 539–542.
10. B. H. Thacker, D. P. Nicolella, S. Kumaresan, N. Yoganandan, and F. A. Pintar, "Probabilistic Finite Element Analysis of the Human Lower Cervical Spine," Mathematical Modeling and Scientific Computing, Vol. 13, No. 1-2, 2001, pp. 12–21.
11. Southwest Research Institute, "NESSUS Version 9," Southwest Research Institute, 2009.
12. Y. T. Wu, "Computational Methods for Efficient Structural Reliability Analysis," AIAA Journal, Vol. 32, No. 8, 1994, pp. 1717–1723.
13. A. Haldar and S. Mahadevan, "Reliability Assessment Using Stochastic Finite Element Analysis," Wiley, 2000.
14. S. Mahadevan, "Probabilistic Finite Element Analysis of Large Structural Systems," Series on stability, vibration and control of systems, Vol. 9, 1997, pp. 1–21.
15. P. J. Laz and M. Browne, "A Review of Probabilistic Analysis in Orthopaedic Biomechanics," Journal of Engineering in Medicine, Vol. 224, 2010.
16. C. Dopico-González, A. M. New, and M. Browne, "Probabilistic Finite Element Analysis of the Uncemented Hip Replacement—Effect of Femur Characteristics and Implant Design Geometry," Journal of Biomechanics Vol. 43, No. 3, 2009, pp. 512–520.
17. C. Dopico-Gonzalez, A. M. New, and M. Browne, "A Computational Tool for the Probabilistic Finite Element Analysis of an Uncemented Total Hip Replacement Considering Variability in Bone-Implant Version Angle," Computer Methods in Biomechanics and Biomedical Engineering, Vol. 13, No. 1, 2010, pp. 1–9.
18. S. Pala, H. Haiderb, P. J. Laza, L. A. Knightc, and P. J. Rullkoetter, "Probabilistic Computational Modeling of Total Knee Replacement Wear " Wear, Vol. 264, No. 7-8, 2008, pp. 701–707.
19. D. Riha, J. Hassan, M. Forrest, and K. Ding, "Development of a Stochastic Approach for Vehicle Fe Models," 2003 ASME International Mechanical Engineering Congress, Washington, D.C., 2003, pp. 37–69.

20. Southwest Research Institute, "Probabilistic Mechanics and Reliability Methods." San Antonio, 2009.
21. M. A. Pérez, J. Grasa, J. M. García-Aznar, J. A. Bea, and M. Doblaré, "Probabilistic Analysis of the Influence of the Bonding Degree of the Stem–Cement Interface in the Performance of Cemented Hip Prostheses," Journal of biomechanics, Vol. 39, No. 10, 2006, pp. 1859–1872.
22. T. H. Jang and S. Ekwaro-Osire, "Random Field Analysis Procedure Applied to Cervical Spine Column," 10th AIAA/ISSMO Multidisciplinary Analysis and Optimization Conference, Albany, New York, 2004.
23. T. H. Jang, S. Ekwaro-Osire, J. BrianGill, and J. Hashemi, "Uncertainty Analysis for Biomechanical Injury of Cervical Spine Column," in 2008 ASME International Mechanical Engineering Congress and Exposition, 2009.
24. A. Rohlmann, A. Mann, T. Zander, and G. Bergmann, "Effect of an Artificial Disc on Lumbar Spine Biomechanics: A Probabilistic Finite Element Study," European Spine Journal Vol. 18, No. 1, 2009, pp. 89–97.
25. X. Dong, T. Guda, and H. Millwater, "Probabilistic Failure Analysis of Bone Using a Finite Element Model of Mineral-Collagen Composites " Journal of Biomechanics, Vol. 42, No. 3, 2009, pp. 202–209.
26. R. Layman, S. Missoum, and J. Geest, "Simulation and Probabilistic Failure Prediction of Grafts for Aortic Aneurysm " Engineering Computations, Vol. 27, No. 1-2, 2010, pp. 84–105.

Chapter 16
A COMPLETE ELECTROCARDIOGRAM (ECG) METHODOLOGY FOR ASSESSMENT OF CHRONIC STRESS IN UNEMPLOYMENT

Wanqing Wu and Jungtae Lee

1 Introduction

Since the nineties, markers of stress and other psychosocial factors are associated with coronary disease [1, 2]. Abundant research implicates mental and physical stress associated with everyday living in the precipitation of sudden cardiac death [3, 4, 5]. Especially, work stress as a huge problem in today's society since about half of work-related illnesses are directly or indirectly related to stress. Several studies have shown a link between the level of work stress and disease [6, 7, 8]. It is well accepted that work is perhaps the greatest contributor to stress in our lives. Job insecurity and rising unemployment have contributed greatly to high stress rates and severe burnout among workers.

Compared to other lifestyle risk factors, stress is different because no consensus exists with respect to either definition or measurement. Inevitably, stress is subjective and it can encompass several aspects from external stressors such as adverse life events, financial problems or job stress to potential reactions such as depression, vital exhaustion, sleeping difficulties or anxiety [9]. From another point of view, Stressors can be physical, mental or both. Physical stress is caused by long-term exposure to negative factors such as an irregular lifestyle, physical overload, environmental toxicity, cigarette or alcohol or drug use, improper diets, etc. Mental stress can be caused by factors such as insecurity, negative emotions, mental overload, confusion, rejection on a social level, family problems, boredom, low self-esteem, etc. Physical and mental stress each elicits physiological responses that are mediated through the autonomic nervous system (ANS). This ANS is both our major defense against stress and the system that demonstrates the principal symptomatic manifestation of stress in its early stages. It is well known that the ANS plays an important part in the overall body's control system for normal functions mediating all the unconscious activities of viscera and organs. Conventionally, the

J. Lee (✉)
Busan, Republic of Korea
e-mail: jtlee@pusan.ac.kr

S. Suh et al. (eds.), *Biomedical Engineering*,
DOI 10.1007/978-1-4614-0116-2_16, © Springer Science+Business Media, LLC 2011

ANS is divided into two parts in a parasympathetic and sympathetic balance: the sympathetic, which activates organs, getting them ready to cope with exercise or other physical stress, and the parasympathetic, which controls background "house-keeping" functions in the body. The balance between these two systems is an indicator of the body's reaction to external and internal demands.

Heart Rate Variability, the beat to beat change in heart rate, is an accurate indicator of autonomic nervous system activity that can provide important insights into a patient's risk of cardiovascular disease. More importantly, it provides a window to observe the heart's ability to respond to normal regulatory impulses that affect its rhythm. It was well known that most of the organs in human body including heart are controlled by ANS, beat to beat variations in heart rate (HR) is regulated by ANS activity. As shown by many studies, HRV signals contain well defined rhythms which include physiological information. Rhythms in the low frequency (LF) range, between 0.04 to 0.15 Hz, are usually considered as markers of sympathetic modulation. The high frequency (HF) range, between 0.15 to 0.4 Hz, can contain the rhythms regulated by parasympathetic activity. Vagal activity is the major contributor to the HF component. Therefore analyzing HRV data can provide a tool to understand the mechanisms of ANS. Moreover, it is a simple and powerful noninvasive methodology having enormous practical advantages with a minimum of technical constraints, which makes it useful everywhere.

Stress involves change of autonomic function and the secretion of several hormones such as cortisol, corticosterone and adrenal catecholamine [10]. Blood pressure and heart rates during stress increase reflecting the predominance of sympathetic nervous system activities [11]. Also, mental stress decreased high frequency component of HRV and increased low frequency component of HRV [12]. HRV decreased in subjects with depression, higher hostility and anxiety. In essence, HRV provides a picture of the interplay between the sympathetic and parasympathetic branches. As such, it reflects the ways in which emotional states affect core physiology including, but not limited to, cardiac function. So it is reasonable to apply ECG methodology for stress detection and analysis which can instruct worker how to reduce their own risk by showing them how their stress states affect their heart health.

Most of us would experience stress at work at some point in time, but we would recognize its subjective nature and variability [13]. Devising an instrument to capture the experience, particularly of chronic work stress, designing a man-machine interface to monitor, analyze and sequentially realize the HRV biofeedback is a challenge for psychologists and sociologists, and studies use a variety of methodology. Differences among the instruments used, analysis method and periods of exposure may account for some of the variations in study findings thus far. The development of a standardized ECG methodology includes a set of hardware, software and biofeedback mechanism that could be used around the world, particularly in newly emerging economies where work stress is likely to be more prevalent, and working hours may be the longest, or in many countries where unemployment is growing in intensity would represent a step forward in the epidemiological investigation of work stress and unemployment strain.

While most papers about HRV and stress focus on the relation between the stress level, assessed by simple questionnaires, and some HRV parameters, the goal of this study is to investigate whether HRV can be used to detect chronic mental and physical stress of unemployment population. In order to perform ECG measurement and HRV analysis, an improved ECG methodology was developed based on the principles of Ambulatory, Simple, High-performance, Robust, and Cost effective. Moreover, 53 healthy (female 9, male 44, 33 volunteers are unemployed drawing the dole between 30 and 40 years, 20 volunteers are undergraduate recruited from local university aged from 20 to 30 years), right-handed volunteers participated in the experiment to compare the variation trend of HRV features in different stimulus mode between unemployment population and normal subjects, accordingly investigated the influence of chronic stress in unemployed population on ANS modulation and psychology-physiological health.

2   ECG Methodology

ECG methodology is an extensive interdisciplinary science. Generally speaking, ECG methodology includes three aspects: hardware platform (data acquisition/ communication), software application (signal processing/monitoring), and analysis method (algorithm/statistical analysis). From the hardware point of view, it correlated with bioengineering, bioinformatics, analog/digital circuit design, printed circuit board (PCB) development, data communication, signal processing and biosensor network, etc. Correspondingly, in terms of software, it related to software engineering, digital signal processing (DSP), man-machine interface (MMI) design, and data storage/management, etc. Put the eyes on physiological signal analysis side, it also referred to multiple disciplines, such as statistics, artificial intelligence, spectral analysis, nonlinear system analysis, HRV analysis and so on. Due to the characteristics of complexity, universality and comprehensiveness, it is difficult to build on a universal standard for ECG methodology. In this paper, we put forward a framework of ECG methodology based on the principles Ambulatory, Simple, High-performance, Robust, and Cost effective. The framework of this ECG methodology illustrated in Figure 1.

2.1   Hardware platform

A number of physiological markers of stress have been identified, including electrodermal activity (EDA), heart rate (HR), various indices of heart rate variability (HRV), blood pressure (BP), muscle tension, and respiration. However, in order to gain acceptance as a method of stress management in daily life, especially for unemployed population, wearable sensors must be minimally obtrusive, so that human beings can perform their tasks with complete freedom, and inconspicuous, to avoid

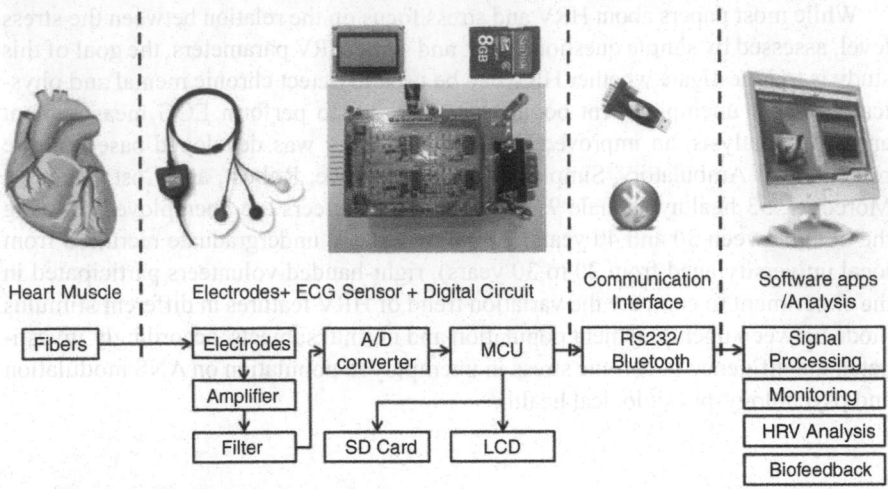

| Heart Muscle | Electrodes+ ECG Sensor + Digital Circuit | Communication Interface | Software apps /Analysis |

```
Fibers ──┬──→ Electrodes ──→  A/D      ──→  MCU  ──→  RS232/    ──→  Signal
         │                    converter                Bluetooth        Processing
         │         ↓                                                 Monitoring
         │     Amplifier                         ↓          ↓
         │         ↓                                                 HRV Analysis
         └──→  Filter         SD Card         LCD                 Biofeedback
```

Figure 1. Framework of ECG Methodology.

anxieties associated with wearing medical devices in public [14]. These usability considerations preclude some of the above measures from being considered as a long-term solution for stress monitoring.

Wearable technology in the biomedical field has also made great progress in recent years, a number of clinical applications of wearable systems emerged in the past few years. They range from simple monitoring of daily activities and physiological signals to integrating miniature sensors to enhance the function of devices utilized by patients. Wearable systems can gather data unobtrusively and reliably over extended periods of time, so the combination of biofeedback techniques with wearable technology is undoubtedly meaningful. It can not only promote the widely usage of the biofeedback technique to enhance health in real life, but also offer a more integrative platform for the deeper research about mechanism of biofeedback technique [15].

Typically, ECG signal that detected by the electrode is very weak with amplitude of 1 mV to 3 mV. Its frequency spectrum lies between 0.05 Hz to 150 Hz. For a full-lead ECG records, 12 leads are derived from 10 electrode locations. Other subsets of 12 leads ECG are used in the situations that do not need as much data recording such as ambulatory ECG (usually 2 leads), intensive care at the bedside (usually 1 or 2 lcads) or in telemetry systems (usually 1 lead). According to the technical requirements, this Bluetooth-enabled hardware platform basically could be divided to six parts: Analog Circuit, Digital Circuit, Signal Condition Circuit, Communication Interface and Power management Circuit, only 1 lead (Lead II) is necessary due to the simplicity and portability factors, however, it was easily expand to standard 12 lead ECG. As most of the energy of ECG waveforms is concentrated in frequency up to 40 Hz only, the cutoff frequency of 5th order Butterworth low-pass filter has set to 40 Hz. This not only simplified the analogy circuit (60 Hz notch filter), but also removed high frequency noise and power line noise (60 Hz in Korea), the total amplifier gain was about 60dB and the Common Mode Reject Ratio (CMRR) was

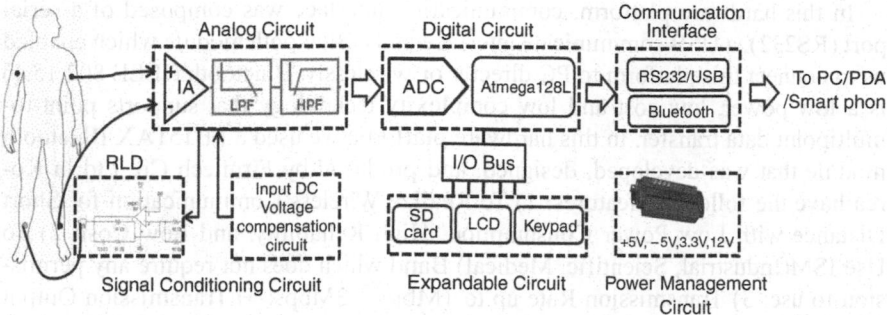

Figure 2. The block diagram of hardware platform.

adjusted to 120dB by introducing Right Leg Driver (RLD) circuit. Figure 2 had shown the block diagram of the hardware platform.

The ECG signal measurement was controlled by Atmega128 which is a low power CMOS 8-bit RISC microcontroller with a throughput of up to 1 MIPS per MHz; Expandable function circuit was optional, it comprised of a Secure Digital Card (SD card), a LCD display module, and a simplified Keypad. The Secure Digital Card (SD card) which allows for easy integration into any design was chosen because it features low power consumption, security, capacity, performance and environmental requirements. And the file system which is fully compatible with the FAT32 specification was designed for this system. Furthermore, the LCD displaying module combined with Simplified keypad provided a man-machine interface for basic interaction function such as input/output, measurement start/stop control, and timer. This hardware platform could be able to configure flexibly according to the aims of application, taken the design principles o into consideration, two schemes (with or without expandable function circuit) was capable of transforming arbitrarily. The expandable circuit interface design and the prototype of hardware platform shown in Figure 3.

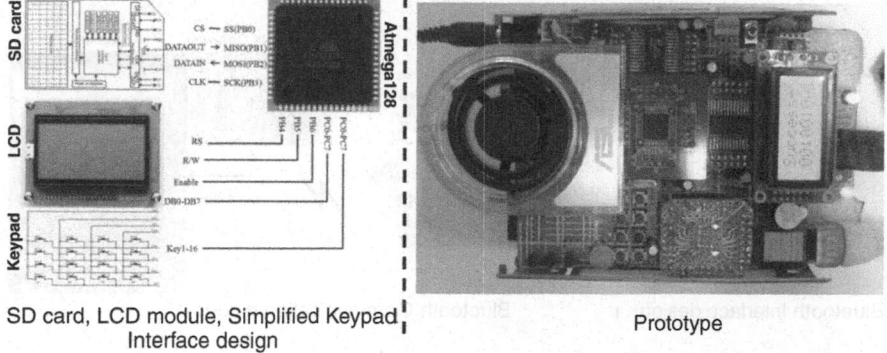

SD card, LCD module, Simplified Keypad
Interface design

Prototype

Figure 3. Interface design of the expandable circuit and the prototype of hardware platform.

In this hardware platform, communication interface was composed of a serial port (RS232), a USB communicate interface and a Bluetooth module which enabled us to connect this platform to PC directly or wirelessly. Bluetooth (IEEE 802.15.1) is a low power, low cost and low complexity technology that supports point-to-multipoint data transfer. In this hardware platform, we used a FB151AX Bluetooth module that was developed, designed, and produced by Firmtech Co. Ltd in Korea have the following features: 1) To Realize Wireless Communication for Short Distance with Low Power Consumption, High Reliability, and Low Cost; 2) To Use ISM(Industrial, Scientific, Medical) Band which does not require any permission to use; 3) Transmission Rate up to 1Mbps ~ 3Mbps; 4) Transmission Output includes 1mW(10 m, Class2) and 100mW(100m Class1); 5) To Guarantee stable wireless communication even under severe noisy environment through adopting the technique of FHSS (Frequency Hopping Spread Spectrum). The parameter of FB151AX was set to 57600bps (Baud rate)/8bit (Data bit)/None (Parity)/1 (Stop bit)/none (Flow control). Communication between Bluetooth modules adopted point-to-point connection (OP_MODE0 in FB151AX), the PC/Smart phone served as master, and the ECG platform acted as slave correspondingly. The ECG data was packed in the Asynchronous Connection Less (ACL) data packets. Interface design of the FB151AS Bluetooth module and the communication mode between hardware platform and PC illustrated in Figure 4.

2.2   Software Application

To develop the PC side software application (named as ECG Viewer, as shown in Figure 6), we used Microsoft Visual Studio .NET Edition with the Nokia Series 60 SDK. As shown in Figure 5, the application realized basic functions include ECG data receive/decoding, Control Command (start/stop control, time control), Real-time monitoring, raw data preprocessing, ECG data storage /Playback and data post-processing.

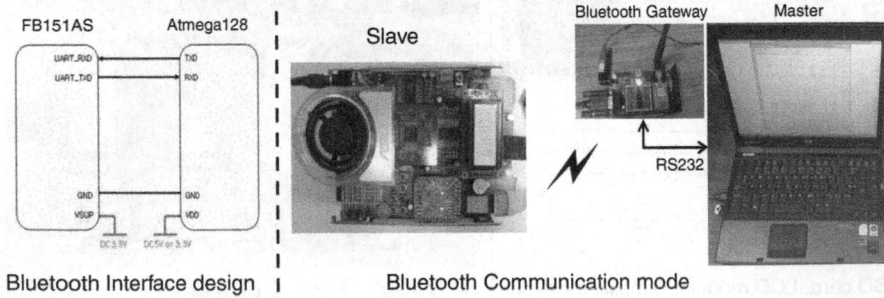

Figure 4. Interface design of the FB151AS Bluetooth module and the communication mode between hardware platform and PC.

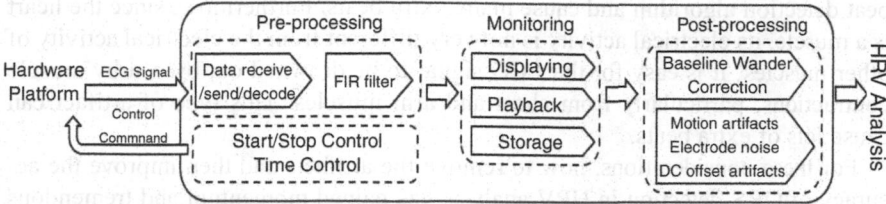

Figure 5. The flowchart of Software design.

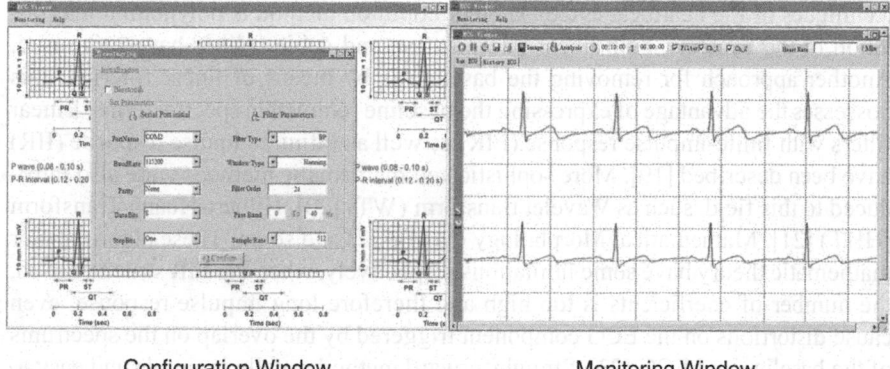

Configuration Window   Monitoring Window

Figure 6. The graphic user interface (GUI) of ECG viewer.

It was well accepted that ECG signal was susceptible to artifacts due to power-line interference, electrode motion, baseline wander, high frequency contents and myoelectric interference [16]. So it was necessary to integrate a 60 Hz notch filter for minimizing the power line interference in real time. This digital filter is a Finite Impulse Response (FIR) band-pass filter with hamming windows. The cut-off frequency was set between 3 Hz to 40 Hz, which is screen out other components such as P wave, T wave, and baseline drift from human motion [17]. The normal ECG spectrum is around 3–40 Hz [18]. The R wave frequency spectrum is higher than 10 Hz, and lower than 40 Hz. The cut off frequency at 40 Hz can protect the noise from power line (60Hz).

When processing ECG signals for HRV analysis in next steps, two kinds of errors can occur: Missed beats and Extra beats. Both types are most often caused by signal distortion which usually aroused by Direct current (DC) offset artifacts, Electrode movement artifacts and Muscle contraction artifacts, etc. The ECG signal normally oscillates around a zero baseline. A direct current offset can make the signal drift up or down on the scale when the impedance between the skin and the electrodes is too high. In some circumstances this can cause missed beats. If the subject moves enough to tug on the extender cable and pull one of the electrodes off the skin you will see very wide signal deviations as the ability to capture the small electric signal is lost. These very wide distortions completely confuse the beat detection algorithm and cause many extra beats. Furthermore, since the heart is a muscle, its electrical activity is not very different from the electrical activity of other muscles. It is easy for the ECG signal to be drowned out by nearby muscle contractions, particularly from chest and arm muscles. This type of artifact can cause lots of extra beats.

For these considerations, how to remove the artifacts and then improve the accuracy of QRS detection in HRV analysis has gained momentum and tremendous amount of studies have been carried out. There were many approaches to the problem of ECG artifacts. Take the baseline wander removal for example, the most straightforward method employs no filtering but relies only on robust averaging techniques of the heartbeat cycle. Another common method is polynomial interpolation of the baseline; both linear interpolation and cubic splines have been used. Another approach for removing the baseline is by means of linear filtering; this possesses the advantage of expressing the baseline removal in spectral terms. Linear filters with finite-impulse response (FIR) as well as infinite impulse response (IIR) have been described [19]. More sophisticated mathematic methods were also introduced to this field, such as Wavelet transform (WT) [20], Hilbert-Huang Transform (HHT) [21], Mathematical Morphology (MM) [22] and so on. These algorithms or mathematic theory have some limitations respectively, when the FIR structure used, the number of coefficients is too high and therefore long impulse response, even cause distortions on the ECG component triggered by the overlap on the spectrums of the baseline and ECG [23]. Straightforward method usually is simple and easy to realize with low efficiency; adversely, the sophisticated mathematic method generally complicated and hard to implement, but high efficiency. The similar methodology applied for others artifacts also have the same worries.

In this paper, we introduce the multi-scale mathematical morphology (3M) filter concept which is a favorable signal analysis tools in many shape-oriented problems into data post-processing of ECG signal due to its rich theoretical framework, low computational complexity, and simple hardware implementation. The uniqueness of mathematical morphology is that it does not require any prior knowledge on the frequency spectrum of the signal under investigation. For ECG signal processing, this feature brings the benefit of avoiding the frequency band overlapping of QRS complexes and other components, such as P/T waves. In the past, the mathematical morphology methods demonstrated the effectiveness of the removal of impulsive noise and background normalization of the ECG signal [24].

3 M is an extension of the single-scale morphology (1M). Both of them share the same basic mathematical morphological operators: dilation, erosion, opening, and closing [25]. The difference between 1 M and 3 M is that the latter applies morphological operators to the signal repeatedly by varying the shape and size of structure elements. The most important operation is to design proper structuring element (SE) which depends on the shape of the signal that is to be preserved, since the opening and closing operations are intended to remove impulses, the SE must be designed so that the waves in the ECG signal are not removed by the process. The values of SE is largely determined by the duration of the major waves and the sampling rate,

denoting the duration of one of the waves as T sec and the sampling rate as S Hz, the length of the SE must be less than T×S [26].

In most applications, opening is used to suppress peaks while closing is used to suppress pits. By using different SE with proper morphology and size according to the target signal information, artifacts can be removed or extracted. In this software application, we used three scales MM to remove Impulsive noise (60 Hz noise), and extracted Baseline wander artifacts, Electrode movement artifacts and Muscle contraction artifacts simultaneously by applying three different SE (SE1,SE2,SE3) to every scale.

The first scale was used to remove impulsive noise and partial EMG noise with SE1. According the character of power line interference, the duration of one cycle is nearly 17 ms (1/60Hz), and the sampling rate of this platform is 512 Hz, so the length of SE1 was set to 5 points (> $17ms \times 512Hz/2$) triangular windows with the values of [0 1 5 1 0]. The baseline signal was extracted in second scale with SE2, and then subtracts it from ECG signal to eliminate the baseline wander. Due to the duration of normal QRS complex is from 50 to 100 ms, the length of SE2 must larger than 51 points (=$100ms \times 512Hz$) for baseline signal extraction. Meanwhile, the hamming window function was used for SE2 because it has similar shape information with ECG waveform. More special, in this application, the length of SE2 could be adjusted manually according to requirement. The last scale was applied to reduce the influence of electrode movement artifacts and muscle contraction artifacts with SE3 which was configured to 30 points triangular wave. The procedures, related window functions and the effect of 3 M filter were demonstrated in Figure 7.

2.3 Analysis Method

QRS detection is very important in HRV analysis; meanwhile, it is difficult, not only because of the physiological variability of the QRS complexes, but also the various types of noise as mentioned above. Within the last decade many new approaches to QRS detection have been proposed, for example, algorithms from the field of artificial neural networks, genetic algorithms, wavelet transforms, filter banks as well as heuristic methods mostly based on nonlinear transforms. Already in the early years of automated QRS detection, an algorithmic structure was developed that is now shared by many algorithms. As shown in Figure 8 it is divided into a preprocessing or feature extraction stage including linear and nonlinear filtering and a decision stage including peak detection and decision logic [27]. The flowchart of improved Adaptive threshold update algorithm for QRS complex detection was shown in Figure 8.

As a relative clean ECG data was achieved through processing by 3 M filter, it was easier for us to detect QRS complex. An adaptive threshold update algorithm has been designed in this application as shown in Figure 8. The threshold update

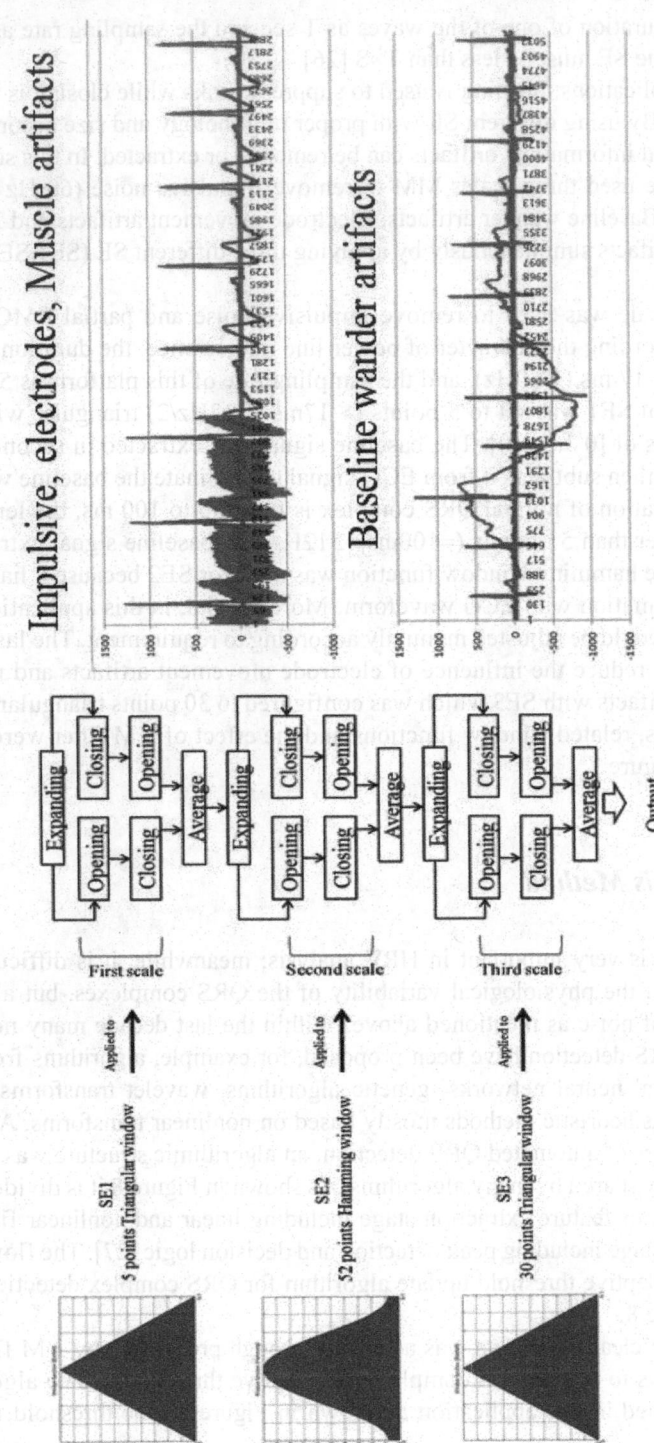

Figure 7. The procedures, related window functions and filter results of 3 M filter.

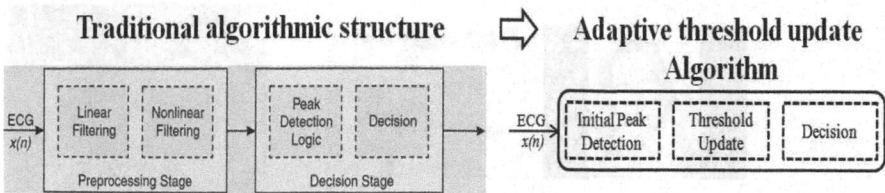

Figure 8. Flowchart of Adaptive QRS complex detection algorithm.

based on the normal cognition that the duration of normal R peak to R peak (RR) varied from 60 bpm to 100 bpm (0.6-1 second). So, if detected interval of RR is larger than 1second, it means a missed peak which can be corrected by decrease initial threshold. On the other hand, if smaller than 0.6 second, it represents extra peak which can be removed by increase initial threshold.

Analysis of HRV consists of a series of measurements of successive RR interval (intervals between QRS complexes of normal sinus depolarisation, or NN interval) variations of sinus origin which provide information about autonomic tone. In 1996 a Task Force of the European Society of Cardiology (ESC) and the North American Society of Pacing and Electrophysiology (NASPE) defined and established standards of measurement, physiological interpretation and clinical use of HRV [28]. Time domain indices, geometric measures and frequency domain indices constitute nowadays the standard clinically used parameters [29]. In this application, we adopted standardized HRV methodology for time domain and frequency domain analysis. In the frequency-domain methods, a power spectrum density (PSD) estimate was calculated for the R-R interval series. The time-domain methods are the simplest to perform since they are applied straight to the series of successive RR interval values. The developed HRV analysis software and related HRV parameters was shown in Figure 9.

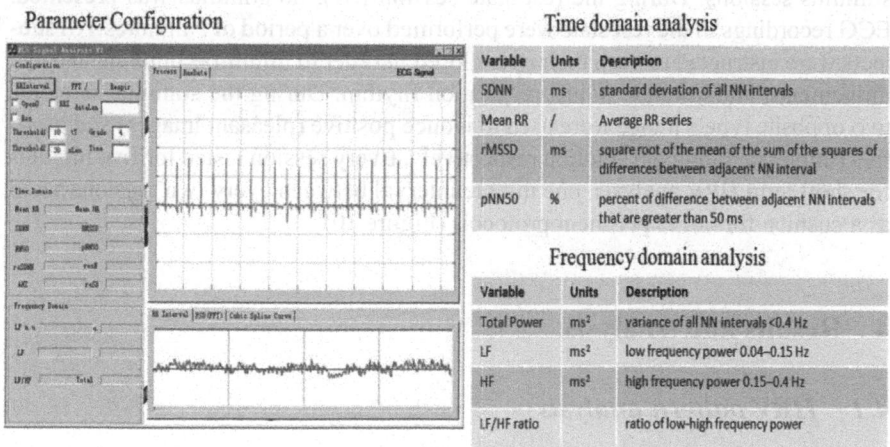

Figure 9. Developed HRV analysis software and related HRV parameters.

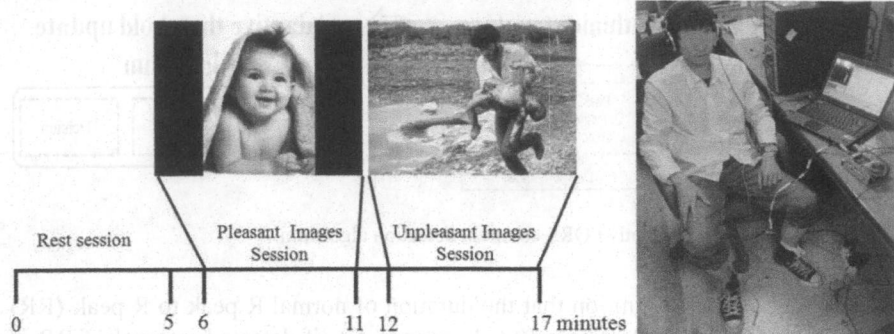

Figure 10. Experiment protocols and Experiment environment.

3   Experiment Protocols

53 healthy (female 9, male 44, 33 volunteers are unemployed drawing the dole between 30 and 40 years, Unemployed group; 20 volunteers are undergraduate recruited from local university aged from 20 to 30 years, Undergraduate group), right-handed volunteers participated in the experiment. We treated the Unemployed group as High Stress group, and served Undergraduate group as Control group. All participants had normal auditory and visual function; none had neurological disorders. This study was approved by the Institutional Review Board (IRB) of Pusan National University Hospital. All of them were given written informed consent to complete the questionnaires and the psycho-physiological protocols.

In this experiment, IAPS (International Affective Picture System) was used to evoke emotion as visual stimulus [30] to investigate the ANS modulation in different pressure groups. All subjects are tested during a rest state session and two stimulus sessions. During the rest state session (S1), no stimulus was presented. ECG recordings in the rest state were performed over a period of 5 minutes. All subjects were instructed to keep their eyes closed in order to minimize blinking and eye movements, and asked to adjust respiration rhythm. During the stimulus sessions, two opposite type's image were used to induce positive (pleasant images) and negative (unpleasant images) emotion respectively. Every session lasted for five minutes for short term HRV analysis, one minute interval between every two sessions acted as a cushion for our experiment protocols (Figure 10).

4   Results and Analysis

4.1   HRV pattern analysis

In HRV pattern analysis, there are two basic patterns: Incoherent HRV pattern which implied low HRV and high risk is characterized by a low peak-to-nadir difference in

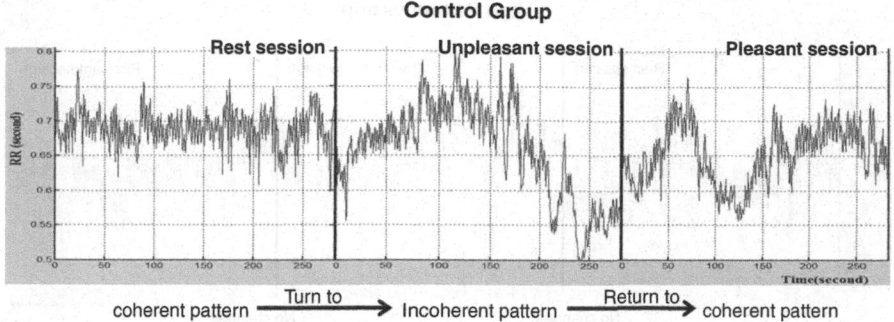

Figure 11. HRV pattern of a subject in Control Group during different emotional states.

the wave form of the heartbeat, and it is typical of states of high sympathetic tone. Chronic sympathetic over activity, as seen in states of depression, anger, anxiety and hostility, trigger increases in parasympathetic tone, reflecting the autonomic system's attempt to achieve homeostasis. In a simple but reasonable analogy, it is like running your car with your foot on the gas and brake at the same time: the conflicting signals produce discordant function, which is reflected in the jagged and incoherent form of the HRV pattern. The more coherent HRV pattern means high HRV and low risk associated with positive emotional states represents a balanced, cohesive ebb and flow of sympathetic and parasympathetic tone. This occurs when both tonalities are at modest to low output, and it is characteristic of low stress, high-satisfaction emotional states [31].

The HRV pattern of a subject in Control Group during different emotional states was shown in Figure 11. It was obvious that the modulatory function of ANS was more powerful, and the reaction capacity of ANS was more rapid when underwent different emotional state. In rest session, there was a coherent pattern of HRV, and turned to Incoherent pattern in Unpleasant session, this phenomenon reflected the autonomic system's attempt to achieve homeostasis when suffered negative emotion. The Pleasant session shifted the HRV back to coherent pattern also indicated the mechanism of ANS modulation in normal subjects.

Correspondingly, Figure 12 illustrated the HRV pattern of High Stress Group in different emotional mode; there was no significant difference among three sessions, the results reflected that the modulatory function of ANS was weaken by long term pressure in High stress group. Compared to Control group, the emotional response of unemployed was slowed due to ANS modulation damage. Negative emotional states such as depression, anger, and anxiety triggered by unemployment have induced Incoherence HRV in High stress group.

4.2   Time domain analysis

The variations in heart rate may be evaluated by a number of methods. Perhaps the simplest to perform are the time domain measures. In this experiment, we ad-

High Stress Group

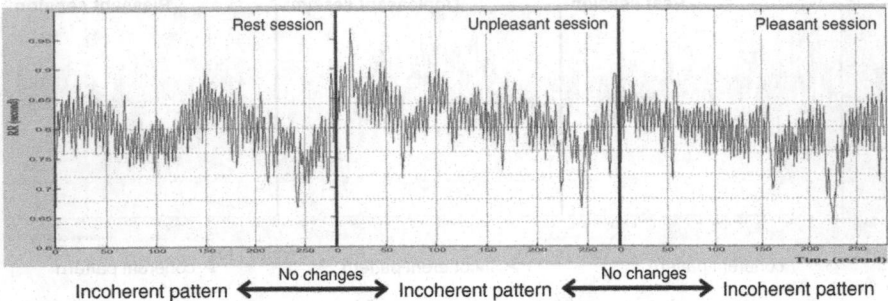

Figure 12. HRV pattern of a subject in High stress Group during different emotional states.

opted three time domain parameters to assess the difference of ANS modulation between Control group and High stress group. The simplest variable to calculate is the standard deviation of the RR intervals (SDNN), that is, the square root of variance which reflects all the cyclic components responsible for variability in the period of recording. Other commonly used statistical variables calculated from segments of the total monitoring period include RMSSD, the square root of the mean squared differences of successive RR intervals, and pNN50, the proportion derived by dividing NN50 by the total number of NN intervals. All of these measurements of short-term variation estimate high-frequency variations in heart rate and thus are highly correlated [28]. The results were shown in Table 1.

Table 1. HRV parameters in High stress group and Control group during three emotional states.

Group	Variable / Session	Time domain Analysis (Mean±STD)			Frequency domain Analysis (Mean±STD)			
		SDNN	rMSSD	pNN50	LF n.u.	HF n.u.	LF/HF ratio	Total power
	Unit	ms	ms	%	ms²	ms²	/	ms²
High stress Group (n=33)	Rest	35± 11.7	27.6± 13.6	10.7± 14.4	45.2± 23.6	54.8± 23.6	1.66± 3.1	1284± 1111
	Unpleasant	41.9± 18.0*	29.9± 17.4	10.2± 14.3	59.9± 22.9*	40.1± 22.9	2.77± 2.9*	2061± 2374*
	Pleasant	41.9± 17.1*	29.6± 16.1	11.1± 15.4	57.9± 20.9*	42.1± 20.8	2.32± 2.8*	1817± 2159*
Control Group (n=20)	Rest	55.9± 14.2	44.7± 12.3	24.4± 12.3	61.1± 16.4	38.9± 16.4	2.2± 1.8	3138± 1920
	Unpleasant	50.4± 16.3*	37.3± 15.2*	16± 13.1*	53.9± 16.7*	46.1± 16.7*	1.56± 1.3*	2435± 1691*
	Pleasant	60± 16.7⁻	40.7± 15.9⁻	20.5± 14.7⁻	66.1± 12.2⁻	33.9± 12.2⁻	2.54± 1.3⁻	3513± 2245⁻

Abbreviations:
STD: standard deviation; *p<0.05(paired t-test): Unpleasant or Pleasant session versus Rest session;⁻p<0.05(paired t-test): Unpleasant versus Pleasant session.

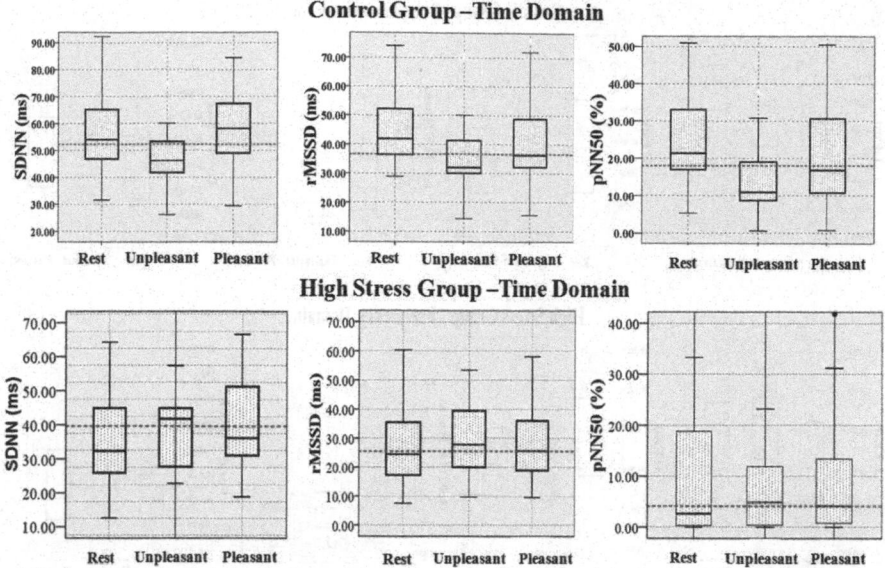

Figure 13. Time domain analysis in Control Group and High Stress Group during three emotional states.

In Control group, the mean SDNN, rMSSD and pNN50 between Rest and Unpleasant session have significant difference ($p < 0.05$), all of them were significantly higher during Rest session (versus Unpleasant session). Accordingly, the mean values in Pleasant session have obvious difference compared to Unpleasant ($p < 0.05$), these results kept in line with HRV pattern analysis which reflected the ANS modulation during different emotional modes in normal subjects (Table 1, Figure 13).

In High stress group (Table 1, Figure 13), there was an abnormal variation in time domain analysis. The mean SDNN was significantly lower in Rest session (35 ± 11.7, versus Unpleasant: 41.9 ± 18.0 or Pleasant: 41.9 ± 17.1 session; $p < 0.05$), and there was no difference between Unpleasant and Pleasant session. Furthermore, the mean rMSSD and pNN50 have slightly increased no matter which types of emotional stimulus acted on subjects compared to Rest session.

4.3   frequency domain analysis

The frequency domain analysis (Table 1, Figure 14) also illustrated the ANS modulatory function in normal subjects (Control group). The mean Normalized LF component of Unpleasant session (53.9 ± 16.7) decreased compared to Rest session (61.1 ± 16.4) and the mean Normalized HF component of them increased correspondingly (46.1 ± 16.7 versus 38.9 ± 16.4). The increased HF n.u. and diminished LF n.u. observed in this study may indicate a shift of sympathy-vagal balance to-

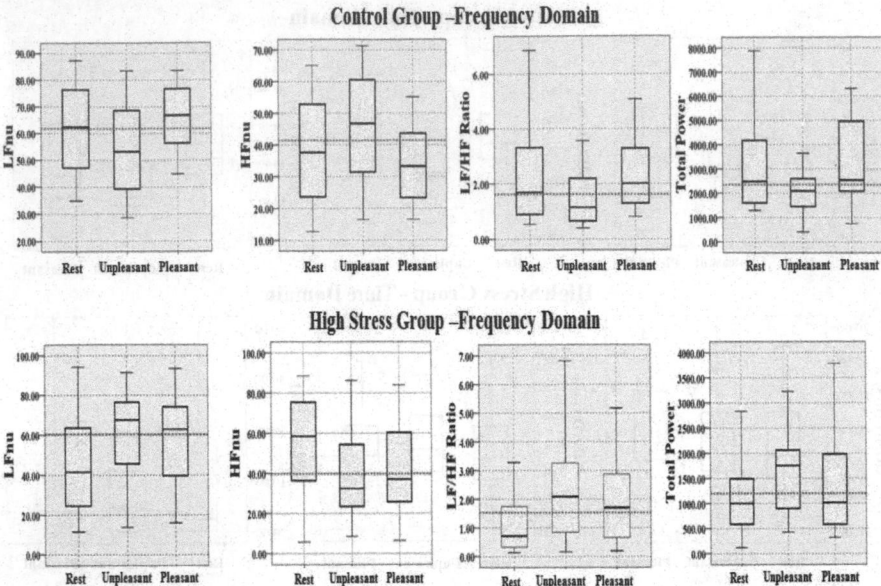

Figure 14. Frequency domain analysis in Control Group and High Stress Group during three emotional states.

wards parasympathetic predominance. In order to resist the negative stimulus, the ANS attempted to achieve homeostasis by increasing the vagal tone, and suppressing the sympathetic activity. However, the mean total power of Unpleasant session (2435±1691) has a significantly decreased (p<0.05) compared to Rest session (3138±1920). These results kept consistent with time domain and HRV pattern analysis. All the mean values of frequency domain in High stress group have a wholly reversed variation trend in contrasted with Control group, no matter Pleasant or Unpleasant session, the sympathy-vagal balance turned to sympathetic predominance which implied the modulatory function and reaction of ANS were weaken due to long term unemployed pressure.

5 Conclusion

This article described a complete ECG methodology to detecting chronic stress in unemployed populations. The approaches relied on estimating the state of the autonomic nervous system from an analysis of heart rate variability (HRV). This methodology consisted of a hardware platform, a software application, and a set of algorithms for ECG signal measurement, signal processing, QRS complex detection, HRV analysis and Biofeedback which based on the principles of Ambulatory, Simple, High-performance, Robust, and Cost effective. With this methodology, emotional shift experiments were performed to evaluate its performances and investigate the mechanism of Autonomic nervous system (ANS) modulation in unemployment with chronic stress. 53 right-handed volunteers were recruited in this experiment. HRV pattern analysis, time domain and frequency domain analysis performed to assess the influence of chronic stress in ANS modulation and response. The experiments of emotional shift have provided us a window to observe the changes of HRV pattern in subjects with chronic stress. Time domain, frequency domain also exhibited a significant difference between normal subjects and unemployed subjects. It was also shown that the modulatory function of ANS declined due to long term unemployed strain.

In conclusion, this methodology was effective for HRV based stress detection and much easier for emotion self-regulatory in unemployed populations. Further work is still needed to use this platform to study the mechanism of interaction between central nervous system (CNS) and autonomic nervous system (ANS), and a wearable mobile biofeedback system based on Wi-Fi PDA also need to be developed.

References

1. H. Hemingway and M. Marmot, "Evidence based cardiology: psychosocial factors in the aetiology and prognosis of coronary heart disease: systematic review of prospective cohort studies," BMJ, vol. 318, pp. 1460–1467, 1999.
2. M. Marmot and S. Stansfeld, Stress and the heart: psychosocial pathways to coronary heart disease. London, NJ: BMJ books, 2001.
3. Fries R, Konig J, Schafers HJ, Bohm M. Triggering effect of physical and mental stress on spontaneous ventricular tachyarrhythmias in patients with implantable cardioverter-defibrillators. Clin Cardiol 2002; 25: 474–8.
4. Steptoe A, Feldman PJ, Kunz S, Owen N, Willemsen G, Marmot M. Stress responsivity and socioeconomic status: a mechanism for increased cardiovascular disease risk? Eur Heart J 2002; 23: 1757–63.
5. Hugo D. Critchley, Peter Taggart, Peter M. Sutton, Diana R. Holdright, Velislav Batchvarov, Katerina Hnatkova, Marek Malik and Raymond J. Dolan. Mental stress and sudden cardiac death: asymmetric midbrain activity as a linking mechanism. Brain, vol.128 (1), pp: 75 – 85, 2005.
6. K. A. Matthews and B. B. Gump. Chronic work stress and marital dissolution increase risk of posttrial mortality in men from the Multiple Risk Factor Intervention Trial. Arch. Intern. Med., vol. 162, pp. 309–315, 2002.
7. M. Kivimaki, P. Leino-Arjas, R. Luukkonen, H. Riihimaki, J. Vahtera and J. Kirjonen. Work stress and risk of cardiovascular mortality: prospective cohort study of industrial employees. BMJ, vol. 325, pp. 857–861, 2002.
8. C. Welin, A. Rosengren, H. Wedel and L. Wilhelmsen. Myocardial infarction in relation to work, family and life events," Cardiovasc. Risk Factors, vol. 5, pp. 30–38.1, 1995.
9. S. Vandeput, J. Taelman, A. Spaepen and S. Van Huffel. Heart Rate Variability as a Tool to Distinguish Periods of Physical and Mental Stress in a Laboratory Environment. Proc. of the 6th International Workshop on Biosignal Interpretation (BSI), New Haven, Connecticut, pp. 187–190. Jun. 2009.
10. Van de Kar LD, Blair ML. Forebrain pathways mediating stress-induced hormone secretion. Front Neuroendocrinol. Vol.20 (1), pp: 41–48. 1999.
11. Tiina Ritvanen, Veikko Louhevaara, Pertti Helin, Sari Vaisanen, Osmo Hanninen. Responses of the autonomic nervous system during periods of perceived high and low work stress in younger and older female teachers. Applied Ergonomics, Vol.37, pp.311–318, 2006.

12. Bernardi, L., Wdowczyk-Szulc, J., Valenti, C., Castoldi, S., Passino, C., Spadacini, G., Sleight, P. Effect of controlled breathing, mental activity, and mental stress with or without verbalization on heart rate variability. Journal of the American College of Cardiology. Vol.35 (6), pp: 1462–1469, 2000.

13. John Yarnell. Stress at work—an independent risk factor for coronary heart disease? Eur Heart J. 2008 Mar. Vol.29 (5), pp: 579–80, 2008.

14. R. Fensli, et al. Sensor acceptance model – measuring patient acceptance of wearable sensors. Methods of Information in Medicine, vol. 47 (1), pp: 89–95, 2008.

15. Zhengbo Zhang, Weidong Wang, Buqing Wang, Hao Wu, Hongyun Liu, Yukai Zhang. A prototype of wearable respiration biofeedback platform and its preliminary evaluation on Cardiovascular Variability. ICBBE'09. IEEE, Beijing, pp: 1–4, 2009.

16. Dipali Bansal, Munna Khan, A. K. Salhan. A Review of Measurement and Analysis of Heart Rate Variability. International Conference on Computer and Automation Engineering (iccae 2009), pp.243–246, 08–10, Mar, 2009.

17. Fei Zhang, Yong Lian. Electrocardiogram QRS Detection Using Multiscale Filtering Based on Mathematical Morphology. EMBS'07, IEEE, Lyon, pp.3196–3199, 2007.

18. Chu, C.-H.H., Delp, E.J. Impulsive noise suppression and background normalization of electrocardiogram signals using morphological operators. IEEE Transactions on Biomedical Engineering. Vol.36 (2), pp: 262–273, 1989.

19. L. Sörnmo. Time-varing digital filtering of ECG baseline wanders. Medical and Biological Engineering and Computing. Vol. 31 (5), pp: 503–508, September, 1993.

20. Mikhled Alfaouri and Khaled Daqrouq. ECG Signal Denoising By Wavelet Transform Thresholding. American Journal of Applied Sciences, Vol.5 (3), pp: 276–281, 2008.

21. Tang Jing-tian; Zou Qing; Tang Yan; Liu Bin; Zhang Xiao-kai. Hilbert-Huang Transform for ECG De-noising. ICBBE 2007.pp: 664–667, 6–8 July, 2007.

22. Yan Sun, Kap Luk Chan, Shankar Muthu Krishnan. ECG signal conditioning by morphological filtering. Journal of computers in biology and medicine, Vol.32 (6), pp: 465–479, November, 2002.

23. Behzad Mozaffary, Mohammad A. Tinati. ECG Baseline Wander Elimination using Wavelet Packets. World Academy of Science, Engeering and Technology. Vol.3, pp: 14–16, 2005.

24. Fei Zhang, Yong Lian. QRS Detection Based on Multiscale Mathematical Morphology for Wearable ECG Devices in Body Area Networks. IEEE Transactions on Biomedical Circuits and Systems. Vol.3 (4), pp: 220–228. Aug, 2009

25. Wanqing Wu, Jungtae Lee. Development of Full-Featured ECG System for Visual Stress Induced Heart Rate Variability (HRV) Assessment. IEEE International Symposium on Signal Processing and Information Technology, Luxor, Egypt. December 15–18, 2010.

26. CHEE-HUNG HENRY. CHU, E.J. Delp, Impulsive noise suppression and background normalization of electrocardiogram signals using morphological operators, IEEE Trans. Biomed. Eng. Vol.36 (2), pp: 262–273, 1989.

27. Kohler, B.-U., Hennig, C., Orglmeister, R. The principles of software QRS detection. IEEE Engineering in Medicine and Biology Magazine, Vol.21 (1), pp: 42–57, February, 2002.

28. Task Force of the European Society of Cardiology and the North American Society of Pacing and Electrophysiology. Heart rate variability. Standards of measurement, physiological interpretation, and clinical use. Circulation; Vol.93, pp: 1043–1065, 1996.

29. Sztajzel J. Heart rate variability: a noninvasive electrocardiographic method to measure the autonomic nervous system. SWISS MED WKLY. Vol.134 (35-36): pp: 514 – 522, 2004.

30. Lang, P.J. Bradley, M.M., and Cuthbert, B.N. NIMH Center for the Study of Emotion and Attention, International Affective Picture System (IAPS). Technical Manual and Affective Ratings. 1995.

31. Lee Lipsenthal, MD. Heart Rate Variability and Emotional Shifting: Powerful Tools for Reducing Cardiovascular Risk. Holistic Primary Care. Vol.5 (4), pp: 2–4, 2004.

Chapter 17
COHERENCE OF GAIT AND MENTAL WORKLOAD

S. M. Hsiang, T. Karakostas, C.-C. Chang, and S. Ekwaro-Osire

1 Introduction

We propose a spectral analytical approach to evaluate the coordination of gait and mental task. Our hypothesis is that when the mental workloads and the gait control are coherent, the capability to recover from or prepare for a perturbation is enhanced. Such analysis is critical for conducting safe foot patrols or walking inspection jobs, for examples, police, foot soldier, immigration officer, meter reader, and healthcare workers [1], where slips, trips and falls (STL) constitute the majority of workplace accidents. Falls, fatal in extremely severe cases, can cause various injuries from head concussion, neck whiplash, paralysis, back pain, broken hip, to torn anterior cruciate ligament (ACL). Even if there is no fall in some situations, trying to catch the balance can cause muscle sprains and strains to muscles, which may have long term and broader impacts to aging process and quality of life due to this accident.

To prevent STL, most studies emphasize tribology, cleaning protocol and precautious measures such as lighting, anti-slip shoe sole design. Few studies address the dual task coordination, particularly the coherence between the gait and the given mental work load, assuming there are physical (rhythmic) component in mental activity and cognitive component in walking. The research gap is partly due to the general assumption that human gaits are cyclic and feed-forward [5]. If a walker can perceive, anticipate and prepare for the incoming hazardous stimuli, at least one to two steps ahead, his gait is less likely to be perturbed. The challenge is quantify the effect of the entrainment between the two activities. When the difference, between the anticipation and environmental stimuli, is so significant that he does not have enough joint moment or time to bring the center of mass inside the base of support, a slip, trip and fall mishap is more likely to happen.

S. Ekwaro-Osire (✉)
Lubbock, TX, USA
e-mail: stephen.ekwaro-osire@ttu.edu

S. Suh et al. (eds.), *Biomedical Engineering*,
DOI 10.1007/978-1-4614-0116-2_17, © Springer Science+Business Media, LLC 2011

2   Coherence Tuning

Since few occupational tasks involve walking only, to identify the underlying mechanism for coordination and adaptation, many research approaches are proposed, such as psychophysics focusing on the perception of the walking conditions; physiology muscle strength; and biomechanics on toe or heel clearance and reaction time. In either aforementioned approaches of gait and mental workload, the critical issue of finding an index of remaining capacity to prepare for a perturbation is not solved. In fact the problem potentially involves three steps: (1) Index and map the coordination; (2) Find the most robust region to the perturbation; (3) Tuning the coherence to the robust region.

Conventionally, coherence or covariability are considered intrinsic within normal human biomechanics and can be depicted as the normal random processes that occur in motor control across multiple repetitions of a task, such as the bipedal gait. However, some studies suggest that many biological behaviors, including chaotic phenomena observed in clinical settings [2], described as stochastic or pendulum in nature are actually the result of nonlinear dynamics [3].

Haken et al. [4] proposed a model to account for the observations on human bimanual coordination that revealed basic features of self-organization which refers to the pattern formation and change in a nonequilibrium system that is open to the exchange of energy and information with its environment. In Haken et al. [4] they found that

1. Only two stable states exist: in-phase motion and antiphase motion.
2. As the frequency increases, the amplitude of motion decreases.
3. As the frequency increases past a critical frequency, antiphase motion abruptly changes to in-phase motion.
4. Beyond this transition, only in-phase motion is possible. That is, for frequencies above the critical frequency, only in-phase motion is stable, while below the critical frequency both in-phase and antiphase motions are stable.

By using these four basics, consider that the two tasks of interest: gait and a mental task, and both can be tracked and indexed by harmonics with amplitude r, and phase ϕ under frequency ω:

$$cx_{gait}(t) = r_{gait}(t) \cos{(\omega t + \phi_{gait}(t))}$$
$$\text{and } x_{task}(t) = r_{task}(t) \cos{(\omega t + \phi_{task}(t))} \tag{1}$$

For either activity:

$$\dot{x}(t) = -r(t)(\omega + \dot{\phi}(t)) \cdot \sin{(\omega t + \phi(t))}$$
$$+ \dot{r}(t) \cdot \cos{(\omega t + \phi(t))} \tag{2}$$

For nearly simple harmonic, they require that

$$-r(t)\dot{\phi}(t) \sin{(\omega t + \phi(t))} + \dot{r}(t) \cos{(\omega t + \phi(t))} = 0 \tag{3}$$

Assume that the gait and the metal task can be coupled with following hybrid model of the Van der Pol and Rayleigh differential equations:

$$
\begin{cases}
\ddot{x}_{gait} - \alpha \dot{x}_{gait} + \beta \dot{x}_{gait}^3 + \gamma x_{gait}^2 \dot{x}_{gait} + \omega^2 x_{gait} = \\
\quad - (\dot{x}_{gait} - \dot{x}_{task}) \left[a - b(\dot{x}_{gait} - \dot{x}_{task})^2 - c(x_{gait} - x_{task})^2 \right] \\
\ddot{x}_{task} - \alpha \dot{x}_{task} + \beta \dot{x}_{task}^3 + \gamma x_{task}^2 \dot{x}_{task} + \omega^2 x_{task} = \\
\quad - (\dot{x}_{task} - \dot{x}_{gait}) \left[a - b(\dot{x}_{task} - \dot{x}_{gait})^2 - c(x_{task} - x_{gait})^2 \right]
\end{cases}
\tag{4}
$$

where the coupling coefficients a, b, and c are assumed to be "small" relative to α, β, and γ, yielding a weak coupling (e.g. $\alpha = 0.5$, $\beta = 0.001$, $\gamma = 0.3875$, $a = 0.049$, $b = 0$, $c = 0.036$). To solve the two equations, we need to integrate both over a period $2\pi\omega$, and set equal to 0 and solve for r_{gait}, r_{task}, and $(\phi_{gait} - \phi_{task})$:

$$
\dot{r}_{gait}(t) \approx \frac{\omega}{2\pi} \int_t^{t+\frac{\omega}{2\pi}} \dot{r}_{gait}(\tau) d\tau =
$$

$$
\frac{1}{8}
\begin{bmatrix}
r_{task} \left(4a - (c + 3b\omega^2) \left(3r_{gait}^2 + r_{task}^2 \right) \right) \cos \Delta\phi \\[2mm]
+ r_{gait}^2 \left(\begin{array}{l} 4(\alpha - a) + r_{gait}^2 (c - \gamma + 3(b - \beta)\omega^2) + \\ r_{task}^2 (c + 3b\omega^2)(2 + \cos 2\Delta\phi) \end{array} \right)
\end{bmatrix}
\tag{5}
$$

$$
\dot{r}_{task}(t) \approx \frac{\omega}{2\pi} \int_t^{t+\frac{\omega}{2\pi}} \dot{r}_{task}(\tau) d\tau =
$$

$$
\frac{1}{8}
\begin{bmatrix}
r_{gait} \left(4a - (c + 3b\omega^2) \left(3r_{task}^2 + r_{gait}^2 \right) \right) \cos \Delta\phi + \\[2mm]
r_{task}^2 \left(\begin{array}{l} 4(\alpha - a) + r_{task}^2 (c - \gamma + 3(b - \beta)\omega^2) + \\ r_L^2 (c + 3b\omega^2)(2 + \cos 2\Delta\phi) \end{array} \right)
\end{bmatrix}
\tag{6}
$$

$$
\Delta\dot{\phi}(t) \approx \frac{\omega}{2\pi} \int_t^{t+\frac{\omega}{2\pi}} \left(\dot{\phi}_{gait}(\tau) - \dot{\phi}_{task}(\tau) \right) d\tau =
$$

$$
\frac{r_{task}^2 + r_{gait}^2}{8 r_{gait} r_{task}}
\begin{bmatrix}
\left(\begin{array}{l} -4a + \\ (c + 3b\omega^2)\left(r_{gait}^2 + r_{task}^2 \right) \end{array} \right) \sin \Delta\phi - \\[2mm]
r_{gait} r_{task} (c + 3b\omega^2) \sin 2\Delta\phi
\end{bmatrix}
\tag{7}
$$

By tuning the magnitude r and phase difference $\Delta\phi$, we can reach the optimal coherence:

Figure 1. Major parameters of the vertical ground reaction forces for left- and right-foot (WA: Weight Acceptance, MS: Mid-Stance and PO: Push-off).

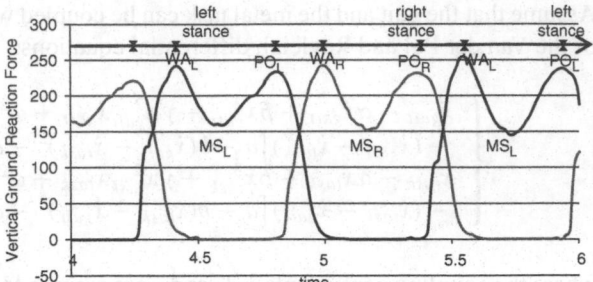

$$\Delta\phi = cos^{-1}\left(\frac{-4a - (c + 3b\omega^2)\,(r_{task}^2 + r_{gait}^2)}{2(c + 3b\omega^2)r_{gait}r_{task}}\right) \tag{8}$$

$$|r_{gait}| = |r_{task}| =$$

$$\frac{2\sqrt{\alpha - a(1 - \cos\Delta\phi)}}{\gamma + 3\beta\omega^2 - (c + 3b\omega^2)\,(3 - 4\cos\Delta\phi + \cos 2\Delta\phi)}$$

3   Experimental Results

Given the model discussed, we propose a new approach to analyze coordination of the kinematic and kinetic variables, where the cross spectral densities, between the ground reaction force and the duration of the mental task, provide the index of coherence. When there are mental workloads, the gait patterns become more automatic and the cross-periodograms show less entropy. As such, the flexibility to prepare for a perturbation is altered. Thirty volunteers without any reported neural, musculoskeletal problems, participated in the study. They walked with harness on instrumented treadmill at an average speed of 2.4±0.3 mph for 10 minutes, and the last 2 minutes (150~240 steps) were collected. Figure 1 show the vertical ground reaction force profile where the basic gait parameters, including weight acceptance (WA), midestance (MS), and push-off (PO) can be identified. Two levels of computations, 1, and 2 digits additions, were introduced to the subjects on a big screen in front of the subjects. The subjects were asked to speak out loud, though the accuracy was not evaluated. Figure 2 shows the overlapping period between the computation and the sum of the vertical forces of both feet.

To match Figures 2 and 1, the MS were identified. Depending on the cognitive proficiency, the computations could be conducted during any period (e.g., WA_L-WA_R, MS_L-MS_R), and there is no clear preferences on the initiation time. Figure 3 shows (a) the amplitude (modulus) of the cross-spectrum estimate against frequency; (b) the coherence squared of gait and computation task, and (c) phase spectrum in radians of gait and computation task. The effect due to the difficulty of computational task is significant to both amplitude and phase. When a bicycle break was applied, to grab and pull a light weight polyurethane rod attached to each subject's shoe heel, causing unexpected tripping effect, the results show that there is a signifi-

Figure. 2. The summation of the left and right ground reaction force during the same period of Figure 1, and the rectangle wave represent the period of computations. Both normalize to unit scale (100%).

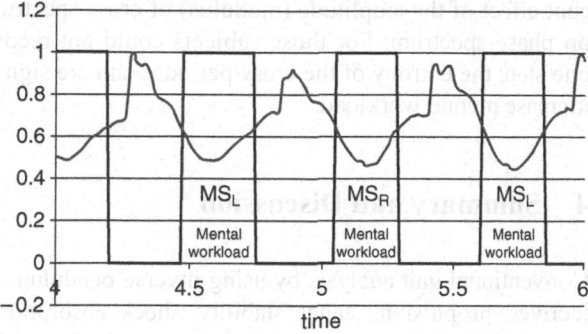

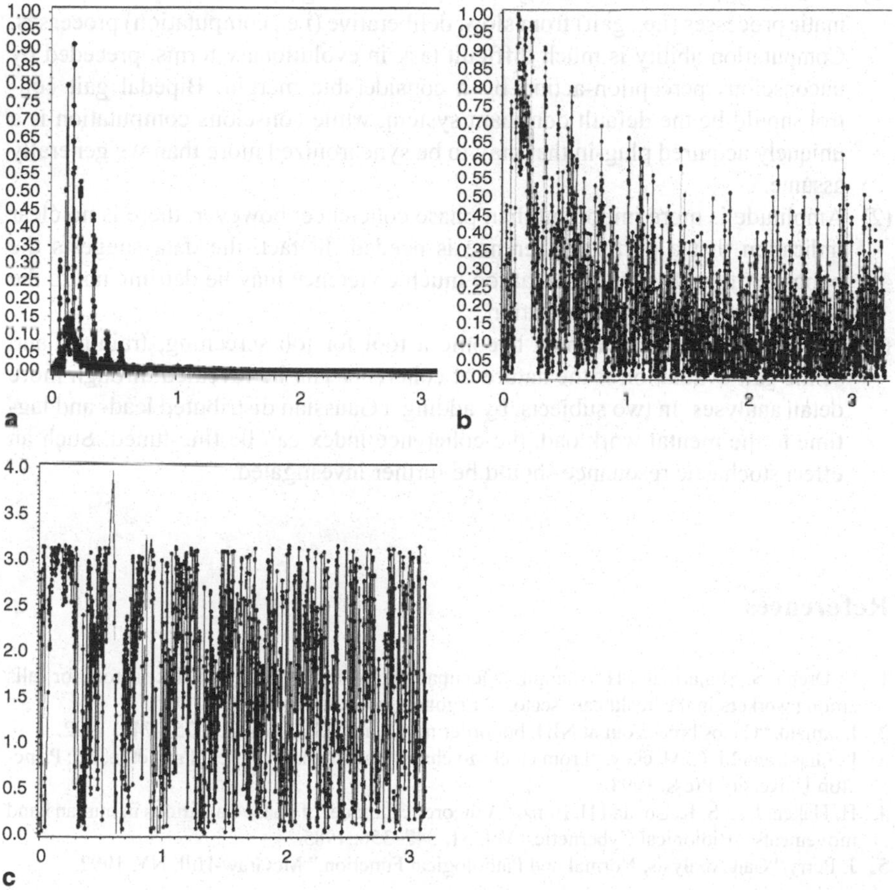

Figure 3. (a) The amplitude of the cross-spectrum estimate against frequency; (b) the coherence squared of gait and computation task; (c) phase spectrum in radians of gait and computation task.

cant effect of the amplitude (modulus) of cross-spectrum, but no significant effect on phase spectrum. For those subjects could not recover the perturbation within one step, the entropy of the cross-periodogram are significantly decayed due to the increase mental workload.

4 Summary and Discussion

Conventional gait analysis by using inverse pendulum model emphasizes four objectives: propulsion, stance stability, shock absorption, and energy conservation (Perry 1992). This study provides a novel tool based on the coordination of gait and a mental task, where the stride-to-stride coherence or covariability analysis may be useful:

(1) As already observed, the study attempts to distinguish and couple fast, automatic processes (i.e., gait) from slow, deliberative (i.e., computation) processes. Computation ability is much difficult task in evolutionary terms, preceded by unconscious perception-action by a considerable margin. Bipedal gait control should be the default dominant system, while conscious computation is a uniquely acquired plug-in that may to be synchronized more than we generally assume.
(2) Amplitude is more important than phase coherence; however, there is no clear indication that a perfect coherence is needed. In fact, the data suggests the opposite of our hypothesis, that too much coherence may be detrimental to the recoverability from a perturbation.
(3) The proposed approach may become a tool for job screening, training, and senile gait evaluation if the nature of coherence can be revealed through more detail analyses. In two subjects, by adding a Gaussian distributed lead- and lag-time to the mental workload, the coherence index can be fine-tuned. Such an effect stochastic resonance should be further investigated.

References

1. S. Drebit, S. Shajari, and H. Alamgir, "Occupational and environmental risk factors for falls among workers in the healthcare sector, " Ergonomics, Vol. 53, 525–536, 2010
2. I. Amato, "Chaos breaks out at NIH, but order may come of it," Science, 257-747, 1992.
3. L. Glass, and M. C. Mackey, "From clocks to chaos: The rhythms of life," Princeton, NJ: Princeton University Press. 1998.
4. H. Haken, J. A. S. Kelso, and H. Bunz, "A theoretical model of phase transitions in human hand movements," Biological Cybernetics, Vol. 51, 347–356, 1985.
5. J. Perry, "Gait Analysis, Normal and Pathological Function," McGraw-Hill, NY. 1992,

Chapter 18
EFFICIENT DESIGN OF LDPC CODE FOR WIRELESS SENSOR NETWORKS

Sang-Min Choi and Byung-Hyun Moon

1 Introduction

In recent years, the idea of wireless sensor networks has produced lots of research, because of wireless sensor networks can be applied widely in many fields. In wireless sensor networks, erroneous transmission can happen by wireless channel noise. Automatic repeat request (ARQ) detects errors using cyclic redundancy check (CRC) and retransmission of data is used as error control scheme for wireless sensor networks. Sensor node requires long lifetime with limited battery, but retransmission reception of data is the primary source of energy consumption and reduces the lifetime of sensor node. Therefore, error control scheme of forward error correction (FEC) for wireless sensor networks is necessary [1].

Low-density parity-check (LDPC) codes were first introduced by Gallager in 1962 and rediscovered by Mackay and Neal in 1996 and come into the spotlight for next generation communication system[2] [3]. The LDPC codes construction can be categorized into Mackay's random construction and sub-block based structured construction. The Mackay's random construction LDPC codes show very good performance. However, this construction is computationally intensive implementation because of large memory requirement. The sub-block based structured construction scheme can be implemented with less complexity than the Mackay's random construction [5].

In this paper, we propose FEC using quasi-cycle (QC) LDPC code for the error control in the wireless sensor networks [4]. The proposed LDPC code has small size parity check matrix and can be implemented easily. The efficient encoding method that is simplified Richardson's encoder is proposed for wireless sensor node application. The proposed encoding method does not require matrix inversion so that this method is suitable for wireless sensor node which has limited computation ability. Also, the bit-flipping (BF) decoding algorithm with hard-decision decoding method is proposed for wireless sensor node application.

B.-H. Moon (✉)
Daegu, Republic of Korea
e-mail: bhmoon@daegu.ac.kr

S. Suh et al. (eds.), *Biomedical Engineering*,
DOI 10.1007/978-1-4614-0116-2_18, © Springer Science+Business Media, LLC 2011

The paper is organized as follows. In Section 2, construction and encoding of LDPC codes is given. In Section 3 the proposed LDPC code construction method for wireless sensor networks is shown. And, the performance of the proposed method is compared with the Mackay's random construction method over AWGN channel is given in Section 4. Finally, the conclusion is made in section 5.

2   Construction and Encoding of LDPC Codes

2.1   QC-LDPC Codes

For wireless sensor network applications, we consider a subclass of LDPC code, QC_LDPC, whose parity-check matrix consists of circulant permutation matrices or the zero matrix.

I_i is the $N_s \times N_s$ permutation matrix which shifts the identity matrix I to the right by i-times for any integer i, $0 \le i \le N_s$. Let I_1 be the $N_s \times N_s$ permutation matrix given by

$$I_1 = \begin{bmatrix} 0 & 1 & 0 & \cdots & 0 \\ 0 & 0 & 1 & \cdots & 0 \\ \vdots & \vdots & \vdots & \ddots & \vdots \\ 0 & 0 & 0 & \cdots & 1 \\ 1 & 0 & 0 & \cdots & 0 \end{bmatrix} \tag{1}$$

Using this notation parity check matrix H can be defined by

$$H = \begin{bmatrix} I_{s_{0,0}} & I_{s_{0,1}} & I_{s_{0,2}} & \cdots & I_{s_{0,(n-1)}} \\ I_{s_{1,0}} & I_{s_{1,1}} & I_{s_{1,2}} & \cdots & I_{s_{1,(n-1)}} \\ I_{s_{2,0}} & I_{s_{2,1}} & I_{s_{2,2}} & \cdots & I_{s_{2,(n-1)}} \\ \vdots & \vdots & \vdots & \ddots & \vdots \\ I_{s_{(m-1),0}} & I_{s_{(m-1),1}} & I_{s_{(m-1),2}} & \cdots & I_{s_{(m-1),(n-1)}} \end{bmatrix} \tag{2}$$

where $S_{i,j}$ is the shift value corresponding position (i, j) sub-block. This value is one of the $\{0, 1, 2, ..., N_s-1\}$ for nonzero sub-block and $Si,j = -1$ for zero matrix. The size of H is $mN_s \times nN_s$.

2.2   Encoding of LDPC Codes

As shown in Figure 1, block type LDPC encoding process has Richardson's encoding algorithm if $M \times N$ Parity check matrix is divided into the form

$$H = \begin{bmatrix} A & B & T \\ C & D & E \end{bmatrix} \tag{3}$$

Figure 1. Parity check matrix for efficient encodding.

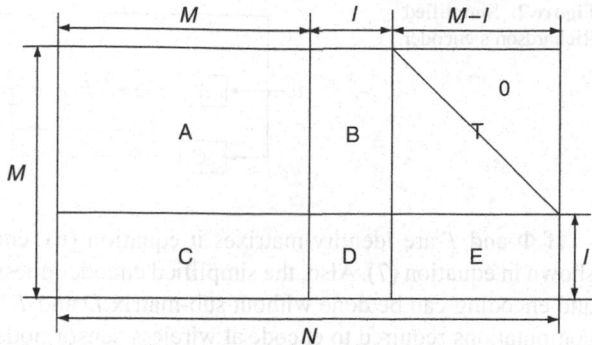

where A is $(M\text{-}l) \times (N\text{-}M)$, B is $(M\text{-}l) \times l$, T is $(M\text{-}l) \times (M\text{-}l)$, C is $l \times (N\text{-}M)$, D is $l \times l$, E is $l \times (M\text{-}l)$. All these matrices are sparse and T is lower triangular with one along the diagonal [5].

Let the codeword $c = \{u \quad p_1 \quad p_2\}$ where u denotes the systematic part, p_1 and p_2 denote the parity part, p_1 has length l, and p_2 has length $(M\text{-}l)$. The codeword c satisfies the following equations.

$$Hc = \begin{bmatrix} A & B & T \\ C & D & E \end{bmatrix} \begin{bmatrix} u^T \\ p_1^T \\ p_2^T \end{bmatrix} = 0 \tag{4}$$

$$Au^T + Bp_1^T + Tp_2^T = 0$$

$$\left(-ET^{-1}A + C\right)u^T + \left(-ET^{-1}B + D\right)p_1^T = 0 \tag{5}$$

Let $\Phi = -ET^{-1}B + D$ and assume that Φ is nonsingular. Then we can obtain parity bits as following equations

$$p_1^T = -\Phi^{-1}(-ET^{-1}A + C)u^T$$
$$p_2^T = -T^{-1}\left(Au^T + Bp_1^T\right) \tag{6}$$

3 LDPC Code for Wireless Sensor Networks

Once sensor node of wireless sensor networks are deployed on fields, it is difficult to replace the battery. Thus, wireless sensor nodes require low power hardware for long lifetime. Since the wireless sensor nodes have small memory size, they must operate with minimum computations to save energy. In order to save energy in the wireless sensor nodes when they encode, the simplified Richardson's encoder is proposed.

Figure 2. Simplified
Richardson's encoder.

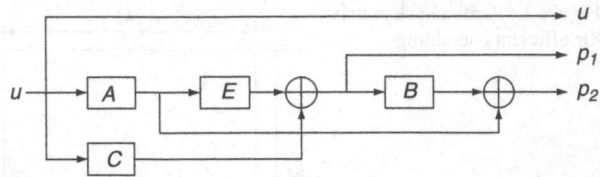

If Φ and T are identity matrixes it equation (6), encoder can be simplified as shown in equation (7). Also, the simplified encoder doesn't require matrix inversion and encoding can be done without sub-matrix D and T. This will reduce amount of computations required to encode at wireless sensor nodes. Figure 2 shows the simplified Richardson's encoder.

$$p_1^T = -(-EA + C)u^T$$
$$p_2^T = -\left(Au^T + Bp_1^T\right) \tag{7}$$

The parity check matrix for this encoder is constructed by 3×6 sub-matrices as shown in equation (8).

$$H = \begin{bmatrix} A_{1,1} & A_{1,2} & A_{1,3} & A_{1,4} & A_{1,5} & A_{1,6} \\ A_{2,1} & A_{2,2} & A_{2,3} & A_{2,4} & A_{2,5} & A_{2,6} \\ A_{3,1} & A_{3,2} & A_{3,3} & A_{3,4} & A_{3,5} & A_{3,6} \end{bmatrix} \tag{8}$$

In equation (8), to make T matrix as identity matrix, let $A_{1,5} = A_{2,6} = I$ and $A_{1,6} = A_{2,5} = I_{-1} = 0$. To make Φ as identity matrix from equation (9), EB should be zero matrix.

$$\begin{aligned} \Phi &= -ET^{-1}B - D \\ &= \begin{bmatrix} A_{3,5} & A_{3,6} \end{bmatrix} \begin{bmatrix} A_{1,4} \\ A_{2,4} \end{bmatrix} + A_{3,4} \\ &= A_{3,5}A_{1,4} + A_{3,6}A_{2,4} + A_{3,4} = I \end{aligned} \tag{9}$$

Also, if we let $A_{3,4} = I$ which corresponds to matrix D as identity matrix, equation (10) is obtained.

$$A_{3,5}A_{1,4} + A_{3,6}A_{2,4} = 0 \tag{10}$$

If $A_{3,5}A_{1,4} = A_{3,6}A_{2,4}$, then Φ become identity matrix. In order to satisfy equation (10), we let $A_{1,4}$ in B matrix and $A_{3,6}$ in E matrix as identity matrix and $A_{3,5}$ and $A_{2,4}$ are equivalent matrix. The construction of the proposed parity check matrix is shown in Figure 3.

In this paper, we propose 96×192 parity check matrix that is made by shifting the 32×32 identity matrix as shown in the following equation (11). The proposed check matrix is shown in Figure 4.

$$H = \begin{bmatrix} I & I & I & I & I & I_{-1} \\ I_1 & I_3 & I_5 & I_7 & I_{-1} & I \\ I_{13} & I_{11} & I_9 & I & I_7 & I \end{bmatrix} \tag{11}$$

Figure 3. Construction of proposed parity check matrix.

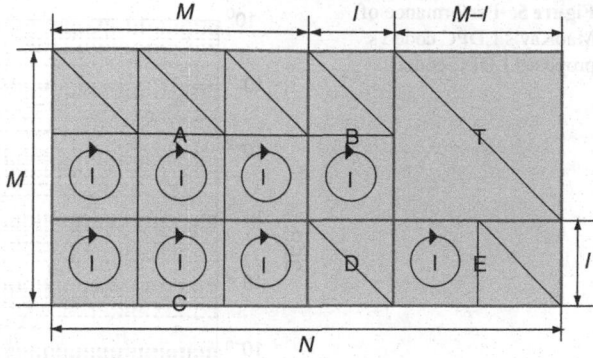

Figure 4. Proposed parity check matrix.

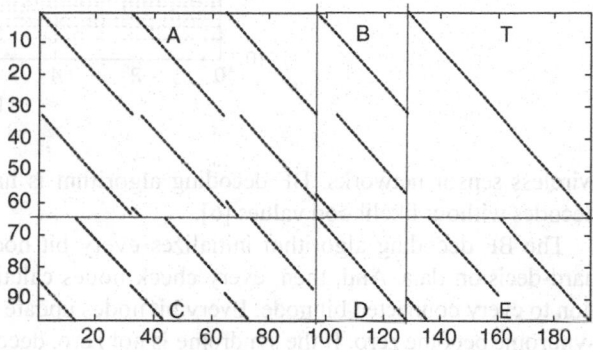

Code rate 1/2 Mackay's random constructed LDPC code with column weight 3 and code length 192 has six cycle-4. However the proposed LDPC code has zero cycle-4 and girth of 6.

4   Simulation Results

The performance of the proposed LDPC code is simulated in the Additive White Gaussian Noise (AWGN) channel. The sum-product algorithm is used to decode LDPC code with the maximum iteration number as 100.

Figure 5 depicts the BER curves of the proposed LDPC code vs the Mackay's random constructed LDPC code. It is shown that the performance of the proposed LDPC code is compared with Mackay's random constructed LDPC code. The proposed LDPC code shows similar result with Mackay's random constructed LDPC code.

For sensor nodes, LDPC decoding by using SPA (Sum-Product Algorithm) seems impossible because of the small size memory, low computation ability of sensor nodes. In this paper, we proposed bit-flipping (BF) decoding algorithm for

Figure 5. Performance of MacKay's LDPC code vs proposed LDPC code.

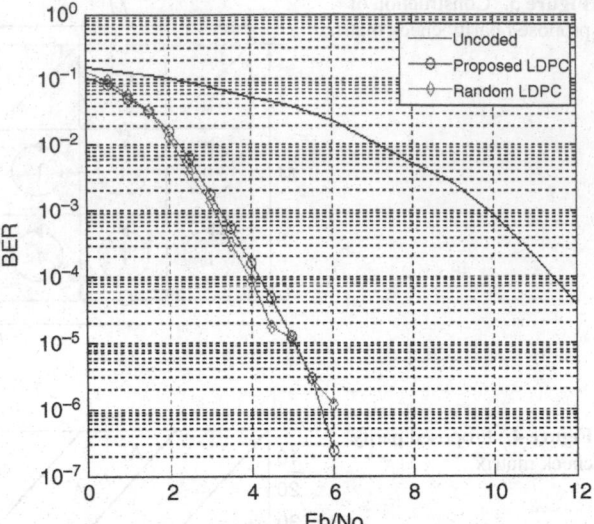

wireless sensor networks. BF decoding algorithm is hard-decision decoding and decodes without likelihood values [6].

The BF decoding algorithm initializes every bit nodes of bipartite graph into hard-decision data. And, then, every check nodes calculate a modulo-2 computation to every connected bit node. Every bit nodes update until check node value and syndrome become zero. If the syndrome is not zero, decoding process continues the next iteration.

Figure 6 depicts the BER curves of the BF decoding algorithm for LDPC code. It is shown that the proposed QC-LDPC code obtains at least 3dB gain over the uncoded case at BER$= 10^{-3}$. And the proposed LDPC code suffers about 1dB loss over Mackay's random constructed LDPC code.

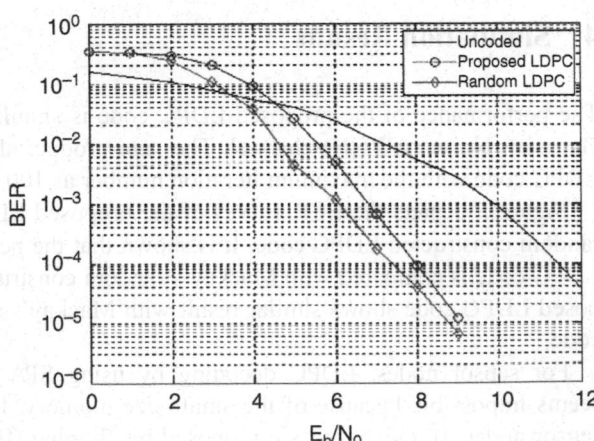

Figure 6. Performance of BF decoding algorithm.

5   Conclusion

In this paper, we propose FEC using QC-LDPC code for the error control in the wireless sensor networks. The proposed LDPC code has small size parity check matrix and easy implementation scheme. The encoding of the proposed QC-LDPC code that constructs sub-matrix T and Φ of Richardson's encoder into identity matrix and requires less computation compare to the Mackay's random constructed LDPC code

The proposed QC-LDPC code with LLR SPA algorithm with 100 iterations is performed. It is shown that the proposed QC-LDPC code showed similar result compare to the Mackay's randomly constructed LDPC code. In order to satisfy the limited power and computation limitation for sensor node application, the proposed code with BF algorithm is tested. The simulation results showed that the proposed QC-LDPC code suffers less that 1dB loss compared to the Mackay's LDPC code.

In the near future, we will implement the proposed QC-LDPC code on the sensor node. And the power consumption of the QC-LDPC code and the Mackay's will be compared.

References

1. Jaein Jeong and Cheng Tien Ee, "Forward error correction in sensor Networks," U.C.Berkeley Technical Report, May, 2003.
2. R. G. Gallager, "Low-Density parity-Check Codes," IRF Trans. On Info. Theory, vol.8, pp.21–28, Jan. 1962.
3. D. J. C. MacKay and R. M. Neal, "Near Shannon linit performance of low density prity check codes," Electron, Lett., vol.32, no.18, pp.1645–1646, Aug. 1996.
4. M. Rossi, M. Sala, "On a class of quasi-cyclic LDPC codes," BCRI Workshop on Coding and Cryptography, Cork, Ireland, 2005.
5. Thomas J, Richardson and R. L. Urbanke, "Efficient Encoding of Low-Density Parity-Check Codes," IEEE Trans. IT, vol.47, pp.638–656, Feb. 2001.
6. Bernhard M. J. Leiner, "LDPC codes – a Brief Tutorial," April 8 2005, http://www.hut.fi/~pat/coding/essays/ldpc.pdf

Chapter 19
TOWARD MULTI-SERVICE ELECTRONIC MEDICAL RECORDS STRUCTURE

Bilal I. Alqudah and Suku Nair

1 Introduction

EMR, or Electronic Medical Records, considered as the modern way of managing patients' records. Providing electronic medical health records can improve the quality of service, increase the level of health care provided, save manpower, and resources. Due to the accelerated development in health care, and the adaption of computer and network technology, concerns like information misuse, privacy violation, and identity theft are evolving rapidly. However, depending on traditional access control, encryption, and physical security, each as independent solution, may not be sufficient in an environment where attacks from inside and outside can occur equally likely. Though, the ubiquitous use of the Internet, Web-based applications, and moving toward EMRs, protecting health records and retaining privacy became a major issue. We are trying to show the importance of data granularity and structure in keeping the information private without relying only on software implementation for electronic medical records, EMR. We are aiming in this work to mitigate privacy breaches, accidental data disclosure, ease information sharing, and minimizing the cost of securing electronic medical records.

The significance of EMR among all community sectors pushes toward more regulations and standardization. This importance came from the amount of personal information stored on it, and its impact if disclosed to the public. However, the consequences of releasing medical information could be job loss, increase of insurance rates, identity theft, sexual crimes, and discrimination based on health problems. According to El Emam *et al.* [9], 63% to 87% of the US populations are identifiable depending on some distinctive characteristics, which provide an indication about the problem extent. Hence, it is not the personal information which identifies the person, the knowledge provided by medical data can identify an individual.

Privacy protection is a problem with twofold. First concern accentuates when data is used within its normal context, such as hospitals, clinics, and insurance com-

B. I. Alqudah (✉)
Dallas, TX, USA
e-mail: balqudah@smu.edu

S. Suh et al. (eds.), *Biomedical Engineering*,
DOI 10.1007/978-1-4614-0116-2_19, © Springer Science+Business Media, LLC 2011

panies where the attack can take place by authorized or unauthorized person [22]. The second concern is the trust and transitive trust when EMRs disclosed for third parties to use in educational, commercial, or research purposes.

Types of privacy contraventions vary according to the context. Despite the record being preserved by the data custodian, an intruder can instigate an active attack, which might cause service interruption or could lead to physical impairment for patients. The passive attack could be performed against the disclosed EMRs, out of its context, to identify a specific victim.

Christian Stingl [22] mentioned that 50% of the attacks targeting the information systems conducted by insiders, which is an *authorized* users who can be classified as *Honest-But-Curious* (HBC). Though if encrypted, having data owners (i.e. patients) away from contributing to their information security impose trust issues like HBC. Since a considerable amount of data stored in one location/form, a lot of information will be disclosed after a single successful attack, it will reveal the patient identity as well as medical information. In other cases where data are kept in databases, with analysis, an attacker can identify the personality of the victim [9].

A patient can be identified by his unique information, or by having a set of activity or health related information, Latanya Sweeney [23] provided an example for identifying William Weld, who was a governor of Massachusetts, after GIC Company released his medical records. By direct matching with the voters lists, the set of data reduced to six people with the same birth date, half of them were men, and he was "the only one in his ZIP code".

El Emam *et al.* [9] provides a comparison between common heuristics to evaluate data importance of HIPAA-specified ID fragments [Table 1] and non-HIPAA specified IDs such as providers' information, this can help us to recognize which data can be ranked as critical, high importance or normal in terms of identifying an individual.

Table. 1 HIPAA regulated Patient identifiers	1. Account Numbers
	2. Name(s) of relative(s)
	3. Biometric identifiers
	4. Names
	5. Certificate/License numbers
	6. Medical Record Number
	7. Dates
	8. Photographs and comparable images
	9. Device identifiers
	10. Postal Address
	11. Email addresses
	12. Social Security Number
	13. Fax numbers
	14. Telephone numbers
	15. Health Plan Numbers
	16. Vehicle identifiers including license plate numbers
	17. IP address numbers
	18. Web URL's
	19. Any other unique identifying number, characteristic, or code

Issues of granting access to some critical information in the electronic medical record, fears from accidental disclosure of the record, sharing information with others or asking for medical opinion without disclosing our private information or our identity, the fear from misusing the information, or being accessible to third parties like insurance companies or employer without our knowledge. All those reasons and more are motivating us to look for practical solutions to keep the data private and gives us the needed flexibility to show or hide our information depending on our needs.

2   Related Work

Due to the ubiquitous use of the Internet and information sharing, it becomes essential to ascertain from having a well-designed access polices, information disclosure roles, standards, regulations, and concrete implementation to retain privacy and protect assets [9], [19], [14], [6], [17], and [3]. However, risk increased gradually by moving toward the electronic medical records and the Internet [24], [3], [21].

In 1992 David Ferraiolo et al. [12] introduced a non-discretionary access control role-based access control,RBAC, which is more appropriate and central than the Discretionary Access Control,DAC, [18], [8]. Hence, it provides a new access control that satisfies the non-military systems by assigning rights to the roles rather than users. Their proposal provided a base for other access control mechanism based on time, location, and access duration.

Bertino et al. [4] introduced the Temporal RBAC, TRBAC, as an extension of RBAC to support periodic role enabling and disabling, which provides more control and compliance with HIPAA constraints. Junzhe Hu et al. [15] introduced further steps to authenticate and grant privileges depending on the context to comply with the HIPPA regulations [14]. This approach, context-aware security, went further than {user, roles, permission, session} in RBAC to refine the authentication criteria. It includes time and place of access, data objects, and operations in addition to other constraints such as context constrains.

Dekker [7] and Ferraiolo et al. [13] discussed the distributed nature of the e-health systems, the increase of access policies, and roles and how they lead to provide a distributed administration model. As distributed databases, and the web-based application became a common trend for development, the need to build trust [25] and enhance the password authentication [26] became a field of interest to provide security management for health systems. Another important aspect in privacy is the accidental or intentional information disclosure where a single click can reveal a large number of records.

Since the health level 7,HL7, clinical documents architecture,CDA, standard gains wide attention by providing a base for integration, exchange, retrieval and sharing of electronic health information, HL7 standards has been adapted in many EMR research [20], [11], [16], [2]. Bhatti et al. [5] proposed a context-aware policy-based approach for federated health-care databases (RHIOs), Elaborating RBAC in policy-based systems, and HL7 standards. Bhatti *et al.* tries to address the requirements for security management in RHIOs from the perspective of database system principles.

3   Problem description

In sections [1, 2] we prefaced for the types of privacy breaches against EMRs. An attack can takes place by analyzing the records, knowing the victim's attributes, and observing some statistical information [11]. For instance, by observing a census division publication for a specific criterion, like number of deaths by HIV in ages within a certain range, combining them with the EMRs released for the same period, the random sample where the victim resides is narrowed. Consequently, identifying a specific victim becomes more easer by confining the set of possibilities as we show in [4].

HL7 presented as a plan text files, well-known architecture, and contains a lot of information in one message. This makes it easy to figure out the contents, relate the files together, and get the big picture. However, current protection techniques, such as access controlling, or expensive encryption are still apart. This separation makes privacy protection relying on the general environment safety and on how many solutions the hospitals are willing to pay for.

In this work, we combine data presentation, existing encryption techniques, and access controlling in one solution to provide services such as identity protection, information sharing, integrity check, and flexibility in access controlling. On the same perspective, we show how we can improve anonymity and privacy threshold depending on data fragmentation and classification.

4   Approach

The last point in table 1 states *"Any other unique identifying number, characteristic, or code"* is considered as identifying information. This point shows the difficulty of bounding all the possible attributes identifies an individual. However, the presence of many non-unique information or non-quasi identifiers can disclose the identity of an individual, or at least it can bind the number of possibilities to a small set of records as shown in relation 4.

Figure [1] shows a sample HL7 message, where we can see that the quasi-identifiers like name, SSN, and address are present on the same file, which makes the EMR identifiable by its nature.

$$\sigma = \{EMR | EMR \text{ is an electronic health record}\}$$
$$\gamma = \{y | y \text{ is a population statistical information}\}$$
$$\beta = \{\overline{EMR} | \overline{EMR} \text{ is a disclosed electronic health record}\}$$
$$V = \{v | v \text{ is a victims' attribute}\}.$$
$$\zeta \ : \text{a set of EMRs where the victim}$$

Figure 1. Sample HL7 message and the classified records.

```
MSH|^~\&|ADT1|MCM|LABADT|MCM|198808181126|SECURITY|ADT^A01|MSG00001|P|2.4
EVN|A01|198808181123
PID|223344_ID||PATID1234^5^M11||JONES^WILLIAM^A^III|Mother Name^Madam^Mothe...
NK1|1|JONES^BARBARA^K|SPO|||||20011105
NK1|1|JONES^MICHAEL^A|FTH
PV1|1|I|2000^2012^01||||004777^LEBAUER^SIDNEY^J.|||SUR||-||1|A0
AL1|1||^PENICILLIN||PRODUCES HIVES RASH
AL1|2||^CAT DANDER
DG1|001|I9|1550^First Diagnosis|MAL NEO LIVER, PRIMARY|19880501103005|F||
DG1|002|I9|1550^Second Diagnosis| Red eyes|19880501103005|F||
PR1|2234|M11|111^CODE151|COMMON PROCEDURES in such cases|198809081123
ROL|45^RECORDER^ROLE MASTER LIST|AD|CP|KATE^SMITH^ELLEN|199505011201
GT1|1122|1519|Guarantor First name^Given name^father Name
IN1|001|A357|1234|BCMD|||||132987
IN2|ID1551001|SSN12345678
ROL|45^RECORDER^ROLE MASTER LIST|AD|CP|KATE^ELLEN|199505011201
```

$$\beta \subseteq \sigma \tag{1}$$

$$\zeta < \beta \tag{2}$$

$$\gamma = \sigma \cap \alpha \cap V \tag{3}$$

$$\zeta \subseteq \gamma \tag{4}$$

Our way of visualizing the approach is; providing a hybrid structure of lightweight techniques in one solution, where the information classified and fragmented to limit access to information category. We use chaining and encryption to protect the information and provide proper integrity check. The proposed architecture allows patients and health-care provider from sharing the information without disclosing patients' identity.

5 Data presentation

The EMR file contains, to mention some, the patient identification data, medical data, insurance information, providers' information. Managing file accessibility, if kept in its original HL7 format, is an application design aspect. However, Information granularity and classification based on their category provides a base on how it will be managed, and broadens accessibility control options.

In this approach, we create a three dimensional data structure or matrix ($Mat3D$). Each cell on $Mat3D$ can hold a block of bytes of a size (S). The principle depends on dividing a single EMR into blocks, say N blocks. Each block contains a homogeneous type of information (i.e., blocks $v_1 \ldots v_7$ contains the medical history of patient P) a parent node will contain all the pointers to one category of information, which is provided by the HL7 standards (Figure 2). The total number of blocks (b) in *all* *EMRs* is given by relation (6) as:

Figure 2. Information classification based on type.

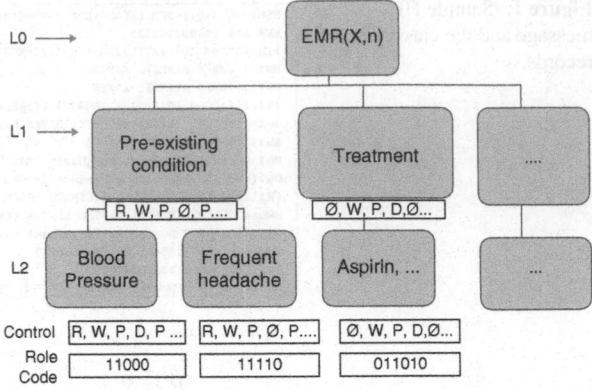

$$N_j = \left[\frac{SizeOf(EMR_j)}{S} \right] \tag{5}$$

$$b = \sum_{j=1}^{n} (N_j) \tag{6}$$

Focusing on one EMR, the blocks will be stored in a 3D structure or matrix, called *Mat3D*, in a random fashion. By aggregating the indices for each block placed in the Mat3D, we will create the access sequence (PR_j). This key will be considered as a part of the quasi-Identifiers of an individual with the same security level, the algorithm in table 2 is used to allocate space for more than one HL7 EMR files.

The condition to continue with the random allocation is to have enough space for the HL7 file we read. The 3D matrix, mat3D, can hold one or more HL7 files. Figure 3 shows the random allocation for twenty HL7-EMR files in a matrix of a size 15^3. As shown in table 2, the key to retrieve the file is to have the sequence of indices as formulated in line 8. The following example shows the result, where each color is a distinguishable EMR file.

Table 2. 3D Data Presentation Algorithm

1- Allocate memory space of *mat3D*[x][y][z] = b
2- *BlockSize* = n, j=0, key = mat3D:getMatrixID()
3- Initiate *Mat3D* with any data or noise
4- If [emptyLocations(mat3D) ≥ sizeOf(F) = BlockSize] → *Continue*; *else*; *goto* 11
5- F [] = getBytes(*HL7Data*), *HL7Data*: file of size n
6- Generate a random x,y,z
7- Mat3D[x][y][z] ← F[j to j⁺ = n]
8- *key* = x||','||y||','||z
9- While F[] has more data go to 4
10- Report key(*key*)
11- for new *HL7Data* go to 3
12- Store *Mat3D*

Figure 3. EMR distribution
in a 3D matrix.

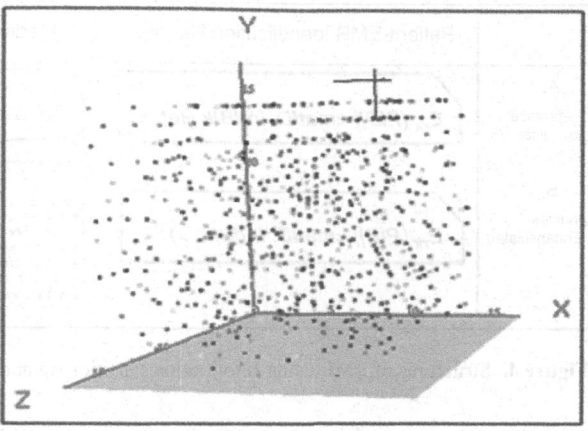

By saving the three dimensional matrix to the storage media (*i.e.* hard drive) as an object or in a liner fashion, the neighboring block for block, say B1, could be non-sensible block, block from different file, or out-of-order block from the same file.

6   Encryption

To provide the required security for the EMR, the user can adopt whatever encryption available for him, and acceptable as a secure algorithm. In our case, we used RSA, and a 512 bit key, table 3 shows the steps of encryption and chaining. The algorithm consists of short data chaining using *XOR* operation, block replacement, RSA encryption, then backward chaining and replacement. This process is in-place process, where no extra memory needed.

As shown by relation (7), the link between the Index Block and the Index Block Reference is encrypted within $\overline{PID}$, which replaces *Block*(0), can be saved in a database, or can be stored in a smartcard.

$$block(0)' = E_{pk}(E_{pk}(block(0)) + IndBlk) \tag{7}$$

Table 3.   Encrypting EMR Blocks Algorithm

Step 0: Strip patient identification data out of the *EMR* and store them in Block(0).

Step 1: Staring from block n : Block(n)=Block(n) $\oplus$ Block(n-1), for all blocks where $n > 0$

Step 2: PID=E_{RSA}(Block(0))

Step 3: Block(i)= Block(i-1) $\oplus$ Block(i), 1< $i \le$ n

Step 4: For each block $\ne$ block number zero, Randomly Store into 3D-matrix (*3DMat*)

Step 5: Generate a sequence of <x,y,z> $\rightarrow$ *IndBlk*, the index block

Step 6: Block(0)= E_{pk} (*PID* || <*matID, IndBlk*>)

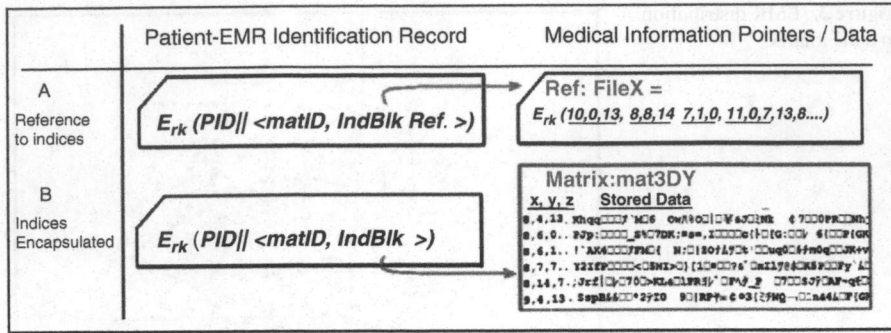

Figure 4. Structures of EMR using Information Partitioning and Classification.

Figure 4 shows the patient identification record/block in two forms, A; which provides a reference to the index block without including the actual coordinates, and B; which provides the indices for the block within the matrix. Structure (B) provides the ability to share a specific portion of the medical record, whereas structure (A) facilitates more than one EMR in a PID block. The same figure shows the matrix and its content after encryption.

Having a closer look to the encrypted blocks within the matrix, we realize the following facts:

- Each block is tied to its location, the initial *xor* operation overloaded the block with information about its neighboring blocks.
- The order of blocks are important to decrypt the file, though, the order is a part of the key.
- Each block is encrypted with different cipher block. Hence, the cipher block used to encrypt block (*i*) is different from the cipher block used to encrypt it adjacent block and so on.
- The structure facilitates integrity check in two ways; any block lost or damaged can be recovered by reversing the *xor* operation. However, the other check

Figure 5. EMR Encoding and Encryption Steps.

Encryption	State	B0/ID	B1	B2	B3	B4	B5
	(S0)	B0/ID	B1	B2	B3	B4	B5
	(S1)	K= E(B0)	a=B1	c=B2+B1	d=B3+B2	e=B4+B3	f=B5+B4
	(S2)	K	a`=a+k	c`=c+a`	d`=d+c`	e`=e+d`	f`=f+e`

Table 4. Accessing EMR

1- D(E(E<B0> | <matID, Coords/CoordsRef*> ← get <MatID, Coords/CoordsRef> +E<B0>

2- OrderedBlocks= fetchBlocks(MatID,Coords)

3- block(i) = block(i) $\oplus$ block(i-1), i=n..2

4- block(j) = block(j) $\oplus$ block(j-1), j= 1..n : *Medical information decrypted*

5- PID = D(<B0>); *Only if the patient wants to disclose his identity*

provided by the first local *xor* and replace process since it serves integrity and obfuscation goals prior to encryption.

The patient identification information consists of the patient quasi-identifier Data (QID), key reference, and the matrix identification number. All encrypted as one block to protect the reference value and the first block from unauthorized usage, figure 5 shows the encryption steps. The steps for legitimate user to retrieve an EMR are shown in table 4. Hence, the user needs to decrypt his EMR one time to be able to retrieve his data without disclosing his identity.

Another variation of this encryption is to perform the same procedure on the tree where each upper level is used to secure the immediate lower level only. Previously, $B(0)$ is the node used to encrypt the whole file. Now, we will use $B(0)$ to secure the next level on the tree only, which is L1. The parent nodes in L1 are used to secure the last level which is L2.

Having the blocks chained, where knowing the correct order is a part of decrypting the file, for unknown file size, the number of permutations is

$$p = \sum_{f=min}^{max} \frac{m!}{(m-f)!} \tag{8}$$

For a brute force attack, the number of tries will be $p/2$. However, an attacker might try to decrypt the block instead of trying all combinations. However, the attacker needs to attack each block in the matrix separately, which means attacking n^3 blocks each as a single block rather than attacking one encrypted file if the EMR was encrypted as a one unit.

In terms of time consumption, a file of size 3 KB needs $2ms$ to be encrypted and $3ms$ for decryption. However, the average time needed to encrypt an EMR of a size 3KB as one unit is $14ms$ and the average time of decryption is $93ms$

7 Access Controlling

As in RBAC, the access A will be granted to the Role r rather than the person. In our model, the access Duration AD, access time AT granted as needed. However, there is a limited set of operations Op can be performed upon an EMR from a certain location L such as ER or ICU in a hospital.

As the EMR classified to three levels, Fig. 2, allows us to grant access at the block level, category, and EMR access. Persona X associated with role r can access a record or partial record R if and only if he has the proper role r, accessing the record from the proper location in the specified time for the granted period, and he can do a specific operation.

$$A_{Xr}(R, \, grant = 1) = (R(X_r) \times L(X_r) \times AD(X_r) \times AT(X_r, \, R) \times$$
$$Op(X_r \rightarrow R) \times level \times grant) - A_{Xr}(\bar{R}, grant = 0) \tag{9}$$
$$where \; grant = \{0, \, 1\}, \; L = \{root, \, Parent, \, Node, \, Block\}$$

Without data classification, access is granted to a role on the whole file, and it is up to the software implementation to limit the amount of information to view. Although, once the EMR is decrypted, it is all accessible. By classifying the information within the EMR, each category can be decrypted separately from other categories, limiting accessibility to that type of data.

As shown in 9, access can be granted to a specific role on leaf information, level 2, such as the names of the medications taken. Also, accessibility can be limited to a category, level 1, such as preexisting conditions. Granting access to the root, EMR, will propagate through the whole file.

In figure 2, the control fields and the role code specifies the operations and the role, where each role will be associated with a set of switches controlling what operations can be performed on the data.

Some data, like critical medications, might need to be excluded from the strict access policies like context access control. In this approach, the context aware policies can be implemented by specifying a control bit on the control vector associated with the record. This bit specifies whether the EMR is governed by context access or not. The system will refer to the proper administration tables to verify the context input if the bit is set.

Patients may need to transfer the files from one provider to another, or share the information without disclosing their identity. This can be achieved by using the public/private key or shared session keys. The patient can encrypt his ID block, which consists of Block (0) encrypted by the patient's private key and used to encrypt the EMR, and the indices using the other hospital public key. The hospital will be able to decrypt the ID block getting Block (0) and the indices required to retrieve the desired information.

The use of smart-cards will bring the combination of what we have and what we know to the front. However, this will make it the responsibility of the patient to keep the whole key secure since the database has no keys to decrypt any part of the data. Though, it lowers the responsibility of the data custodians regarding the number of breaches.

8 Conclusion

The high cost of security and privacy protection solution could prevent many medical service providers from adopting them. However, it is a trade off, more security, more cost. In this work we tried to show the effect of re-engineering the system and how it can provide multiple services, enhances and increase system flexibility and scalability.

As the need to share information without disclosing identity increase, restrict accessibility to the minimum amount of information needed, thinking of system redesigning within the standards can provide more than what it shows. In this work, we have presented the capabilities and the effort needed to achieve high goals. This work can be extended to include communication protocols and more detailed context-aware access controlling.

References

1. DE-ID DATA CORP, "http://www.de-idata.com/" March 2010.
2. O. Ajayi, R. O. Sinnott, and A. Stell, "Dynamic trust negotiation for flexible e-health collaborations," in Mardi Gras Conference, (Baton Rouge, Louisiana, USA), p. 8, 2008.
3. A. Berler, S. Spyrou, E. Monochristou, Y. A. Tolias, G. Konnis, N. Magglaveras, and D. Koutsouris, "Risk assessment in integrated regional healthcare networks," in Interoperability & Security in Medical Information Systems (F. Makedon and J. Ford, eds.), vol. 2, The Electronic Journal for E-Commerce Tools & Applications (eJETA), eJETA.org, May 2007.
4. E. Bertino, P. A. Bonatti, and E. Ferrari, "Trbac: A temporal rolebased access control model," ACM Trans. Inf. Syst. Secur., vol. 4, no. 3, pp. 191–233, 2001.
5. R. Bhatti, K. Moidu, and A. Ghafoor, "Policy-based security management for federated healthcare databases (or rhios)," in HIKM, (Sheraton Crystal City Hotel, Arlington, VA), pp. 41–48, International Workshop on Health Information and Knowledge Management (HIKM 2006), November 2006.
6. Capgemini, "http://www.capgemini.com/," March 2010.
7. M. A. C. Dekker, J. Crampton, and S. Etalle, "Rbac administration in distributed systems," in SACMAT, pp. 93–102, 2008.
8. DoD, "Trusted computer system evaluation criteria," DECEMBER 1985.
9. K. E. Emam, "Heuristics for de-identifying health data," IEEE Security & Privacy, vol. 6, no. 4, pp. 58–61, 2008.
10. K. E. Emam and F. Kamal, "Protecting privacy using k-anonymity," Journal of the American Medical Informatics Association, vol. 15, pp. 627–637, August 2008.
11. E. B. Fernandez and T. Sorgente, "An analysis of modeling flaws in hl7 and jahis," in SAC, pp. 216–223, 2005.
12. D. F. Ferraiolo and D. R. Kuhn, "Role-based access controls," in 15th National Computer Security Conference, (National Institute of Standards and Technology, Technology Administration, U.S. Department of Commerce, Gaithersburg, d. 20899 USA), pp. 554 – 563, 1992. 3
13. D. F. Ferraiolo, D. R. Kuhn, and R. Chandramouli, Role-based access control. Computer Security Series, Artech House, 2003.
14. HHS, "US department of health &human services http://aspe.hhs.gov/admnsimp/." Web site, March 2010.
15. Junzhe and A. C.Weaver, "A dynamic, context-aware security infrastructure for distributed healthcare applications," in Pervasive Security, Privacy and Trust (PSPT 2004), (University of Virginia, Charlottesville, VA 22904), 2004.
16. J. W. Lebak, J. Yao, and S. Warren, "Hl7-compliant healthcare information system for home monitoring," in Proceedings of the 26th Annual International Conference of the IEEE EMBS, (Department of Electrical & Computer Engineering, Kansas State University, Manhattan, KS, USA), 2006.
17. Microsoft, "Connected health framework architecture and design blueprint," March 2009.
18. J. PATRICK R GALLAGHER, "Computer security subsystem interpretation of the trusted computer system evaluation criteria,"September 1988.
19. L. Reed-Fourquet, J. T. Lynch, M. K. Martin, M. Cascio, W.-Y. Leung, and P. P. Ruenhorst, "Managing information privacy & security in healthcare the chime-trust healthcare public key infrastructure and trusted third party services: A case-study," case study, Healthcare Information and Management Systems Society (HIMSS), Jan 2007. CHIME Inc., Wallingford Connecticut.
20. B. Smith and W. Ceusters, "Hl7 rim: An incoherent standard," in Studies in Health Technology and Informatics, vol. 124, (University of Buffalo, NY. USA), p. 133138, 2006.
21. A. Stell, R. Sinnott, and O. Ajayi, "Secure, reliable and dynamic access to distributed clinical data," in Proceedings of the LSGRID2006: Yokohama, (University of Glasgow, National e-Science Centre, Glasgow, G12 8QQ, UK), 2006.

22. C. Stingl and D. Slamanig, "Privacy-enhancing methods for ehealth applications: how to prevent statistical analyses and attacks." IJBIDM, vol. 3, no. 3, pp. 236–254, 2008.

23. L. Sweeney, "k-anonymity: A model for protecting privacy," International Journal of Uncertainty, Fuzziness and Knowledge-Based Systems, vol. 10, pp. 557–570, may 2002.

24. K. T. Win, H. Phung, L. Young, M. Tran, C. Alcock, and K. Hillman, "Electronic health record system risk assessment: a case study from the minet," Health Information Management, vol. 32, pp. 43–48, 2004.

25. Z. Wu and A. C. Weaver, "Dynamic trust establishment with privacy protection for web services," in ICWS, pp. 811–812, 2005.

26. Y. Yang, R. H. Deng, and F. Bao, "Fortifying password authentication in integrated healthcare delivery systems," in ASIACCS, pp. 255–265, 2006

Chapter 20
THE NEW HYBRID METHOD FOR CLASSIFICATION OF PATIENTS BY GENE EXPRESSION PROFILING

Erdal Cosgun and Ergun Karaagaoglu

1 Introduction

Improvements that should not be underestimated have been provided in genetic researches together with the improvements especially in the technology in the recent years. Contribution of Bioinformatics is also important in these improvements. Because the input received should be evaluated most effectively. Methods used for this purpose are generally the Data Mining methods. The subject which we aim to study is the classification of the patients with the gene expression data. Artificial and the real data which public access is available will be used this study. When literature is reviewed a lot of publications exist, especially related with the production of gene expression. The method used in most of the studies is about how to try the suggested method on one or two data sets which have been put in the public use via internet, and then to prove the reliability on artificial data with various scenarios. The most important reason of such an approach is the high cost of genetic researches.

The number of patients is pretty low compared with the number of genes in the microarray inputs within the study. The classical statistical approaches cannot be used because of that reason. Data Mining methods suggest some different dimension reduction methods for these input sets. The most important of these is the Independent Components Analysis (ICA). When studies in the literature are studied [4, 12, 14], it has been observed that the Independent Components Analysis (ICA) has been used on microarray inputs and the results provided have proved to be more accurate compared with the other dimension reduction methods (Multi Spectral Method, Principal Component Analysis). Another approach used in the patient classification with gene expression data is 'Unsupervised Learning' based classification. With the assistance of this method, variables are separated into clusters according to certain distance measurements first, and then classification is done with these clusters. Furthermore, a new method called "RF Clustering has been developed from the"Random Forest (RF)" which is a classification method in some studies.

E. Cosgun (✉)
Ankara, Turkey
e-mail: erdalcosgunn@gmail.com

S. Suh et al. (eds.), *Biomedical Engineering*,
DOI 10.1007/978-1-4614-0116-2_20, © Springer Science+Business Media, LLC 2011

These methods have been used in the generalization of the results and achieving higher true classification rates especially in the studies aiming classification and clustering. Thus, `Unsupervised Learning` sense is combined with the supervised classification method. Both over-fit problems have been overcome and possibility of defining the genes that cause the disease more clearly have been obtained. There are also examples in microarray inputs aiming dimension reduction. The most important problem in these researches related with classification and prediction is how to adapt different methods on algorithms to be able to generalize the results. Because generally data sets used in studies are small in number. Thus, obtained results may not give the same result in every data set. Some generalization methods have been used to get rid of this problem [11, 19] such as Bootstrap, Boosting, and Cross Validation. Thus the reliability of the results of the suggested method is put forward.

Another method used in the analysis of gene expression data is the clustering method. Clustering of the genes has become the starting point of many researches. The main objective in this approach is how to bring the genes which have same characteristics together by using different means of distance measurements. This approach has especially become important in the cancer researches. Because it is considered important to have a little idea about the relations among genes in the treatment of the diseases.

2 Methods

Machine learning methods are often categorized into supervised (outcome labels are used) and unsupervised (outcome label are not used) learning methods [17]. In this study we have combined these two methods based on dimension reduction on publicity microarray data sets` patients class prediction. The most important point while doing the classification of the patients with the assistance of gene expression data is how to choose the method which will be used. It is not realistic to expect a single method to have high true classification rate in every data set. For that reason, the method or the methods which will give the highest performance under different scenarios must be chosen. It is seen that methods which are developed from the combination of several methods have given more generalizable results. Because the bias in the analysis can easily be eliminated in such combined methods. For this reason methods will be brought together when the classification is done with gene expression inputs within the study.

First of these is the Independent Components Analysis (ICA) which is the dimension reduction method, the second is Kohonen Map method which is the clustering method and the last one is the Random Forest Method which is used as the classification method. Gene expression inputs are mostly the inputs in which the number of subjects is far less than the number of the genes. There are thousands of genes belonging to a person. And in these researches a lot of people cannot participate in the studies due to the cost restrictions. It means that such data do not provide the most important assumptions of many statistical methods. Therefore using classical methods mostly lead to obtain wrong or overfit results. Gene expression inputs will primarily be generated the study. Genes will be reduced to a smaller number of factors by the Independent Components Analysis (ICA) to eliminate the problems of being multi-dimensional afterwards. The purpose here is to form a common factor from the genes which have different expression levels. At the second stage, factors will be clustered by Kohonen Map method. The aim here is to bring the factors which have same features together.

The reason for not doing the clustering at the first stage is the fact that the clustering methods may produce incorrect results when the dimensions increase. Consequently similar factors which the independent genes constitute will be clustered in relation with the method which the Independent Components Analysis (ICA) uses. At the last stage, the classification of the patients with RF is targeted by choosing a certain number of clusters among clusters with Bootstrap method 1000 times randomly. The reason of using the Bootstrap method is to eliminate the bias in choosing the clusters which will be used in classification. The reliability of the classification obtained consequently will be higher due to the fact that RF method uses the Bootstrap method in its own algorithms.

The proposed method implemented in the Clementine 12.0 (SPSS Inc., Chicago, IL, USA), STATISTICA 7 (Statsoft Inc.), MATLAB R2009b and R statistical programming language. (Packages: randomForest, fastICA, kohonen, factoMiner, boot) **Source of R packages:**
http://cran.r-project.org/web/packages

2.1   Data Sets

This new classification method trained on two real data sets:

1) Small Round Blood Cell Tumor (SRBCT): The entire data set includes the expression data of 2308 genes. There are totally 63 training samples. The 63 training samples contain 23 Ewing family of tumors (EWS), 20 Rhabdomyosarcoma (RMS), 12 Neuroblastoma (NB), and 8 Burkitt lymphomas (BL). *2) Colon Cancer Data Set:* It consists of 62 samples of colon epithelial cells taken from cancer patients. Each sample contains 2000 gene expression levels. 20 out of 62 samples are normal samples and the remaining are colon cancer samples. Obviously, the best way of proving the performance of the proposed method is using real data sets and compare results with other studies which used same data sets. For further studies, proposal needs to show its performance on simulated data with various kinds of scenarios.

2.2   Independent Component Analysis (ICA)

The goal of ICA is to find a linear representation of non-Gaussian data so that the components are statistically independent, or as independent as possible [32]. In mi-

Figure 1. Theoretical framework of ICA algorithms on microarray gene expression data. (Ref. 32-page 3)

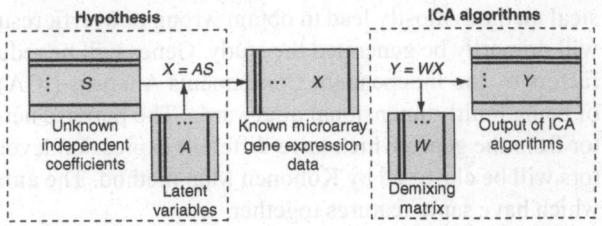

croarray experiments, we observe n random variables $X_1 \ldots\ldots X_n$ which are modeled as linear combinations of n random signal $S_1 \ldots\ldots S_n$:

$$X_i = a_{i1}s_1 + a_{i2}s2 + \ldots\ldots + a_{in}s_n, \quad \text{for all } i = 1, \ldots.n \tag{1}$$

Where the $a_{ij}, i,j = 1, \ldots\ldots, n$ are some real coefficients. By definition, the s_i are statistically mutually independent. [18] But this assumption is an unpractical assumption in many applications like microarray experiments. Implicit in the work of gene expression analysis today is the assumption that no gene in the human genome is expressed completely independently of other genes. [32] So it is hard to provide this assumption in real life.

For explaining clearly, interpret the model with vector-matrix notation. Figure 1 represents the vector- matrix for microarray experiments.

$$X = A^*S + N \tag{2}$$

where X is the matrix of acquired signals xi(t) and S is the matrix of source signals si(t). The coefficients of A determine the contribution of the individual sources $s_i(t)$ in the measured signals and N represents the Gaussian noise. The aim of blind source separation comes down to the determination of the source signals $s_i(t)$ from the measured signals $x_i(t)$. These $s_i(t)$ can be estimated as

$$S = W^*X \tag{3}$$

with W the unmixing matrix. However, since both coefficients and sources are unknown, it is generally impossible to determine them without imposing additional constraints. Therefore, several possible assumptions about the sources have been proposed in order to obtain a unique decomposition. The most well-known is the constraint of statistical independence as imposed in ICA. As indicated in the intro-

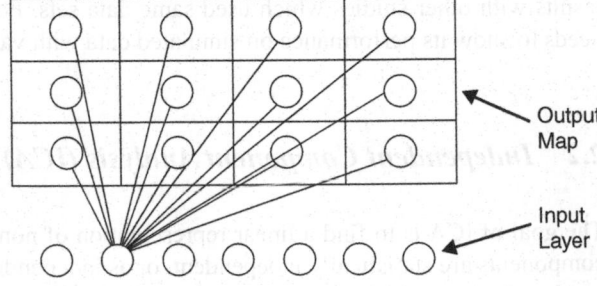

Figure 2. Structure of Kohonen Map.

duction, ICA is a signal processing technique that recovers independent sources from a set of simultaneously recorded signals that result from a linear mixing of the source signals (Comon et al., 1994).

ICA has different algorithms i.e. minimum mutual information (MMI), FastICA and maximum non-Gaussianity. But fastICA is has more general usage than others for microarray data analysis. In FastICA, maximizing negentropy is used as the contrast function since negentropy is an excellent measurement of nongaussianity [32]. For further information [32].

2.3   Kohonen Map (KM)

Kohonen models (Kohonen, 2001) are a special kind of neural network model that performs unsupervised learning. This type of network can be used to cluster the dataset into distinct groups when you don't know what those groups are at the beginning. Records are grouped so that records within a group or cluster tend to be similar to each other, and records in different groups are dissimilar [31].

Kohonen developed the KM network between 1979 and 1982 based on the earlier work of Willshaw and Malsburg [27]. It is designed to capture topologies and hierarchical structures of higher dimensional input spaces. Unlike most neural network applications, the KM performs unsupervised training, i.e., during the learning (training) stage, KM processes the input units in the network and adjusts their weights primarily based on the lateral feedback connections [22].

2.4   KM learning algorithm [Lippmann, 1987]

1. Initialize weights to small random values.
2. Choose input randomly from dataset.
3. Compute **distance** to all processing elements.
4. Select winning processing element j with minimum distance.
5. **Update weight** vectors to processing element j and its neighbors using following learning law. The learning law moves weight vector toward input vector.
6. Go to step 2 or stop iteration when enough inputs are presented.

2.4.1   Distance

Kohonen Map algorithm uses Euclidean Distance for finding the closer input neuron to output neuron.

$$d_{ij} = \sqrt{\sum_k (x_{ik} - w_{jk})^2} \qquad (4)$$

where is the value of the k_{th} input field for the i_{th} record, and is the weight for the k_{th} input field on the j_{th} output unit [31].

2.4.2 Neighborhoods

The neighborhood function is based on the Chebychev distance, which considers only the maximum distance on any single dimension:

$$d_c(x, y) = \max_i |x_i - y_i| \qquad (5)$$

where is the location of unit x on dimension i of the output grid, and is the location of another unit y on the same dimension [31].

2.4.3 Weight Updates

For the winning output node, and its neighbors if the neighborhood is > 0, the weights are adjusted by adding a portion of the difference between the input vector and the current weight vector. The magnitude of the change is determined by the learning rate parameter (eta). The weight change is calculated as

$$\Delta w_j = \eta \cdot (w_j - i_j) \qquad (6)$$

where is the weight corresponding to input unit j for the output unit being updated, and is the jth input unit [31].

2.5 Random Forest

Random Forests are a combination of tree predictors such that each tree depends on the values of a random vector sampled independently and with the same distribution for all trees in the forest. The generalization error for forests converges a.s. to a limit as the number of trees in the forest becomes large. The generalization error of a forest of tree classifiers depends on the strength of the individual trees in the forest and the correlation between them [15].

1. Draw a bootstrap sample from the data. Call those not in the bootstrap sample the "out-of-bag" data.
2. Grow a "random" tree, where at each node, the best split is chosen among m randomly selected variables. The tree is grown to maximum size and not pruned back.
3. Use the tree to predict out-of-bag data.
4. Use the predictions on out-of-bag data to form majority votes.
5. Repeat, N times and collect an ensemble of N trees. Prediction of test data is done by majority votes from predictions from the ensemble of trees. [28]

Figure 3. Random Forest
Training Procedure.

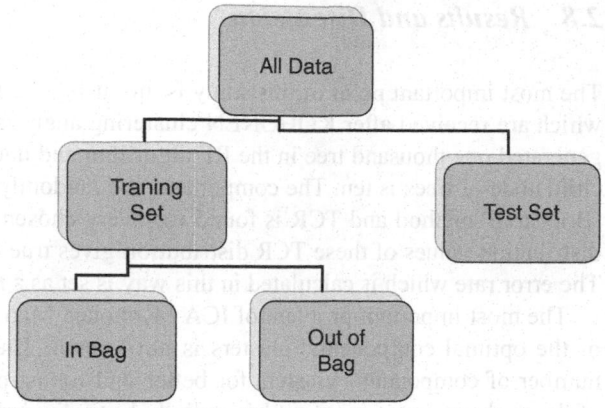

RF determines the relative importance of each gene, through various methods, such as calculation of the Gini Index, which assesses the importance of the variable and carries out accurate variable selection [29].

2.6   Bootstrap Idea

The original sample represents the population from which it was drawn. So resamples from this sample represent what we would get if we took many samples from the population. The bootstrap distribution of a statistic, based on many resamples, represents the sampling distribution of the statistic, based on many samples [25]. For this study, we choose components with bootstrap as an input variable for classification. Therefore, results are more generalize than using one sample classification.

2.7   Support Vector Machines (SVMs)

SVMs are a relatively new computational learning methods based on the statistical learning theory presented by Vapnik (1999). In SVMs, original input space mapped into a high-dimensional dot product space called a feature space, and in the feature space the optimal hyper plane is determined to maximize the generalization ability of the classifier. The maximal hyper plane is found by exploiting the optimization theory, and respecting insights provided by the statistical learning theory [23]. In this study SVMs is only used for performance comparison with proposed method.

2.8   Results and Discussion

The most important point of this study is, not using the Kohonen Clusters directly, which are received after KOHONEN clustering analysis, in the RF algorithm. We generated one thousand tree in the RF algorithm and number of minimum cases in child node of trees is ten. The components are randomly chosen 1000 times by the 'Bootstrap' method and TCR is found for every chosen sub-sample. The standard distribution values of these TCR distribution gives true error rate of classification. The error rate which is calculated in this way is set as a robust value.

The most important problem of ICA / Kohonen Map is the fact that the number of the optimal components/ clusters is not known. Therefore we chose different number of component / clusters for better and robust prediction. In the first step of the analysis, every input set has initially been divided into 25 to 50 components by the ICA. (Parameters of ICA are shown in Table 1) These components are then divided into 5,10,20 clusters with KOHONEN MAP method and clusters which were chosen among these with Bootstrap method were put into RF algorithm as input. (Fig. 4)

According to the results shown in Table 1, correlation between number of components and TCR is negative. 25 components` TCR results are higher than 50 components`. On the other hand, correlation between TCR and number of clusters is not so clear for these data sets. 10 cluster results have higher TCR than 5 or 20. So optimal classification structure for this study is 25 ICA components – 10 Kohonen map clusters.

Proposed method with its results are shown in Table 2. According to that, the highest TCR has been obtained with the suggested method for 2 data sets. `Only RF` or `only SVMs` [10-fold cross validation- Radial based kernel function] has lower TCR than new method. (Table 3) Specially, new method outperforms `only

Table 1. Chosen parameters of ICA.

Parameter	Value
Tolerance	0.0001
Alpha	1
Max. iteration number	200

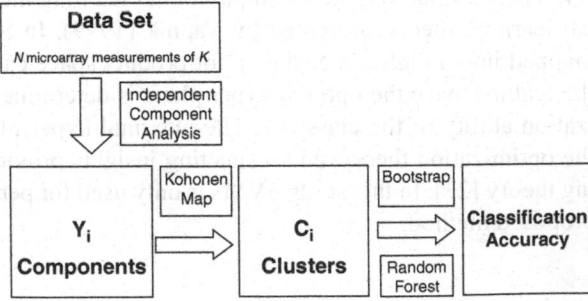

Figure 4. Procedure of Proposal Method.

Table 2. TCR results of proposal method.

ICA NUMBER OF COMPONENTS	KOHONEN MAP NUMBER OF CLUSTERS	TCR (%)	
		SRBCT	COLON CANCER
25	5	86.20	88.12
	10	88.71	90.61
	20	83.10	88.90
50	5	80.12	85.12
	10	82.81	87.61
	20	78.12	84.10

Table 3. TCR results of SVMs and RF method.

METHOD	SRBCT (%)	COLON CANCER (%)
ONLY SVMs*	81.30	86.97
ONLY RF	76.10	75.80

*SVMs [10-fold cross validation- Radial based kernel function, Best Kernel Function results.

RF` results. It is important because main objective of this study to increase the TCR with combined methods.

Obviously as anticipated, every new method added in the analysis has increased the classification performance. The most important reason of receiving this result is to choose the methods consciously and to take the examples in the literature into consideration.

Another important reason of suggesting this method is, a lot of the genes' low probability of being important in microarray input sets respecting the classification. Starting at this point, components in which all of these genes take place at a certain load have been constituted with ICA, and similar components have been brought together in clusters. Thus irrelevant genes have been brought together. Then the best loading combinations have been provided with the ability to make the classification with bootstrap.

3   Conclusion

The new unsupervised based classification technique studied in this paper, we have presented an experimental study in which I have compared some of the most commonly used classification method for microarray data sets: SVMs. I have applied SVMs and proposal method on two publicly available data sets, and have compared how these methods have performed in patients' classification prediction. Proposal methods outperform the `only SVMs` and `only RF` for each data sets. (Table 3) Results have revealed the importance of `bootstrap clustering` after dimension reduction in accurately classifying new samples. The integrated dimension reduction and clustering methods with classification algorithm are the most effective ways of prediction of class label and finding important genes.

At the second stage of this study, it is planned how to make the classification in synthetic and real data sets with other machine learning methods apart from RF, SVMs (i.e. Naïve Bayes, Neural Networks), use different clustering methods with KM, (i.e. Hierarchical Clustering, K-Means), try different number of component/ loading, and compare the performances with an area under Receiver Operating Characteristic (ROC) curve.

Acknowledgements We thank Dr. Murat M. Tanik, Dr. David Allison and Dr. Christine W. Duarte for their helpful feedback. We appreciate Abidin Cosgun's language control.

References

1. Ron Wehrens, Lutgarde M. C. Buydens, Self and Super-Organizing Maps in R: The Kohonen Package, Journal of Statistical Software, Volume. 25, 2007, Issue 5.
2. Pablo Tamayo et al., Interpreting Patterns of Gene Expression with Self-Organizing Maps : Methods and Application to hematopoietic differentiation. Proc.Natl. Acad.Sci.,Volume 96, 1999, 2907–2912
3. Rudolph S. Parrish, Horace J. Spencer, Ping Xu,Distribution Modeling and Simulation of Gene Expression Data, Computational Statistics and Data Analysis,2009
4. Su-In Lee, Serafim Batzoglou, An Application of Independent Component Analysis to Microarrays, Genome Biology,2003,4:R76
5. Ka Yee Yeung, Mario Medvedovic and Roger E. Bumgarner, Clustering Gene Expression Data With Repeated Measurements, Genome Biology, 2003, 4:R74
6. Hae-Sang Park, Chi- Hyuck Jun, Joo-Yeon Yoo, Classifying Genes According To Predefined Patterns By Controlling False Discovery Rate, Expert Systems with Applications, Volume: 36, 2009, 11753–11759
7. Jiawei Han, How Can Data Mining Help Bio-Data Analysis?, Workshop on Data Mining in Bioinformatics,2002
8. Ruffino, F. Muselli, M. Valentini, G., Biological Specifications for a Synthetic Gene Expression Data Generation Model, Lecture Notes In Computer Science, NUMB 3849,2006, 277–283
9. Pekka Ruusuvuori et al., Microarray Simulator as Educational Tool, Proceedings of the 29th Annual International Conference of The IEEE EMBS, 2007, 5919–5922
10. Xin Jin, Rongfang Bie, Random Forest and PCA for Self-Organizing Maps Based Automatic Music Genre Discrimination, Conference on Data Mining,2006, 414–417
11. Samir A Saidi at al., Independent Component Analysis Of Microarray Data In The Study Of Endometrial Cancer, Oncogene, 2004, 23, 6677–6683
12. A. Hyvärinen, E. Oja, Independent Component Analysis: Algorithms and Application, Neural Networks, 13(4-5), 2000, 411–430
13. J.V. Stone, A Brief Introduction to Independent Component Analysis in Encyclopedia of Statistics in Behavioral Science, Volume 2, pp. 907–912, Editors Brian S. Everitt & David C. Howell, John Wiley & Sons, Ltd, Chichester,2005, ISBN 978-0-470-86080-9
14. International Journal of Innovative Computing, Information and Control ICIC International, Independent Component Analysis For Classification Of Remotely Sensed Images, Volume 2, Number,2006, 31349–4198,
15. Breiman, Leo, "Random Forests". Machine Learning, 2001,45 (1): 5–32.

16. Mehdi Pirooznia, Jack Y Yang, Mary Qu Yang and Youping Deng, A comparative study of different machine learning methods on microarray gene expression data, BMC Genomics, Volume 9, 2008, S13

17. Tao Shi and Steve Horvath, Unsupervised Learning with Random Forest Predictors. Journal of Computational and Graphical Statistics. Volume 15, Number 1, 2006, 118–138 (21)

18. Aapo Hyvärinen, Juha Karhunen, Erkki Oja, Independent Component Analysis, Copyright by John Wiley & Sons, Inc, 2001

19. Dhammika Amaratunga, Javier Cabrera and Yung-Seop Lee, Enriched Random Forests, Bioinformatics, Vol. 24, 2008, Pages 2010–2014

20. Yeo Lee Chin, Safaai Deris, A Study On Gene Selection And Classification Algorithms For Classification Of Microarray Gene Expression Data, Jurnal Teknologi,2005, 43(D): 111–124

21. Ng Ee Ling, Yahya Abu Hasan, Classification On Microarray Data, Proceedings of the 2nd IMT-GT Regional Conference on Mathematics, Statistics and Applications,2006

22. Kohonen, T., Self-organization and Associative Memory. Springer, Berlin, 1984

23. Achmad Widodo et al., Combination of independent component analysis and support vector machines for intelligent faults diagnosis of induction motors, Expert Systems with Applications 32, 2007, 299–312

24. Katrien Vanderperren et al.,Removal of BCG artifacts from EEGrecordings inside the MR scanner: A comparison of methodological and validation-related aspects, NeuroImage 50, 2010, 920–934.

25. Tim Hesterberg, David S. Moore, Shaun Monaghan, Ashley Clipson, Rachel Epstein, Bootstrap Methods And Permutation Tests -Companion Chapter 18 To The Practice Of Business Statistics, W. H. Freeman and Company New York, 2003

26. Federico Marini, Jure Zupan, Antonio L. Magr, Class-modeling using Kohonen artificial neural networks, Analytica Chimica Acta,544, 2005, 306–314

27. M.B. Wilk, S.S. Shapiro,The joint assessment of normality of several independent samples, Technometrics 10,1968, 825–839.

28. Course Notes of `Exploring/Data Mining Pharmaceutical Data` by Birol Emir (PFIZER) – Prof. Javier Cabrera, 10 MAY 2009, Pre-conference Course of IBS-EMR 2009, ISTANBUL, TURKEY

29, Torri A, Beretta O, Ranghetti A, Granucci F, Ricciardi Castagnoli P, et al., Gene Expression Profiles Identify Inflammatory Signatures in Dendritic Cells. PLoS ONE 5(2):,2010, e9404. doi:10.1371/journal.pone.0009404

30. Hyvärinen, A. and E. Oja., A fast fixed-point algorithm for independent component analysis. Neural Comput. 9, 1997, 1483–1492.

31. Clementine® 12.0 Algorithms Guide, Copyright © 2007 by Integral Solutions Limited.

32. Wei Kong, Charles R. Vanderburg, Hiromi Gunshin, Jack T. Rogers, Xudong Huang,A review of independent component analysis application to microarray gene expression data, BioTechniques, 45, 2008, 501–520, doi 10.2144/000112950

33. Corinna Cortes and V. Vapnik, "Support-Vector Networks", Machine Learning, 20, 3, 1995, 273–297

Chapter 21
NEW MUSCULOSKELETAL JOINT MODELING PARADIGMS

Ibrahim Esat and Neriman Ozada

1 Introduction

There are some fundamental difficulties in modeling human musculoskeletal biomechanics. These difficulties relate to the skeletal joints. Some difficulties are simply mathematical but some relate to understanding and modeling of muscles, bones, tendon and cartilage. The theories used in industry and in the field in modeling joints are derivations from the engineering modeling schemes. For example it is common to use standard engineering joints to represent human joints such as revolute, prismatic, spherical etc. The truth is that none of these schemes are adequate in modeling real human joints. Furthermore such schemes make the modeling of certain diagnostics impossible, such as joint laxity. The chapter presents some theories developed in order to describe human joint in its anatomical fidelity and argues that representation in terms of joint flexibility enables various joint parameters to be investigated which would not be possible otherwise.

The chapter shows the main elements of mathematical modeling of a multi body system connected by flexible elements representing a joint made of contact springs and tissue holding the joint together. However as well as offering opportunities the scheme presents serious challenges. In the absence of axes of rotation, centre of rotation which are present in standard engineering joints need to be obtained in order the new scheme to be viable. The chapter also gives the derivation of some of these properties such as the centre of stiffness and axes of stiffness. Equally the formulation allows, by obtaining the static deflection, to calculate the centre and axes of rotation.

The integral properties also overcome two major difficulties in modeling joints as flexible structures. The flexible representation is prone to create ill conditioned multi body systems. As the contact stiffnesses may be substantially higher than the connecting soft tissue resulting in stiff systems. The second problem is the solution efficiency. Using variable step length in time integration is probably the only safe

I. Esat (✉)
Uxbridge, Middlesex, UK
e-mail: Ibrahim.Esat@brunel.ac.uk

S. Suh et al. (eds.), *Biomedical Engineering*,
DOI 10.1007/978-1-4614-0116-2_21, © Springer Science+Business Media, LLC 2011

way to ensure required accuracy, but this result in extremely slow solution. Integral representation speed up the solution many fold.

2 A theory of flexible joint modeling

To develop a multi body system flexibly connected, bodies in Figure 1 are considered.

Expressing the displacement of two points on body i and j,

$$\mathbf{d}_i = \mathbf{x}_i + \theta_i \times \mathbf{r}_i$$
$$\mathbf{d}_j = \mathbf{x}_j + \theta_j \times \mathbf{r}_j \tag{1}$$

Taking the displacements on i and j and converting them to the global axes, multiplying the relative motion between them with the stiffness matrix, also described in the global axis, the force vector between bodies can be obtained. Taking moments due to this force, and force vector itself assembled together, the final equations of motion for each body can be written. The full derivation and the list of notation can be found in [1, 2].

$$\mathbf{T}_i' \mathbf{M}_i \mathbf{T}_i'^T \ddot{\mathbf{U}}_i + \sum_p \left(\mathbf{T}_i' \mathbf{A}_i^T \mathbf{T}_i^T \mathbf{k}_p \mathbf{T}_i \mathbf{A}_{ip} \mathbf{T}_i'^T \right) \mathbf{U}_i -$$
$$\sum_p \left(\mathbf{T}_i' \mathbf{A}_i^T \mathbf{T}_i^T \mathbf{k}_p \mathbf{T}_j \mathbf{A}_{jp} \mathbf{T}_j'^T \right) \mathbf{U}_j = 0$$
$$\mathbf{T}_j' \mathbf{M}_j \mathbf{T}_j'^T \ddot{\mathbf{U}}_j - \sum_p \left(\mathbf{T}_j' \mathbf{A}_j^T \mathbf{T}_j^T \mathbf{k}_p \mathbf{T}_i \mathbf{A}_{ip} \mathbf{T}_i'^T \right) \mathbf{U}_i + \tag{2}$$
$$\sum_p \left(\mathbf{T}_j' \mathbf{A}_j^T \mathbf{T}_j^T \mathbf{k}_p \mathbf{T}_j \mathbf{A}_{jp} \mathbf{T}_j'^T \right) \mathbf{U}_j = 0$$

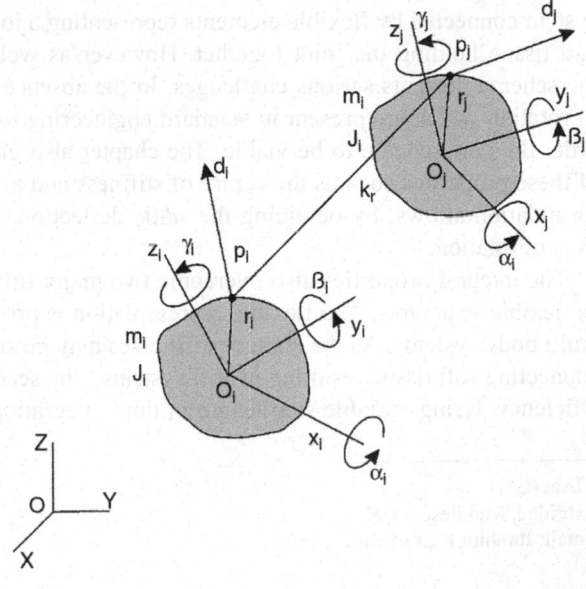

Figure 1. Bodies i and j.

These formulations now describe multi rigid bodies connected together by flexible springs.

In human joint there are joint related measures which can be used in joint diagnostics. One such measure is the joint laxity. This measure is widely used and it is quick and cheap way of getting an idea if a joint has a problem, especially with respect to cartilage and particularly ligament damage. Although, widely used the method needs scientific grounding. And the joint formulation presented here can be used to achieve this. The laxity primarily is a measure of joint stiffness. Having already defined joint stiffness in tensorial form, it is easy to extract various invariants related to the stiffness of the joint.

2.1   Centre and principal axes of stiffness

The calculation of the centre of the rotation can be defined as the centre, relative to which any rotation will not generate a resultant force. In vectorial terms this can be expressed as:

$$\sum_i k_i(\theta \times (\mathbf{r}_i - \rho_i)) = 0 \tag{3}$$

Where k_i is the stiffness of the i_{th} tissue in the global axis (this is expressed in global axis for convenience before assembling the formulation for the sake of simplicity). θ is the arbitrary rotation vector and $\mathbf{r}_i$ is the position of the i_{th} tissue attachment on body i (remembering that body j body is fixed in space) all measured relative to the local body axes (with axes parallel to the global axes, which does not need any transformation). ρ is the position vector of the centre of stiffness. The equation now can be simplified to,

$$\sum_{alltissues} k_i(\theta \times \mathbf{r}_i) - k_i(\theta \times \rho) = 0 \tag{4}$$

Now replacing the vector equation with its matrix equivalents,

$$\sum_{alltissues} \left[\begin{bmatrix} k_{xxi} & k_{xyi} & k_{xzi} \\ k_{yxi} & k_{yyi} & k_{yzi} \\ k_{zxi} & k_{zyi} & k_{zzi} \end{bmatrix} \begin{bmatrix} 0 & z_{pi} & -y_{pi} \\ -z_{pi} & 0 & x_{pi} \\ y_{pi} & -x_{pi} & 0 \end{bmatrix} \right] \begin{Bmatrix} \theta_{xi} \\ \theta_{yi} \\ \theta_{zi} \end{Bmatrix} =$$

$$\sum_{alltissues} \left[\begin{bmatrix} k_{xxi} & k_{xyi} & k_{xzi} \\ k_{yxi} & k_{yyi} & k_{yzi} \\ k_{zxi} & k_{zyi} & k_{zzi} \end{bmatrix} \right] \begin{bmatrix} 0 & \rho_z & -\rho_y \\ -\rho_z & 0 & \rho_x \\ \rho_y & -\rho_x & 0 \end{bmatrix} \begin{Bmatrix} \theta_{xi} \\ \theta_{yi} \\ \theta_{zi} \end{Bmatrix} \tag{5}$$

There are many combinations of setting equations to calculate three unknowns $\{\rho_x \quad \rho_y \quad \rho_z\}$, for example, a set of equations for solution is given in Eqn. (6), apart from fact that it is ill conditioned, not having the terms k_{xx} or k_{zz} means that if stiffnesses of a system happen to be only in the principal directions where the global axes coincide with these axes means that all cross stiffnesses will disappear.

$$
\begin{bmatrix} 0 & \sum k_{yzi} & -\sum k_{yyi} \\ -\sum k_{yzi} & 0 & \sum k_{yxi} \\ \sum k_{yyi} & -\sum k_{yxi} & 0 \end{bmatrix} \begin{Bmatrix} \rho_x \\ \rho_y \\ \rho_z \end{Bmatrix} = \begin{Bmatrix} \sum \left(-z_{pi}k_{yyi} + y_{pi}k_{yzi} \right) \\ \sum \left(z_{pi}k_{yxi} - x_{pi}k_{yzi} \right) \\ \sum \left(-y_{pi}k_{yxi} + x_{pi}k_{yyi} \right) \end{Bmatrix} \tag{6}
$$

Thus there will be no solution. Therefore it is important that the equations are selected in such a way that a viable solution always exist (that is even if all cross stiffness elements disappear due to choice of axes systems.

$$
-\rho_z \sum k_{yyi} + \rho_y \sum k_{yzi} = \sum \left(-z_{pi}k_{yyi} + y_{pi}k_{yzi} \right) \tag{7}
$$

$$
-\rho_z \sum k_{zyi} + \rho_y \sum k_{zzi} = \sum \left(-z_{pi}k_{zyi} + y_{pi}k_{zzi} \right)
$$

$$
\begin{bmatrix} \sum k_{yzi} & -\sum k_{yyi} \\ \sum k_{zzi} & -\sum k_{zyi} \end{bmatrix} \begin{Bmatrix} \rho_y \\ \rho_z \end{Bmatrix} = \begin{Bmatrix} \sum \left(-z_{pi}k_{yyi} + y_{pi}k_{yzi} \right) \\ \sum \left(-z_{pi}k_{zyi} + y_{pi}k_{zzi} \right) \end{Bmatrix} \tag{8}
$$

$$
\begin{Bmatrix} \rho_y \\ \rho_z \end{Bmatrix} = \frac{1}{-\sum k_{yzi} \sum k_{zyi} + \sum k_{zzi} \sum k_{yyi}}
$$

$$
\begin{bmatrix} -\sum k_{zyi} & \sum k_{yyi} \\ -\sum k_{zzi} & \sum k_{yzi} \end{bmatrix} \begin{Bmatrix} \sum \left(-z_{pi}k_{yyi} + y_{pi}k_{yzi} \right) \\ \sum \left(-z_{pi}k_{zyi} + y_{pi}k_{zzi} \right) \end{Bmatrix} \tag{9}
$$

The formulation above ensures that there is always a solution provided, k_{zz}, k_{yy} exist. And from,

$$
-\rho_y \sum k_{yxi} + \rho_x \sum k_{yyi} = \sum \left(-y_{pi}k_{yxi} + x_{pi}k_{yyi} \right)
$$

$$
\rho_x = \frac{1}{\sum k_{yyi}} \left(\rho_y \sum k_{yxi} + \sum \left(-y_{pi}k_{yxi} + x_{pi}k_{yyi} \right) \right) \tag{10}
$$

Where, again numerator and de-nominator contain diagonal stiffness elements which ensure that there is always a feasible solution. Once this is calculated, the Eigenvalues of the lower 3×3 matrix of the re assembled stiffness matrix gives the principal joint stiffnesses and the eigen vectors gives the axes of articulation. Summarising the procedure, initially there is a stiffness matrix as:

$$
\begin{bmatrix} \mathbf{k_p} & \mathbf{k_p}\mathbf{R_{pi}} \\ \mathbf{R_{pi}^T}\mathbf{k_p} & \mathbf{R_{pi}^T}\mathbf{k_p}\mathbf{R_{pi}} \end{bmatrix} \tag{11}
$$

From the matrix formulation above, the centre of stiffness is given by,

$$
\sum_p \left[\mathbf{k_p}\mathbf{R_{pi}} \right] = \left[\sum_p \mathbf{k_p} \right] \rho \tag{12}
$$

The torsional stiffnesses and the principal axes of rotation are given by:

$$\left[\mathbf{R}_{pi}^T \mathbf{k}_p \mathbf{R}_{pi} - \lambda \mathbf{I} \right]_\rho \mathbf{X} = 0 \tag{13}$$

Where, λ gives principal stiffnesses and $\mathbf{X}$ gives principal axes of stiffness. For static analysis nothing special is developed, static deflection analysis is standard capability of the Musculoskeletal Joint Modeller (MJM) software developed for the analysis.

2.2 Axis of rotation as an Eigenvalue problem

The objective function can be formed by adding all i, j combinations (double summation sign is removed from the following equations for simplicity) of all surface normals (n_i and n_j), shown in the Figure 2, extracted at each node point of the mesh. u is the unit vector along the axes of rotation, the error function (or objective function) to be minimised can be formed as.

$$E = [u \cdot n_i \times n_j]^2 + \lambda[|u| - 1] \tag{14}$$

Expanding and performing the dot product,

$$
\begin{aligned}
E = \big[& u_x \left(n_{iy} n_{jz} - n_{iz} n_{jy} \right) + u_y \left(n_{iz} n_{jx} - n_{ix} n_{jz} \right) \\
& + u_z \left(n_{ix} n_{jy} - n_{iy} n_{jx} \right) \big]^2 + \lambda \left[|u| - 1 \right]
\end{aligned} \tag{15}
$$

and minimizing the error

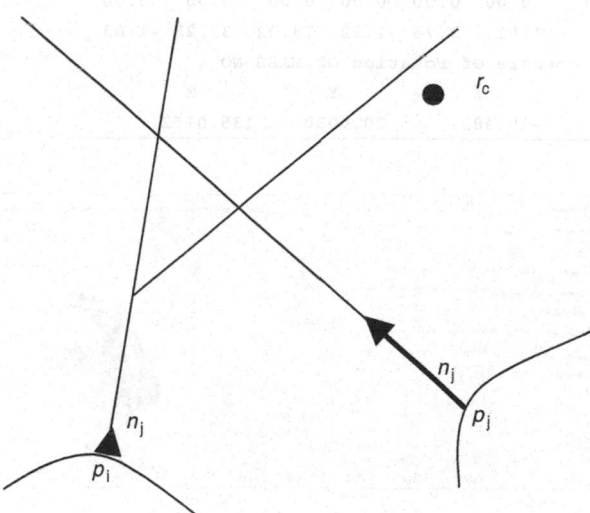

Figure 2. Two representative articulating surface and normals.

$$\frac{\partial E}{\partial u_x} = 0, \quad \frac{\partial E}{\partial u_y} = 0, \quad \frac{\partial E}{\partial u_z} = 0 \tag{16}$$

$$a = (n_{iy}n_{jz} - n_{iz}n_{jy}); \quad b = (n_{iz}n_{jx} - n_{ix}n_{jz}); \quad c = (n_{ix}n_{jy} - n_{iy}n_{jx}) \tag{17}$$

The axes of rotation can be converted to eigenvalue problem

$$\left\{ \begin{bmatrix} a^2 & ab & ac \\ ba & b^2 & bc \\ ca & cb & c^2 \end{bmatrix} + \lambda \begin{bmatrix} 1 & 0 & 0 \\ 0 & 1 & 0 \\ 0 & 0 & 1 \end{bmatrix} \right\} \left\{ \begin{matrix} u_x \\ u_y \\ u_z \end{matrix} \right\} = \left\{ \begin{matrix} 0 \\ 0 \\ 0 \end{matrix} \right\} \tag{18}$$

3 Results

Static deflection of elbow joint under an applied load is given in Table 1. The elbow system shown in Figure 3 analysed by using the formulations presented in the methods section. For centre of rotation calculation the static deflection results are used. MJM is the musculoskeletal joint modeling software which incorporated all theories presented in this chapter [2]. The centre of rotation from small deflection results is shown in Table 1. Eigenvalue version of the results is not presented here.

Table 1. Deflections under static loads.

Project Name	: TwoBoneOneJointMuscleWrap				
DEFLECTIONS					
mass	X	Y	Z	ALFA	BETA	GAMA
1	0.00	0.00	0.00	0.00	0.00	0.00
2	-4.51	3.76	-1.22	28.02	33.25	-1.03

```
   centre of rotation OF MASS NO =
           X           Y           Z
      -19.4837     20.6030     135.0162
```

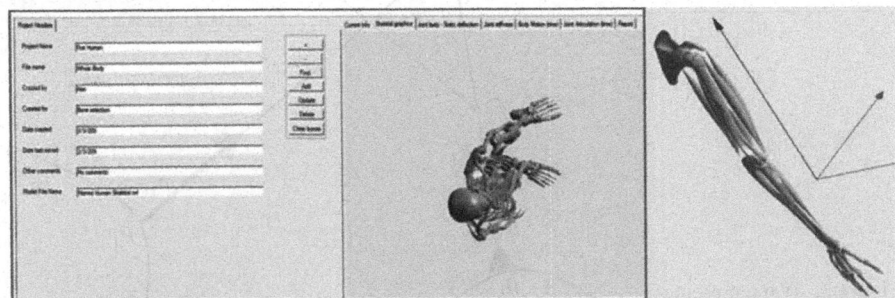

Figure 3. MJM and elbow modelling on MJM.

Table 2. Joint stiffness and related results.

```
Centre of stiffness of the joint
between body  1 and 2
-------------------------------------------
X Coordinate = -34.44
Y Coordinate = 2.86
Z Coordinate = 134.92
PRINCIPAL STIFFNESSES
R for real and I for imaginary
-------------------------------------------
R   305.967282234803   I   0
R   17.6352419578363   I   0
R   305.526200056914   I   0
PRINCIPAL AXES
--------------
   1.000e 00   2.925e-01   2.937e-01
   3.177e-02   0.000e 00   8.876e-01
   0.000e 00   8.702e-01   1.000e 00
```

3.1   Centre of stiffnes and Axis of stiffness

The principal stiffnesses and the principal axes of stiffness are given in the Table 2. Again the centre of stiffness and the centre of rotation are very close to each other, especially the z axis position.

4   Some Experimental Devices to Study Centre of Rotation and the Centre of Stiffness

A number of devices (only some presented here) developed to measure joint laxity, centre of rotation and he axes of rotation. The Stewart platform measure full centre kinematics (Figure 4 and Figure 5), the device in Figure 6 produces data suitable to calculate centre of stiffness and the device in Figure 7 is used to measure laxity.

5   Discussion and Conclusions

The chapter presents the basic idea that a joint can be presented in terms of its stiffness characteristics. Full derivation is not given but references are given for a fuller treatment. Following this a full mathematical treatment of the centre and axis of stiffness of a joint is obtained. The investigation primarily investigates whether

Figure 4. Stewart platform.

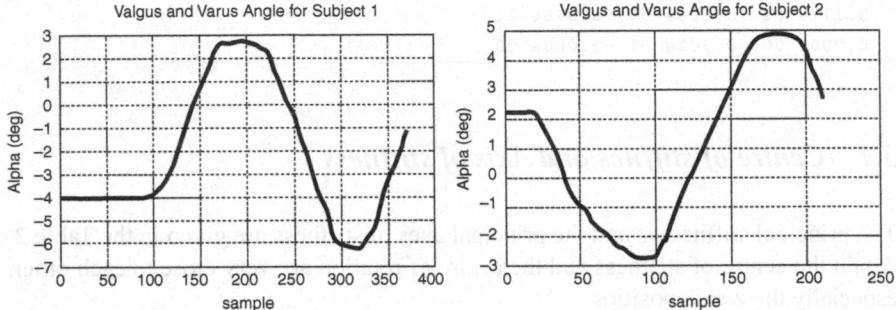

Figure 5. Stewart Platform results.

4 **Some Experimental Devices to Study Centre of Rotation and the Centre of Stiffness**

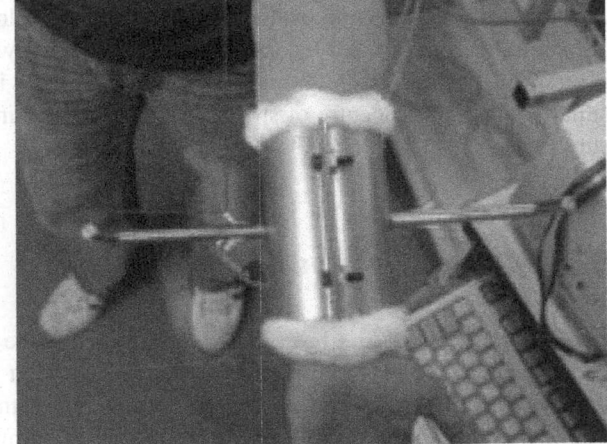

Figure 6. Measuring the centre of stiffness.

Figure 7. Laxity measurement experiment.

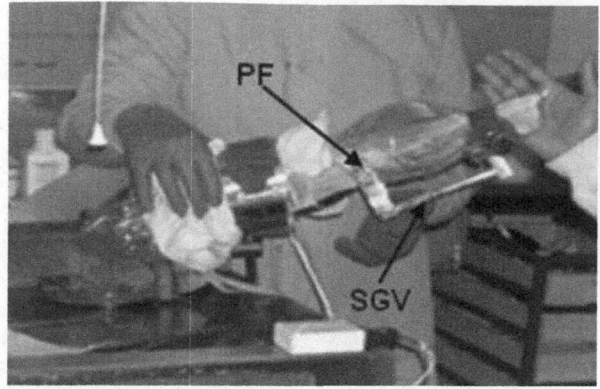

there is a relationship between the centre of rotation and the centre of stiffness, to that end the centre of rotation is obtained from the static deflection using the stiffness definition obtained. Also presented, further work has also been carried out to show that the axis of rotation can also be expressed as an Eigen value problem. The early results indicate that the invariants investigated can be effective in joint characterization for diagnostics and implant design. To that end the formulations can help placing some concepts "such as joint laxity" on a more scientific ground.

References

1. Ibrahim I. Esat and Neriman Ozada, Articular human joint modelling,Robotica, Volume 28, Issue 02, March 2010, pp. 321–339.
2. Neriman Ozada, Human joint modeling, PhD Thesis, Brunel University,England, June 2009.

Chapter 22
THE ROLE OF SOCIAL COMPUTING FOR HEALTH SERVICES SUPPORT

Jerry A. Higgs, Varadraj P. Gurupur, and Murat M. Tanik

1 Introduction

Rapid technological advances, regulatory issues, and an aging population introduces increasingly complex and difficult to tackle problems into the field of health sciences and services [1, 2]. It has been widely accepted that there is a need for a comprehensive approach to resolve the fragmented solutions within health sciences and services. Therefore, there is a growing recognition to develop methods, theories, and conceptual models that integrate interdisciplinary perspectives into the systems developed for health sciences and services [3].

Collaboration via social computing provides the computational means to substitute the ad-hoc approaches of knowledge exchange across disciplines and specializations [4, 5]. Furthermore, this allows social computing to be used as a new paradigm to control the exchange of information providing an opportunity for expanded collaboration and knowledge exchange. The second generation of web-based communities and hosted services facilitates an ideal mix of technologies for collaborating and sharing among participants. These communities of users could be scientists and doctors in health sciences and services as well as other support professionals in the health related professions including the general public benefiting from these services.

For the uninitiated, it is prudent that we begin with a definition of the terms Medicine 2.0 and Health 2.0 [6]:

1. Medicine 2.0 is the science of maintaining and/or restoring human health through the study, diagnosis and treatment of patients utilizing advanced internet-based services, including web-based community sites, blogs, wikis, social bookmarking, folksonomies (tagging) and Really Simple Syndication (RSS), to collaborate, exchange information and share knowledge. Physicians, nurses, medical students and health researchers who consume web media can actively participate in the creation and distribution of content, helping to customize information and technology for their own purposes.

J. A. Higgs (✉)
Birmingham, AL, USA
e-mail: jhiggs@uab.edu

S. Suh et al. (eds.), *Biomedical Engineering*,
DOI 10.1007/978-1-4614-0116-2_22, © Springer Science+Business Media, LLC 2011

Figure 1. The areas of collaboration that make up the knowledge base.

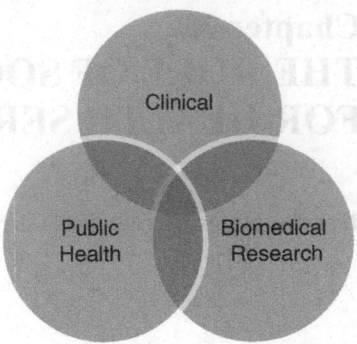

2. Health 2.0, a new concept of healthcare, also utilizes advanced internet-based services but is focused on healthcare value (meaning outcome/price). Patients, physicians, providers and payers use competition at the medical condition level over the full cycle of care as a catalyst for improving safety, efficiency and quality of healthcare delivery.

Providing these definitions is less about differentiating the goals of Medicine 2.0, Health 2.0, but more about understanding their purpose as mechanisms to deliver optimal medical outcomes through individualized care. In this chapter, we are proposing a collaborations engine that fosters cooperation among members of the health sciences community (Figure 1).

This new type of collaboration empowers the entire health and medical community with the ability for multidisciplinary or even transdisciplinary engagement [30]. Public health scientists, hospitals, clinics, specialists, pharmaceutical researchers and medical researchers would all have the ability to share and collaborate. The creation of such an environment would foster innovative even transformative medicine. In spite of all this, currently we do not have a mechanism to facilitate this type of collaboration. Social computing by culminating many aspects of Web 2.0 would potentially introduce an organic development to address this need.

2   Understanding the effects of Social Computing on Medical Science

Social computing and networking has caused a dramatic evolution in the way people collaborate and interact via the Internet. Essentially, social computing represents the collection of technologies that gather, process, compute, and visualize social information [7]. This new social structure has emerged creating an array of loosely integrated social components. In fact, integration of social computing in biomedical research (Science 2.0) and scholarly publishing have acted as enablers, bridging

Figure 2. Medicine 2.0/ Health 2.0 map (with some current exemplary applications and services). (Gunther Eysenbach. Medicine 2.0: Social Networking, Collaboration, Participation, Apomediation, and Openness. J Med Internet Res 2008 (in press) http://dx.doi.org/10.2196/jmir.1030. DOI:10.2196/jmir.1030).

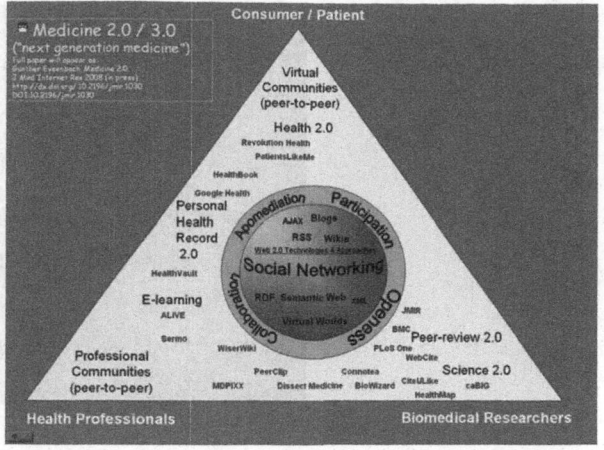

the principles of participation and collaboration across different points along the continuum of knowledge production and dissemination [8]. With medicine becoming one of the most information-intensive sectors of the economy—the National Library of Medicine (NLM)'s bibliographic database MEDLINE [9] added over 1 million new publications in 2010. Most of the new activity can probably be attributed to Medicine 2.0 and Health 2.0. However, Medicine 2.0 and Health 2.0 are more consumer and patient centric (Figure 2).

We are proposing an application that involves modeling complex connections between various aspects of health and medical sciences. Although, our application focuses more on the connections among the scientists the outcome of a new, better health system, which emphasized collaboration, participation and innovation are all embodied in the potential benefits gained [8]. As noted by the Institute of Medicine's vision of a "learning healthcare system" which emphasizes data fluidity as information moves transparently through levels of stewardship. We see an expansion of this methodology by creating an environment that empowers scientist with the ability to combine the advances in new molecular, genomic, and biomedical techniques. Biomedical scientists are also engaging in experiments to connect their data resources in ways that will accelerate discovery. These efforts originated from individual research efforts and clinical practices where biomedical data is available in hundreds of public and private databases, which are all accessible due to new technologies and the internet [10]. The problem is finding innovative ways to take advantage of this influx of advanced techniques and increased access to data.

In other words, the problem is not the lack of technology or innovations but a cohesive integration of these advances and a systematic mapping to the recipients in a cost effective modality. Social computing, being by its very nature suitable for organic growth, provides an ideal medium for sharing data, ideas and computational

Figure 3. A tag cloud (a typical Web 2.0 phenomenon in it) presentingWeb 2.0 themes. (http://en.wikipedia.org/wiki/Tag_cloud).

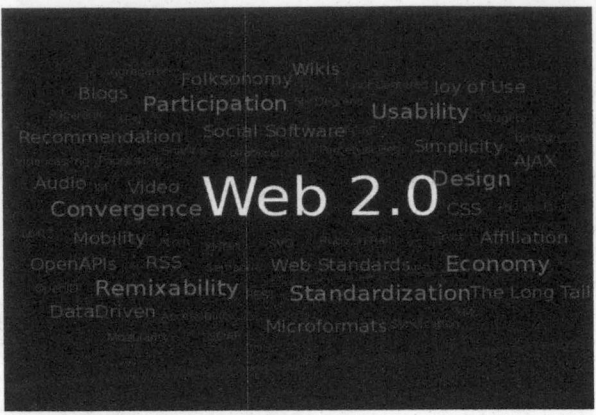

techniques. This allows useful patterns and knowledge base to become transferable commodity.

Today's challenge, then, is to harness the power of health IT to create a collaboration engine for the health sciences. In other words, the issue to be concentrated is not to invent new point technologies but to develop innovative approaches for communication systems among these point technologies and people. A forward looking approach dictates that a collaboration engine should be built on the Web 2.0 methodologies (Figure 3), yet scalable enough to incorporate the semantics of Web 3.0. This is the logical next step as we investigate ways to add meaning to the data available through the Internet. In the industrial age of medicine, healthcare was a highly structured commodity offered to patients in a reactive, mass-produced way [11, 12, and 13]. In the information age, medicine could evolve to become predictive, personalized, preemptive, and participative [14]. In this new era of medicine, it will take the work of many scientific disciplines in an integrated way to effect the changes needed for promoting ongoing advances without causing unanticipated peripheral impairment. Innovation is required to create the architectures needed to promote social participation. This is why a paradigm shift is needed, which can be initiated with the adoption of a social computing model that enables the social aspects of Web 2.0 and other social computing technologies to be viewed as individual systems. These systems would then facilitate the integration of a variety of technologies while providing unprecedented user experience.

In other words, this problem can be solved by applying a system of systems framework, which treats each component as a separate system. Each system communicates as a set of defined processes with specific inputs and outputs. This creates a collection of social computing systems that are easily reusable and scalable reducing the complexity involved in integrating these systems. The Transformative Software Engineering Framework defines a process for building a system as a collection of sub- or meta- systems. In fact, to investigate the interactions, the Transformative Software Engineering Framework was used to engineer a collaboration engine prototype known as HealthScienceMed (HScMed).

3 Brief Description of the Importance of Web 2.0 on Medical Science

By using technologies like XHTML, DHTML, CSS, JavaScript, and JSON developers have been able to provide users with a more advanced browsing experience. Coupled with an XMLHttpRequest object, browsers begin to mirror desktop applications. By converting WebPages to Web 2.0 sites, asynchronous HTTP requests are an effective technique in bringing seemingly static pages to life. However, the usual XMLHttpRequest-based AJAX (Asynchronous JavaScript and XML) clients suffer from the limitation of only being able to communicate to the server from where they are downloaded. This becomes problematic for deployment environments that span multiple domains. In addition, developers end up writing browser-specific code since each of the main browsers implements this XML request object differently. However, for our prototype HealthScienceMed (HScMed) we chose an approach based on JavaScript Object Notation (JSON) that, in the spirit of Web 2.0, makes it easy to build mashup applications without the cross-domain and cross-browser limitations of AJAX [15].

The way people interact on the Web has been changing continuously and drastically. Many Web applications are now being developed with participation, collaboration, and openness. These characteristics define Web 2.0, as a participatory information sharing, interoperability, and user-centered design medium for collaboration on the World Wide Web (Figure 3) [16]. Client-side technologies like JavaScript, which were relegated to doing menial client-side tricks/validations, have come to prominence and are playing an important role in delivering a richer experience to Web developers and users [15].

4 HScMed Application Prototype

Social computing has the promise of complementing the traditional role of doctors and scientists with that of e-physicians and e-scientists. In this new role, information can be shared on a global scale, giving physicians and scientists the ability to make informed decisions, while making social connections with one another based on their common interests and goals. To facilitate these collaborations we have developed HealthScienceMed (HScMed) prototype, which promotes scientific collaboration in a manner that preserves the scientific process and scientists' interests while deliberately promoting (using Web 2.0 technologies) the removal of geographic and social barriers [17] (Figure 4). HScMed provides a centrist means for scientists to collaborate and share resources throughout the scientific development process. In developing HScMed, social driven networking is combined with data driven categorization facilitate interdisciplinary and even transdisciplinary collaborations based on both shared interests [17, 20, 21] (Figure 5).

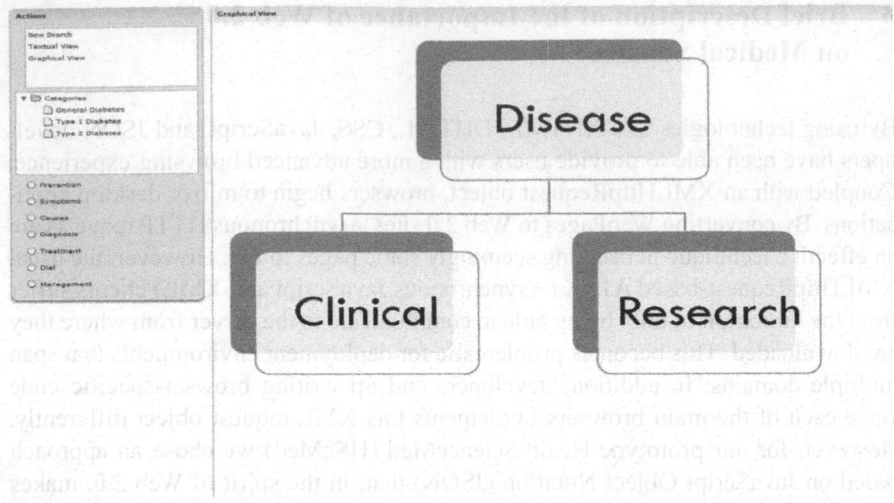

Figure 4. HScMed displaying a graphical search hierarchy for Diabetes.

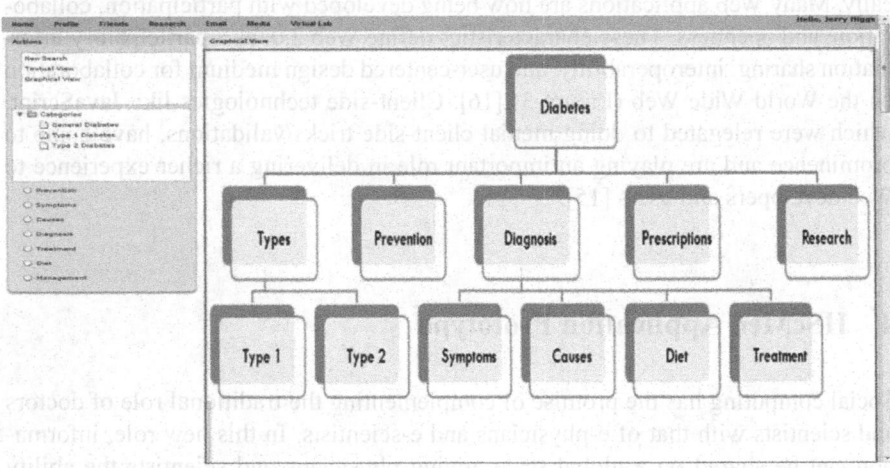

Figure 5. Graphical disease search.

The data/resource is the catalyst between some researchers while the social profile is the method for others. This has the potential to dramatically improve resource sharing and transdisciplinary scientific collaboration by bringing together both conventional and unconventional collaborations [17, 18, and 19]. HScMed have the ability to search through listings of projects (Figure 6) that are listed for various scientists. Once selected, scientists are able to view full details of provided project information (Figure 7).

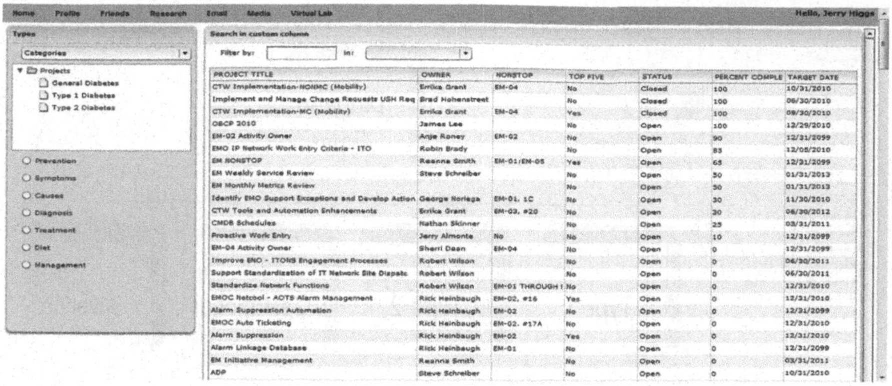

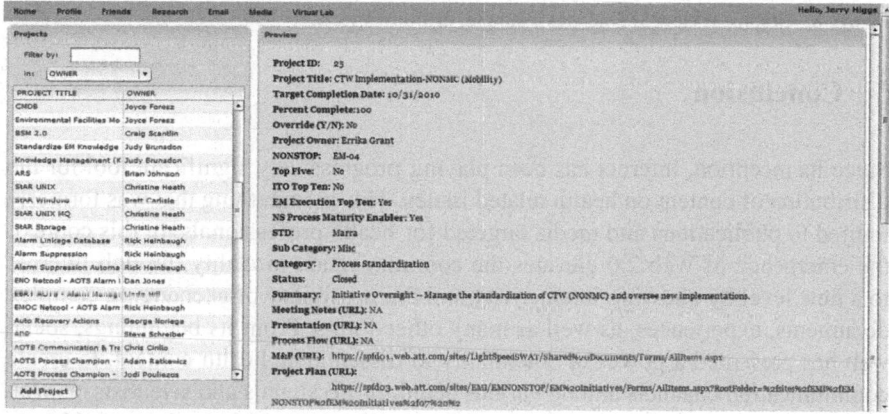

Figure 6. Projects search displayed in grid format.

Figure 7. Detailed project view.

HScMed also allows searches based on a specific disease (Figure 8), making data driven collaborations simple and efficient. Health sciences data integration is important in biomedical research for achieving a comprehensive understanding of various phenomena. The Semantic Web is a technology that accomplishes data integration over heterogeneous datasets distributed on the Web. On the Semantic Web, each data item is identified by a globally unique identifier, *i.e.* a URI, and a relationship between two data items is described as a statement including a triple of subject, predicate and object. It provides a Representational State Transfer (REST) Web service that retrieves Semantic Web data described in the light-weight JSON format. Because JSON is supported by most popular programming languages, the service interface is defined independently in a programming language used for implementation.

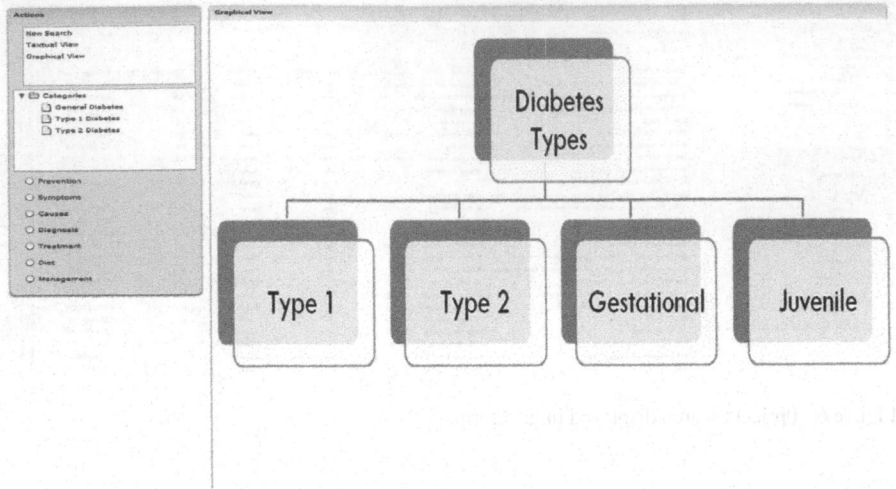

Figure 8. Graphical view of Diabetes example, drilldown view.

5 Conclusion

Since its inception, Internet has been playing progressively significant tool for the distribution of content on health related issues. This is something that was formerly limited to publications and media targeted for health professionals. In this context, the emergence of Web 2.0 elevates the communication modality through Internet to a new level by allowing interactivity and the possibility of exchanging complex documents, experiences, as well as many other forms. From its beginnings, social web has presented a powerful potentiality to disseminate health content, to create communication channels among various health professionals and scientists, as well as public health institutions. HScMed is an early stage platform that provides an experimental prototype to investigate the nature of these collaborations so that we can develop advanced integrated systems with promising results.

References

1. Kessel FS, Rosenfield PL, Anderson NB, eds. Expanding the boundaries of health and social science: case studies in interdisciplinary innovation. New York: Oxford University Press, 2003.
2. Committee on Facilitating Interdisciplinary Research, National Academy of Sciences, National Academy of Engineering, Institute of Medicine. Facilitating interdisciplinary research. Washington DC: The National Academies Press, 2005. http://www.nap.edu/catalog.php?record_id=11153#toc.
3. Hiatt RA, Breen N. The social determinants of cancer: a challenge for transdisciplinary science. Am J Prev Med 2008;35(2S):S141–S150.
4. Hall, KL. The Collaboration Readiness of Transdisciplinary Research Teams and Centers. American Journal of Preventive Medicine, 2008.

5. Klein JT. Evaluation of interdisciplinary and transdisciplinary research: a literature review. Am J Prev Med 2008;35(2S):S116–S123.

6. Walter, J. Medicine and the Second Generation of Internet-based Services. [Online]. Available: http://blog.highlighthealth.info/medicine-20/medicine-20-10-medicine-and-the-second-generation-of-internet-based-services/.

7. "The 2011 International Workshop on Social Computing, Network, and Services (Social-ComNet 2011)." [Online]. Available: http://www.ftrai.org/socialcomnet2011/.

8. "JMIR: Journal of Medical Internet Research." [Online]. Available: http://www.jmir.org/2008/3/e22/.

9. "MEDLINE: U.S. National Library of Medicine (NLM), National Institutes of Heath." [Online]. Available: http://www.nlm.nih.gov/databases/databases_medline.html.

10. Chen, H, Fuller, SS, Friedman, C, Hersh, W. Knowledge Management, Data Mining, and Text Mining in Medical Informatics. Integrated Series in Information Systems, Volume 8, Unit 1, 2005.

11. Stokols D, Fuqua J, Gress J, et al. Evaluating transdisciplinary science. Nicotine Tob Res 2003;5(1S):S21–S39.

12. Stokols D, Misra S, Moser RP, Hall KL, Taylor BK. The ecology of team science: understanding contextual influences on transdisciplinary collaboration. Am J Prev Med 2008;35(2S):S96–S115.

13. Stokols D, Harvey R, Gress J, Fuqua J, Phillips K. In vivo studies of transdisciplinary scientific collaboration: lessons learned and implications for active living research. Am J Prev Med 2005;28(2S2):202–13.

14. Hughes B, Joshi I, Wareham J. Health 2.0 and Medicine 2.0: Tensions and Controversies in the Field, *Journal of Medical Internet Research*, **10**(3): e23

15. Raina, A, Jimenez, J. Dynamic Webpages with JSON: A JSON-based design approach lacks AJAX's cross-domain restrictions. "JAVAWORLD." [Online]. Available: http://www.javaworld.com/javaworld/jw-11-2006/jw-1115-json.html?fsrc=rss-index.

16. "Wikipedia: The Free Encyclopedia." [Online]. Available: http://en.wikipedia.org/wiki/Web_2.0.

17. Cheung, K, Mitchell, A, Greenberg, S, Miller, P. OrphanData.org: Enabling Transdisciplinary Scientific Collaboration Using Web 2.0. Medicine 2.0 (Proceedings). Vol. 2, 2009. [Online]. Available: http://www.medicine20congress.com/ocs/public/conferences/1/schedConfs/2/Med_2.0.09_Proceedings.pdf.

18. IEEE, *Message from the SocialCom 2009 General Chairs*, 2009.

19. Segaran, T, "Programming Collective Intelligence",O'Reilly Meida publishing, 2007.

20. Yanziang, X, Tiejian L, Haibo H. "Social Computing Research Map" IEEE, 2010.

21. Stoica, A and Prieur, C, "Structure of Neighborhoods in a Large Social Network", in press: SocialCom2009.

22. Microsoft Research [Online]. Available: http://research.microsoft.com/en-us/groups/scg/

23. IBM Research [Online]. Available: https://researcher.ibm.com/researcher/view_project.php?id=1782

24. Hewlett Packard Research [Online]. Available: http://www.hpl.hp.com/research/idl/

25. "Wikipedia: The Free Encyclopedia." [Online]. Available: http://en.wikipedia.org/wiki/System_of_systems.

26. "Wikipedia: The Free Encyclopedia." [Online]. Available: http://en.wikipedia.org/wiki/System_of_systems_engineering.

27. "Purdue University, College of Engineering, Systems of Systems (SoS)," [Online]. Available: https://engineering.purdue.edu/Engr/Research/Initiatives/SoS/

28. "The Conversation Prison." [Online]. Available: http://www.briansolis.com/2008/08/introducing-conversation-prism/

29. Gurupur, VP and Tanik, MM, "A System for Building Clinical Research Applications using Semantic Web-Based Approach", UAB, 2010.

30. Gurupur, VP and Tanik, MM, "Abstract Software Design Framework: A Semantic Service Composition Approach", IEEE, 2009.